Basic Structured Grid Generation
with an introduction to unstructured grid generation

Basic Structured Grid Generation

with an introduction to unstructured grid generation

M. Farrashkhalvat and J.P. Miles

BUTTERWORTH
HEINEMANN

OXFORD AMSTERDAM BOSTON LONDON NEW YORK PARIS
SAN DIEGO SAN FRANCISCO SINGAPORE SYDNEY TOKYO

Butterworth-Heinemann
An imprint of Elsevier Science
Linacre House, Jordan Hill, Oxford OX2 8DP
200 Wheeler Rd, Burlington MA 01803

First published 2003

British Library Cataloguing in Publication Data
A catalogue record for this book is available from the British Library

Library of Congress Cataloguing in Publication Data
A catalogue record for this book is available from the Library of Congress

ISBN 0 7506 5058 3

Typeset by Laserwords Private Limited, Chennai, India
Printed and bound in Great Britain by MPG Books Ltd, Bodmin, Cornwall

Contents

Preface

Over the past two decades, efficient methods of grid generation, together with the power of modern digital computers, have been the key to the development of numerical finite-difference (as well as finite-volume and finite-element) solutions of linear and non-linear partial differential equations in regions with boundaries of complex shape. Although much of this development has been directed toward fluid mechanics problems, the techniques are equally applicable to other fields of physics and engineering where field solutions are important. *Structured* grid generation is, broadly speaking, concerned with the construction of co-ordinate systems which provide co-ordinate curves (in two dimensions) and co-ordinate surfaces (in three dimensions) that remain coincident with the boundaries of the solution domain in a given problem. Grid points then arise in the interior of the solution domain at the intersection of these curves or surfaces, the grid cells, lying between pairs of intersecting adjacent curves or surfaces, being generally four-sided figures in two dimensions and small volumes with six curved faces in three dimensions.

It is very helpful to have a good grasp of the underlying mathematics, which is principally to be found in the areas of differential geometry (of what is now a fairly old-fashioned variety) and tensor analysis. We have tried to present a reasonably self-contained account of what is required from these subjects in Chapters 1 to 3. It is hoped that these chapters may also serve as a helpful source of background reference equations.

The following two chapters contain an introduction to the basic techniques (mainly in two dimensions) of structured grid generation, involving algebraic methods and differential models. Again, in an attempt to be reasonably inclusive, we have given a brief account of the most commonly-used numerical analysis techniques for interpolation and for solving algebraic equations. The differential models considered cover elliptic and hyperbolic partial differential equations, with particular reference to the use of forcing functions for the control of grid-density in the solution domain. For solution domains with complex geometries, various techniques are used in practice, including the multi-block method, in which a complex solution domain is split up into simpler sub-domains. Grids may then be generated in each sub-domain (using the sort of methods we have presented), and a matching routine, which reassembles the sub-domains and matches the individual grids at the boundaries of the sub-domains, is used. We show a simple matching routine at the end of Chapter 5.

A number of variational approaches (preceded by a short introduction to variational methods in general) are presented in Chapter 6, showing how grid properties such as smoothness, orthogonality, and grid density can be controlled by the minimization of an appropriate functional (dependent on the components of a fundamental metric tensor). Surface grid generation has been considered here in the general context of harmonic maps. In Chapter 7 time-dependent problems with moving boundaries are considered. Finally, Chapter 8 provides an introduction to the currently very active area of *unstructured* grid generation, presenting the fundamentals of Delaunay triangulation and advancing front techniques.

Our aim throughout is to provide a straightforward and compact introduction to grid generation, covering the essential mathematical background (in which, in our view, tensor calculus forms an important part), while steering a middle course regarding the level of mathematical difficulty. Mathematical exercises are suggested from time to time to assist the reader. In addition, the companion website (www.bh.com/companions/0750650583) provides a series of easy-to-follow, clearly annotated numerical codes, closely associated with Chapters 4, 5, 6, and 8. The aim has been to show the application of the theory to the generation of numerical grids in fairly simple two-dimensional domains, varying from rectangles, circles and ellipses to more complex geometries, such as C-grids over an airfoil, and thus to offer the reader a basis for further progress in this field. Programs involve some of the most frequently used and familiar stable numerical techniques, such as the Thomas Algorithm for the solution of tridiagonal matrix equations, the Gauss-Seidel method, the Conjugate Gradient method, Successive Over Relaxation (SOR), Successive Line Over Relaxation, and the Alternating Direction Implicit (ADI) method, as well as Transfinite Interpolation and the marching algorithm (a grid generator for hyperbolic partial differential equations). The programming language is the standard FORTRAN 77/90.

Our objective in this book is to give an introduction to the most important aspects of grid generation. Our coverage of the literature is rather selective, and by no means complete. For further information and a much wider range of references, texts such as Carey (1997), Knupp and Steinberg (1993), Thompson, Warsi, and Mastin (1985), and Liseikin (1999) may be consulted. Unstructured grid generation is treated in George (1991). A very comprehensive survey of modern developments, together with a great deal of background information, is provided by Thompson, Soni, and Weatherill (1999).

The authors would like to express their gratitude to Mr. Thomas Sippel-Dau, LINUX Service Manager at Imperial College of Science, Technology and Medicine for help with computer administration.

M. Farrashkhalvat

J.P. Miles

1

Mathematical preliminaries – vector and tensor analysis

1.1 Introduction

In this chapter we review the fundamental results of vector and tensor calculus which form the basis of the mathematics of structured grid generation. We do not feel it necessary to give derivations of these results from the perspective of modern differential geometry; the derivations provided here are intended to be appropriate to the background of most engineers working in the area of grid generation. Helpful introductions to tensor calculus may be found in Kay (1988), Kreyzig (1968), and Spain (1953), as well as many books on continuum mechanics, such as Aris (1962). Nevertheless, we have tried to make this chapter reasonably self-contained. Some of the essential results were presented by the authors in Farrashkhalvat and Miles (1990); this book started at an elementary level, and had the restricted aim, compared with many of the more wide-ranging books on tensor calculus, of showing how to use tensor methods to transform partial differential equations of physics and engineering from one co-ordinate system to another (an aim which remains relevant in the present context). There are some minor differences in notation between the present book and Farrashkhalvat and Miles (1990).

1.2 Curvilinear co-ordinate systems and base vectors in E^3

We consider a general set of curvilinear co-ordinates x^i, $i = 1, 2, 3$, by which points in a three-dimensional Euclidean space E^3 may be specified. The set $\{x^1, x^2, x^3\}$ could stand for cylindrical polar co-ordinates $\{r, \theta, z\}$, spherical polars $\{r, \theta, \varphi\}$, etc. A special case would be a set of rectangular cartesian co-ordinates, which we shall generally denote by $\{y_1, y_2, y_3\}$ (where our convention of writing the integer indices as subscripts instead of superscripts will distinguish cartesian from other systems), or sometimes by $\{x, y, z\}$ if this would aid clarity. Instead of $\{x^1, x^2, x^3\}$, it may occasionally be clearer to use notation such as $\{\xi, \eta, \varsigma\}$ without indices.

The position vector $\mathbf{r}$ of a point P in space with respect to some origin O may be expressed as

$$\mathbf{r} = y_1\mathbf{i}_1 + y_2\mathbf{i}_2 + y_3\mathbf{i}_3, \tag{1.1}$$

where $\{\mathbf{i}_1, \mathbf{i}_2, \mathbf{i}_3\}$, alternatively written as $\{\mathbf{i}, \mathbf{j}, \mathbf{k}\}$, are unit vectors in the direction of the rectangular cartesian axes. We assume that there is an invertible relationship between this background set of cartesian co-ordinates and the set of curvilinear co-ordinates, i.e.

$$y_i = y_i(x^1, x^2, x^3), \quad i = 1, 2, 3, \tag{1.2}$$

with the inverse relationship

$$x^i = x^i(y_1, y_2, y_3), \quad i = 1, 2, 3. \tag{1.3}$$

We also assume that these relationships are differentiable. Differentiating eqn (1.1) with respect to x^i gives the set of *covariant base vectors*

$$\mathbf{g}_i = \frac{\partial \mathbf{r}}{\partial x^i}, \quad i = 1, 2, 3, \tag{1.4}$$

with background cartesian components

$$(\mathbf{g}_i)_j = \frac{\partial y_j}{\partial x^i}, \quad j = 1, 2, 3. \tag{1.5}$$

At any point P each of these vectors is tangential to a co-ordinate curve passing through P, i.e. a curve on which one of the x^is varies while the other two remain constant (Fig. 1.1). In general the $\mathbf{g}_i$s are neither unit vectors nor orthogonal to each other. But so that they may constitute a set of basis vectors for vectors in E^3 we demand that they are not co-planar, which is equivalent to requiring that the scalar triple product $\{\mathbf{g}_1 \cdot (\mathbf{g}_2 \times \mathbf{g}_3)\} \neq 0$. Furthermore, this condition is equivalent to the requirement that the *Jacobian* of the transformation (1.2), i.e. the determinant of the matrix of partial derivatives $(\partial y_i/\partial x^j)$, is non-zero; this condition guarantees the existence of the inverse relationship (1.3).

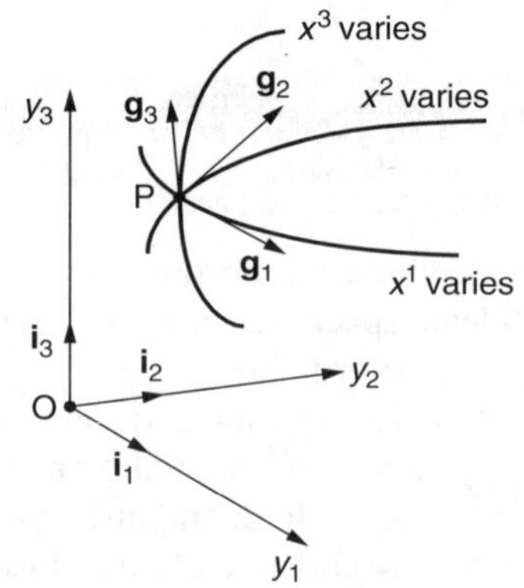

Fig. 1.1 Covariant base vectors at a point P in three dimensions.

Given the set $\{\mathbf{g}_1, \mathbf{g}_2, \mathbf{g}_3\}$ we can form the set of *contravariant base vectors* at P, $\{\mathbf{g}^1, \mathbf{g}^2, \mathbf{g}^3\}$, defined by the set of scalar product identities

$$\mathbf{g}^i \cdot \mathbf{g}_j = \delta^i_j \tag{1.6}$$

where δ^i_j is the Kronecker symbol given by

$$\delta^i_j = \begin{cases} 1 & \text{when } i = j \\ 0 & \text{when } i \neq j \end{cases} \tag{1.7}$$

Exercise 1. Deduce from the definitions (1.6) that the $\mathbf{g}^i$s may be expressed in terms of vector products as

$$\mathbf{g}^1 = \frac{\mathbf{g}_2 \times \mathbf{g}_3}{V}, \quad \mathbf{g}^2 = \frac{\mathbf{g}_3 \times \mathbf{g}_1}{V}, \quad \mathbf{g}^3 = \frac{\mathbf{g}_1 \times \mathbf{g}_2}{V} \tag{1.8}$$

where $V = \{\mathbf{g}_1 \cdot (\mathbf{g}_2 \times \mathbf{g}_3)\}$. (Note that V represents the volume of a parallelepiped (Fig. 1.2) with sides $\mathbf{g}_1$, $\mathbf{g}_2$, $\mathbf{g}_3$.)

The fact that $\mathbf{g}^1$ is perpendicular to $\mathbf{g}_2$ and $\mathbf{g}_3$, which are tangential to the co-ordinate curves on which x^2 and x^3, respectively, vary, implies that $\mathbf{g}^1$ must be perpendicular to the plane which contains these tangential directions; this is just the tangent plane to the co-ordinate surface at P on which x^1 is constant. Thus $\mathbf{g}^i$ must be *normal* to the co-ordinate surface x^i = constant.

Comparison between eqn (1.6), with the scalar product expressed in terms of cartesian components, and the chain rule

$$\frac{\partial x^i}{\partial y_1}\frac{\partial y_1}{\partial x^j} + \frac{\partial x^i}{\partial y_2}\frac{\partial y_2}{\partial x^j} + \frac{\partial x^i}{\partial y_3}\frac{\partial y_3}{\partial x^j} = \frac{\partial x^i}{\partial y_k}\frac{\partial y_k}{\partial x^j} = \frac{\partial x^i}{\partial x^j} = \delta^i_j \tag{1.9}$$

for partial derivatives shows that the background cartesian components of $\mathbf{g}^i$ are given by

$$(\mathbf{g}^i)_j = \frac{\partial x^i}{\partial y_j}, \quad j = 1, 2, 3. \tag{1.10}$$

In eqn (1.9) we have made use of the *summation convention*, by which repeated indices in an expression are automatically assumed to be summed over their range

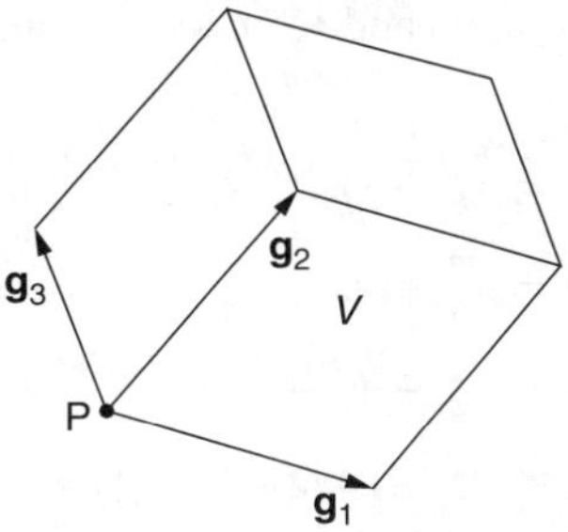

Fig. 1.2 Parallelepiped of base vectors at point P.

of values. *(In expressions involving general curvilinear co-ordinates the summation convention applies only when one of the repeated indices appears as a subscript and the other as a superscript.)* The comparison shows that

$$\frac{\partial x^i}{\partial y_1}\mathbf{i}_1 + \frac{\partial x^i}{\partial y_2}\mathbf{i}_2 + \frac{\partial x^i}{\partial y_3}\mathbf{i}_3 = \nabla x^i = \mathbf{g}^i, \tag{1.11}$$

where the gradient operator ∇, or grad, is defined in cartesians by

$$\nabla = \mathbf{i}_1\frac{\partial}{\partial y_1} + \mathbf{i}_2\frac{\partial}{\partial y_2} + \mathbf{i}_3\frac{\partial}{\partial y_3} = \mathbf{i}_k\frac{\partial}{\partial y_k}. \tag{1.12}$$

For a general scalar field φ we have

$$\nabla\varphi = \mathbf{i}_k\frac{\partial\varphi}{\partial y_k} = \mathbf{i}_k\frac{\partial\varphi}{\partial x^j}\frac{\partial x^j}{\partial y_k} = \left(\mathbf{i}_k\frac{\partial x^j}{\partial y_k}\right)\frac{\partial\varphi}{\partial x^j} = \mathbf{g}^j\frac{\partial\varphi}{\partial x^j}, \tag{1.13}$$

making use of a chain rule again and eqn (1.11); this gives the representation of the gradient operator in general curvilinear co-ordinates.

1.3 Metric tensors

Given a set of curvilinear co-ordinates $\{x^i\}$ with covariant base vectors $\mathbf{g}_i$ and contravariant base vectors $\mathbf{g}^i$, we can define the covariant and contravariant *metric tensors* respectively as the scalar products

$$g_{ij} = \mathbf{g}_i \cdot \mathbf{g}_j \tag{1.14}$$

$$g^{ij} = \mathbf{g}^i \cdot \mathbf{g}^j, \tag{1.15}$$

where i and j can take any values from 1 to 3. From eqns (1.5), (1.10), for the background cartesian components of $\mathbf{g}_i$ and $\mathbf{g}^i$, it follows that

$$g_{ij} = \frac{\partial y_k}{\partial x^i}\frac{\partial y_k}{\partial x^j} \tag{1.16}$$

and

$$g^{ij} = \frac{\partial x^i}{\partial y_k}\frac{\partial x^j}{\partial y_k}. \tag{1.17}$$

If we write (x, y, z) for cartesians and (ξ, η, ς) for curvilinear co-ordinates, we have the formulas

$$\begin{aligned}
g_{11} &= x_\xi^2 + y_\xi^2 + z_\xi^2 \\
g_{22} &= x_\eta^2 + y_\eta^2 + z_\eta^2 \\
g_{33} &= x_\varsigma^2 + y_\varsigma^2 + z_\varsigma^2 \\
g_{12} &= g_{21} = x_\xi x_\eta + y_\xi y_\eta + z_\xi z_\eta
\end{aligned} \tag{1.18}$$

$$g_{23} = g_{32} = x_\eta x_\varsigma + y_\eta y_\varsigma + z_\eta z_\varsigma$$

$$g_{31} = g_{13} = x_\varsigma x_\xi + y_\varsigma y_\xi + z_\varsigma z_\xi$$

where a typical partial derivative $\frac{\partial x}{\partial \xi}$ has been written as x_ξ, and the superscript 2 now represents squaring.

Exercise 2. For the case of spherical polar co-ordinates, with $\xi = r, \eta = \theta, \varsigma = \varphi$, and

$$x = r\sin\theta\cos\varphi, \quad y = r\sin\theta\sin\varphi, \quad z = r\cos\theta$$

show that

$$\begin{pmatrix} g_{11} & g_{12} & g_{13} \\ g_{21} & g_{22} & g_{23} \\ g_{31} & g_{32} & g_{33} \end{pmatrix} = \begin{pmatrix} 1 & 0 & 0 \\ 0 & r^2 & 0 \\ 0 & 0 & r^2\sin^2\theta \end{pmatrix}, \tag{1.19}$$

where (r, θ, φ) take the place of (ξ, η, ς).

Formulas for g^{ij} are, similarly,

$$\begin{aligned} g^{11} &= \xi_x^2 + \xi_y^2 + \xi_z^2 \\ g^{22} &= \eta_x^2 + \eta_y^2 + \eta_z^2 \\ g^{33} &= \varsigma_x^2 + \varsigma_y^2 + \varsigma_z^2 \\ g^{12} &= g^{21} = \xi_x\eta_x + \xi_y\eta_y + \xi_z\eta_z \\ g^{23} &= g^{32} = \eta_x\varsigma_x + \eta_y\varsigma_y + \eta_z\varsigma_z \\ g^{31} &= g^{13} = \varsigma_x\xi_x + \varsigma_y\xi_y + \varsigma_z\xi_z. \end{aligned} \tag{1.20}$$

The metric tensor g_{ij} provides a measure of the distance ds between neighbouring points. If the difference in position vectors between the two points is $\mathbf{dr}$ and the infinitesimal differences in curvilinear co-ordinates are dx^1, dx^2, dx^3, then

$$ds^2 = \mathbf{dr} \cdot \mathbf{dr} = \left(\sum_{i=1}^{3} \frac{\partial \mathbf{r}}{\partial x^i} dx^i\right) \cdot \left(\sum_{j=1}^{3} \frac{\partial \mathbf{r}}{\partial x^j} dx^j\right) = \frac{\partial \mathbf{r}}{\partial x^i} \cdot \frac{\partial \mathbf{r}}{\partial x^j} dx^i\, dx^j = g_{ij}\, dx^i\, dx^j, \tag{1.21}$$

making use of the summation convention. As previously remarked, the summation convention may be employed in generalized (curvilinear) co-ordinates only when each of the repeated indices appears once as a subscript and once as a superscript.

We can form the 3×3 matrix L whose row i contains the background cartesian components of $\mathbf{g}_i$ and the matrix M whose row i contains the background cartesian components of $\mathbf{g}^i$. We may write, in shorthand form,

$$L = \begin{pmatrix} \mathbf{g}_1 \\ \mathbf{g}_2 \\ \mathbf{g}_3 \end{pmatrix}, \quad M^T = \begin{pmatrix} \mathbf{g}^1 & \mathbf{g}^2 & \mathbf{g}^3 \end{pmatrix} \tag{1.22}$$

and $L_{ij} = \dfrac{\partial y_j}{\partial x^i}$, $M_{ij} = \dfrac{\partial x^i}{\partial y_j}$; it may be seen directly from eqn (1.6) that

$$LM^T = I, \tag{1.23}$$

where I is the 3×3 identity matrix. Thus L and M^T are mutual inverses. Moreover

$$\det L = \{\mathbf{g}_1 \cdot (\mathbf{g}_2 \times \mathbf{g}_3)\} = V \tag{1.24}$$

as previously defined in eqn (1.8). Since $M^T = L^{-1}$, it follows that

$$\det M = \{\mathbf{g}^1 \cdot (\mathbf{g}^2 \times \mathbf{g}^3)\} = V^{-1}. \tag{1.25}$$

It is easy to see that the symmetric matrix arrays (g_{ij}) and (g^{ij}) for the associated metric tensors are now given by

$$(g_{ij}) = LL^T, \quad (g^{ij}) = MM^T. \tag{1.26}$$

Since $M^T = L^{-1}$ and $M = (L^T)^{-1}$, it follows that

$$(g^{ij}) = (g_{ij})^{-1}. \tag{1.27}$$

In component form this is equivalent to

$$g_{ik}g^{jk} = \delta_i^j. \tag{1.28}$$

From the properties of determinants it also follows that

$$g = \det(g_{ij}) = (\det L)^2 = V^2, \tag{1.29}$$

$$\det(g^{ij}) = g^{-1}, \tag{1.30}$$

and

$$V = \{\mathbf{g}_1 \cdot (\mathbf{g}_2 \times \mathbf{g}_3)\} = \sqrt{g}, \tag{1.31}$$

where g must be a positive quantity.

Thus in place of eqn (1.8) we can write

$$\mathbf{g}^1 = \frac{1}{\sqrt{g}}\mathbf{g}_2 \times \mathbf{g}_3, \quad \mathbf{g}^2 = \frac{1}{\sqrt{g}}\mathbf{g}_3 \times \mathbf{g}_1, \quad \mathbf{g}^3 = \frac{1}{\sqrt{g}}\mathbf{g}_1 \times \mathbf{g}_2. \tag{1.32}$$

From eqn (1.27) and standard 3×3 matrix inversion, we can also deduce the following formula:

$$g^{ij} = \frac{1}{g}\begin{pmatrix} G_1 & G_4 & G_5 \\ G_4 & G_2 & G_6 \\ G_5 & G_6 & G_3 \end{pmatrix}, \tag{1.33}$$

where the co-factors of (g_{ij}) are given by

$$G_1 = g_{22}g_{33} - (g_{23})^2, \quad G_2 = g_{11}g_{33} - (g_{13})^2, \quad G_3 = g_{11}g_{22} - (g_{12})^2$$

$$G_4 = g_{13}g_{23} - g_{12}g_{33}, \quad G_5 = g_{12}g_{23} - g_{13}g_{22}, \quad G_6 = g_{12}g_{13} - g_{23}g_{11}. \tag{1.34}$$

The cofactors of the matrix L in eqn (1.22) are the various background cartesian components of $(\mathbf{g}_j \times \mathbf{g}_k)$, which may be expressed, with the notation used in eqn (1.18), as

$$\begin{aligned} \alpha_1 &= y_\eta z_\varsigma - y_\varsigma z_\eta, & \alpha_2 &= x_\varsigma z_\eta - x_\eta z_\varsigma, & \alpha_3 &= x_\eta y_\varsigma - x_\varsigma y_\eta \\ \beta_1 &= y_\varsigma z_\xi - y_\xi z_\varsigma, & \beta_2 &= x_\xi z_\varsigma - x_\varsigma z_\xi, & \beta_3 &= x_\varsigma y_\xi - x_\xi y_\varsigma \\ \gamma_1 &= y_\xi z_\eta - y_\eta z_\xi, & \gamma_2 &= x_\eta z_\xi - x_\xi z_\eta, & \gamma_3 &= x_\xi y_\eta - x_\eta y_\xi \end{aligned} \tag{1.35}$$

so that

$$\mathbf{g}_2 \times \mathbf{g}_3 = \alpha_1\mathbf{i} + \alpha_2\mathbf{j} + \alpha_3\mathbf{k}, \quad \mathbf{g}_3 \times \mathbf{g}_1 = \beta_1\mathbf{i} + \beta_2\mathbf{j} + \beta_3\mathbf{k}, \quad \mathbf{g}_1 \times \mathbf{g}_2 = \gamma_1\mathbf{i} + \gamma_2\mathbf{j} + \gamma_3\mathbf{k} \tag{1.36}$$

and

$$\mathbf{g}^1 = \frac{1}{\sqrt{g}}(\alpha_1\mathbf{i}+\alpha_2\mathbf{j}+\alpha_3\mathbf{k}), \quad \mathbf{g}^2 = \frac{1}{\sqrt{g}}(\beta_1\mathbf{i}+\beta_2\mathbf{j}+\beta_3\mathbf{k}), \quad \mathbf{g}^3 = \frac{1}{\sqrt{g}}(\gamma_1\mathbf{i}+\gamma_2\mathbf{j}+\gamma_3\mathbf{k}). \tag{1.37}$$

Since $M = L^{-1}$, we also have, in the same notation, the matrix elements of M:

$$\begin{aligned} \xi_x &= \alpha_1/\sqrt{g}, & \xi_y &= \alpha_2/\sqrt{g}, & \xi_z &= \alpha_3/\sqrt{g} \\ \eta_x &= \beta_1/\sqrt{g}, & \eta_y &= \beta_2/\sqrt{g}, & \eta_z &= \beta_3/\sqrt{g} \\ \varsigma_x &= \gamma_1/\sqrt{g}, & \varsigma_y &= \gamma_2/\sqrt{g}, & \varsigma_z &= \gamma_3/\sqrt{g}. \end{aligned} \tag{1.38}$$

Exercise 3. Using eqn (1.29) and standard determinant expansions, derive the following formulas for the determinant g:

$$\begin{aligned} g &= g_{11}G_1 + g_{12}G_4 + g_{13}G_5 = (\alpha_1 x_\xi + \beta_1 x_\eta + \gamma_1 x_\varsigma)^2 \\ &= g_{22}G_2 + g_{12}G_4 + g_{23}G_6 = (\alpha_2 y_\xi + \beta_2 y_\eta + \gamma_2 y_\varsigma)^2 \\ &= g_{33}G_3 + g_{13}G_5 + g_{23}G_6 = (\alpha_3 z_\xi + \beta_3 z_\eta + \gamma_3 z_\varsigma)^2. \end{aligned} \tag{1.39}$$

From eqn (1.32) it follows that

$$g^{ip} = \mathbf{g}^i \cdot \mathbf{g}^p = \frac{1}{g}(\mathbf{g}_j \times \mathbf{g}_k) \cdot (\mathbf{g}_q \times \mathbf{g}_r),$$

where $\{i, j, k\}$ and $\{p, q, r\}$ are in cyclic order $\{1, 2, 3\}$. Using the standard Lagrange vector identity

$$(\mathbf{A} \times \mathbf{B}) \cdot (\mathbf{C} \times \mathbf{D}) = (\mathbf{A} \cdot \mathbf{C})(\mathbf{B} \cdot \mathbf{D}) - (\mathbf{A} \cdot \mathbf{D})(\mathbf{B} \cdot \mathbf{C}), \tag{1.40}$$

we have

$$\begin{aligned} g^{ip} &= \frac{1}{g}\{(\mathbf{g}_j \cdot \mathbf{g}_q)(\mathbf{g}_k \cdot \mathbf{g}_r) - (\mathbf{g}_j \cdot \mathbf{g}_r)(\mathbf{g}_k \cdot \mathbf{g}_q)\} \\ &= \frac{1}{g}(g_{jq}g_{kr} - g_{jr}g_{kq}). \end{aligned} \tag{1.41}$$

For example,

$$g^{13} = \frac{1}{g}(g_{21}g_{32} - g_{22}g_{31}).$$

1.4 Line, area, and volume elements

Lengths of general infinitesimal line-elements are given by eqn (1.21). An element of the x^1 co-ordinate curve on which $dx^2 = dx^3 = 0$ is therefore given by $(ds)^2 = g_{11}(dx^1)^2$. Thus arc-length along the x^i-curve is

$$ds = \sqrt{g_{ii}}\, dx^i \tag{1.42}$$

(with no summation over i).

A line-element along the x^1-curve may be written $\frac{\partial \mathbf{r}}{\partial x^1} dx^1 = \mathbf{g}_1\, dx^1$, and similarly a line-element along the x^2-curve is $\mathbf{g}_2\, dx^2$. The infinitesimal vector area of the parallelogram of which these two line-elements form the sides is the vector product $(\mathbf{g}_1\, dx^1 \times \mathbf{g}_2\, dx^2)$, which has magnitude

$$dA_3 = |\mathbf{g}_1 \times \mathbf{g}_2|\, dx^1\, dx^2. \tag{1.43}$$

Again by the Lagrange vector identity we have

$$\begin{aligned}|\mathbf{g}_1 \times \mathbf{g}_2|^2 &= (\mathbf{g}_1 \times \mathbf{g}_2)\cdot(\mathbf{g}_1 \times \mathbf{g}_2) = (\mathbf{g}_1\cdot\mathbf{g}_1)(\mathbf{g}_2\cdot\mathbf{g}_2) - (\mathbf{g}_1\cdot\mathbf{g}_2)(\mathbf{g}_1\cdot\mathbf{g}_2)\\ &= g_{11}g_{22} - (g_{12})^2.\end{aligned}$$

Hence $dA_3 = \sqrt{g_{11}g_{22} - (g_{12})^2}\, dx^1\, dx^2$, giving the general expression

$$dA_i = \sqrt{g_{jj}g_{kk} - (g_{jk})^2}\, dx^j\, dx^k = G_i\, dx^j\, dx^k, \tag{1.44}$$

using eqn (1.34), where i, j, k must be taken in cyclic order $1, 2, 3$, and again there is no summation over j and k.

The parallelepiped generated by line-elements $\mathbf{g}_1\, dx^1$, $\mathbf{g}_2\, dx^2$, $\mathbf{g}_3\, dx^3$, along the co-ordinate curves has infinitesimal volume

$$dV = \mathbf{g}_1\, dx^1 \cdot (\mathbf{g}_2\, dx^2 \times \mathbf{g}_3\, dx^3) = \{\mathbf{g}_1 \cdot (\mathbf{g}_2 \times \mathbf{g}_3)\} dx^1\, dx^2\, dx^3.$$

By eqn (1.31) we have

$$dV = \sqrt{g}\, dx^1\, dx^2\, dx^3. \tag{1.45}$$

1.5 Generalized vectors and tensors

A vector field $\mathbf{u}$ (a function of position $\mathbf{r}$) may be expressed at a point P in terms of the covariant base vectors $\mathbf{g}_1, \mathbf{g}_2, \mathbf{g}_3$, or in terms of the contravariant base vectors $\mathbf{g}^1$, $\mathbf{g}^2$, $\mathbf{g}^3$. Thus we have

$$\mathbf{u} = u^1\mathbf{g}_1 + u^2\mathbf{g}_2 + u^3\mathbf{g}_3 = u^i\mathbf{g}_i \tag{1.46}$$

$$= u_1\mathbf{g}^1 + u_2\mathbf{g}^2 + u_3\mathbf{g}^3 = u_i\mathbf{g}^i, \tag{1.47}$$

where u^i and u_i are called the *contravariant* and *covariant* components of $\mathbf{u}$, respectively. Taking the scalar product of both sides of eqn (1.46) with $\mathbf{g}^j$ gives

$$\mathbf{u}\cdot\mathbf{g}^j = u^i\mathbf{g}_i\cdot\mathbf{g}^j = u^i\delta_i^j = u^j.$$

Hence

$$u^i = \mathbf{u} \cdot \mathbf{g}^i, \tag{1.48}$$

and, similarly,

$$u_i = \mathbf{u} \cdot \mathbf{g}_i. \tag{1.49}$$

A similar procedure shows, incidentally, that

$$\mathbf{g}_i = g_{ij}\mathbf{g}^j, \tag{1.50}$$

and

$$\mathbf{g}^i = g^{ij}\mathbf{g}_j. \tag{1.51}$$

We then easily deduce that

$$u^i = g^{ij}u_j \tag{1.52}$$

and

$$u_i = g_{ij}u^j. \tag{1.53}$$

These equations may be interpreted as demonstrating that the action of g^{ij} on u_j and that of g_{ij} on u^j are effectively equivalent to 'raising the index' and 'lowering the index', respectively.

It is straightforward to show that the scalar product of vectors **u** and **v** is given by

$$\mathbf{u} \cdot \mathbf{v} = u^i v_i = u_i v^i = g_{ij}u^i v^j = g^{ij}u_i v_j \tag{1.54}$$

and hence that the magnitude of a vector **u** is given by

$$|\mathbf{u}| = \sqrt{g_{ij}u^i u^j} = \sqrt{g^{ij}u_i u_j}. \tag{1.55}$$

It is important to note the special transformation properties of covariant and contravariant components under a change of curvilinear co-ordinate system. We consider another system of co-ordinates $\overline{x}^i$, $i = 1, 2, 3$, related to the first system by the transformation equations

$$\overline{x}^i = \overline{x}^i(x^1, x^2, x^3), \quad i = 1, 2, 3. \tag{1.56}$$

These equations are assumed to be invertible and differentiable. In particular, differentials in the two systems are related by the chain rule

$$\mathrm{d}\overline{x}^i = \frac{\partial \overline{x}^i}{\partial x^j}\,\mathrm{d}x^j, \tag{1.57}$$

or, in matrix terms,

$$\begin{pmatrix} \mathrm{d}\overline{x}^1 \\ \mathrm{d}\overline{x}^2 \\ \mathrm{d}\overline{x}^3 \end{pmatrix} = A \begin{pmatrix} \mathrm{d}x^1 \\ \mathrm{d}x^2 \\ \mathrm{d}x^3 \end{pmatrix}, \tag{1.58}$$

where we assume that the matrix A of the transformation, with i-j element equal to $\partial \overline{x}^i / \partial x^j$, has a determinant not equal to zero, so that eqn (1.58) may be inverted. We define the Jacobian J of the transformation as

$$J = \det A. \tag{1.59}$$

Exercise 4. Show that if we define the matrix B as that whose i-j element is equal to $\partial x^j/\partial \overline{x}^i$, then

$$AB^T = I \tag{1.60}$$

and

$$\det B = J^{-1}. \tag{1.61}$$

We obtain new covariant base vectors, which transform according to the rule

$$\overline{\mathbf{g}}_i = \frac{\partial \mathbf{r}}{\partial \overline{x}^i} = \frac{\partial \mathbf{r}}{\partial x^j}\frac{\partial x^j}{\partial \overline{x}^i} = \frac{\partial x^j}{\partial \overline{x}^i}\mathbf{g}_j, \tag{1.62}$$

with the inverse relationship

$$\mathbf{g}_i = \frac{\partial \overline{x}^j}{\partial x^i}\overline{\mathbf{g}}_j. \tag{1.63}$$

In background cartesian components, eqn (1.62) may be written in matrix form as

$$\overline{L} = BL, \tag{1.64}$$

where $\overline{L}$ is the matrix with i-j component given by $\partial y_j/\partial \overline{x}^i$, and from eqn (1.23) and eqn (1.60) we deduce that

$$\overline{M} = AM, \tag{1.65}$$

where $\overline{M}$ is the matrix with i-j component $\partial \overline{x}^i/\partial y_j$.

The new system of co-ordinates has associated metric tensors given, in comparison with eqn (1.26), by

$$(\overline{g}_{ij}) = \overline{L}\,\overline{L}^T, \quad (\overline{g}^{ij}) = \overline{M}\,\overline{M}^T, \tag{1.66}$$

so that the corresponding determinant $\overline{g} = \det(\overline{g}_{ij})$ is given by

$$\overline{g} = (\det \overline{L})^2.$$

Hence $\det \overline{L} = \sqrt{\overline{g}}$, $\det L = \sqrt{g}$, and $\det \overline{L} = \det B \det L = J^{-1} \det L$ from eqn (1.64). Thus we have

$$J = \sqrt{\frac{g}{\overline{g}}}. \tag{1.67}$$

Equation (1.65) yields the relation between corresponding contravariant base vectors:

$$\overline{\mathbf{g}}^i = \frac{\partial \overline{x}^i}{\partial x^j}\mathbf{g}^j. \tag{1.68}$$

Expressing $\mathbf{u}$ as a linear combination of the base vectors in the new system gives

$$\mathbf{u} = \overline{u}^i\overline{\mathbf{g}}_i = \overline{u}_i\overline{\mathbf{g}}^i. \tag{1.69}$$

We now easily obtain, using eqn (1.62), the transformation rule for the covariant components of a vector:

$$\overline{u}_i = \mathbf{u}\cdot\overline{\mathbf{g}}_i = \mathbf{u}\cdot\left(\frac{\partial x^j}{\partial \overline{x}^i}\mathbf{g}_j\right) = \frac{\partial x^j}{\partial \overline{x}^i}\mathbf{u}\cdot\mathbf{g}_j = \frac{\partial x^j}{\partial \overline{x}^i}u_j, \tag{1.70}$$

or, in matrix form,

$$\begin{pmatrix} \overline{u}_1 \\ \overline{u}_2 \\ \overline{u}_3 \end{pmatrix} = B \begin{pmatrix} u_1 \\ u_2 \\ u_3 \end{pmatrix}. \tag{1.71}$$

The set of components $\frac{\partial \varphi}{\partial x^j}$ (where φ is a scalar field) found in eqn (1.13) can be said to constitute a covariant vector, since by the usual chain rule they transform according to eqn (1.70), i.e.

$$\frac{\partial \varphi}{\partial \overline{x}^i} = \frac{\partial x^j}{\partial \overline{x}^i} \frac{\partial \varphi}{\partial x^j}.$$

Exercise 5. Show that the transformation rule for contravariant components of a vector is

$$\overline{u}^i = \frac{\partial \overline{x}^i}{\partial x^j} u^j, \tag{1.72}$$

or

$$\begin{pmatrix} \overline{u}^1 \\ \overline{u}^2 \\ \overline{u}^3 \end{pmatrix} = A \begin{pmatrix} u^1 \\ u^2 \\ u^3 \end{pmatrix}. \tag{1.73}$$

Note the important consequence that the scalar product (1.54) is an *invariant* quantity (a true scalar), since it is unaffected by co-ordinate transformations. In fact

$$\overline{u}^i \overline{v}_i = \left(\frac{\partial \overline{x}^i}{\partial x^j} u^j \right) \left(\frac{\partial x^k}{\partial \overline{x}^i} v_k \right) = \left(\frac{\partial \overline{x}^i}{\partial x^j} \frac{\partial x^k}{\partial \overline{x}^i} \right) u^j v_k = \delta^k_j u^j v_k = u^j v_j.$$

From eqns (1.64), (1.65), and (1.66), we obtain the transformation rules:

$$(\overline{g}_{ij}) = \overline{L}\,\overline{L}^T = BLL^T B^T = B(g_{ij})B^T, \tag{1.74}$$

$$(\overline{g}^{ij}) = \overline{M}\,\overline{M}^T = AMM^T A^T = A(g^{ij})A^T. \tag{1.75}$$

In fact g_{ij} is a particular case of a *covariant tensor of order two*, which may be defined here as a set of quantities which take the values T_{ij}, say, when the curvilinear co-ordinates x^i are chosen and the values $\overline{T}_{ij}$ when a different set $\overline{x}^i$ are chosen, with a transformation rule between the two sets of values being given in co-ordinate form by

$$\overline{T}_{ij} = \frac{\partial x^k}{\partial \overline{x}^i} \frac{\partial x^l}{\partial \overline{x}^j} T_{kl} \tag{1.76}$$

with summation over k and l, or in matrix form

$$\overline{T} = BTB^T. \tag{1.77}$$

Similarly, g^{ij} is a particular case of a *contravariant tensor of order two*. This is defined as an entity which has components T^{ij} obeying the transformation rules

$$\overline{T}^{ij} = \frac{\partial \overline{x}^i}{\partial x^k} \frac{\partial \overline{x}^j}{\partial x^l} T^{kl} \tag{1.78}$$

or, equivalently,

$$\overline{T} = ATA^T. \tag{1.79}$$

We can also define *mixed* second-order tensors $T_i^{\cdot j}$ and $T_{.i}^{j}$, for which the transformation rules are

$$\overline{T}_i^{\cdot j} = \frac{\partial x^k}{\partial \overline{x}^i}\frac{\partial \overline{x}^j}{\partial x^l} T_k^{\cdot l} \tag{1.80}$$

$$\overline{T} = BTA^T, \tag{1.81}$$

and

$$\overline{T}^{i}_{\cdot\cdot j} = \frac{\partial \overline{x}^i}{\partial x^k}\frac{\partial x^l}{\partial \overline{x}^j} T^k_{.l} \tag{1.82}$$

$$\overline{T} = ATB^T. \tag{1.83}$$

Exercise 6. Show from the transformation rules (1.80) and (1.82) that the quantities $T^k_{.k}$ and $T_k^{.k}$ are invariants.

Given two vectors **u** and **v**, second-order tensors can be generated by taking products of covariant or contravariant vector components, giving the covariant tensor $u_i v_j$, the contravariant tensors $u^i v^j$, and the mixed tensors $u^i v_j$ and $u_i v^j$. In this case these tensors are said to be *associated*, since they are all derived from an entity which can be written in absolute, co-ordinate-free, terms, as $\mathbf{u} \otimes \mathbf{v}$; this is called the *dyadic product* of **u** and **v**. The dyadic product may also be regarded as a linear operator which acts on vectors **w** according to the rule

$$(\mathbf{u} \otimes \mathbf{v})\mathbf{w} = \mathbf{u}(\mathbf{v} \cdot \mathbf{w}), \tag{1.84}$$

an equation which has various co-ordinate representations, such as

$$(u_i v_j)w^j = u_i(v_j w^j).$$

It may also be expressed in the following various ways:

$$\mathbf{u} \otimes \mathbf{v} = u_i v_j \mathbf{g}^i \otimes \mathbf{g}^j = u^i v^j \mathbf{g}_i \otimes \mathbf{g}_j = u^i v_j \mathbf{g}_i \otimes \mathbf{g}^j = u_i v^j \mathbf{g}^i \otimes \mathbf{g}_j \tag{1.85}$$

with summation over i and j in each case.

In general, covariant, contravariant, and mixed components $T_{ij}, T^{ij}, T_i^{\cdot j}, T^i_{.j}$, are associated if there exists an entity **T**, a linear operator which can operate on vectors, such that

$$\mathbf{T} = T_{ij}\mathbf{g}^i \otimes \mathbf{g}^j = T^{ij}\mathbf{g}_i \otimes \mathbf{g}_j = T_i^{\cdot\cdot j}\mathbf{g}^i \otimes \mathbf{g}_j = T^i_{\cdot\cdot j}\mathbf{g}_i \otimes \mathbf{g}^j. \tag{1.86}$$

Thus the action of **T** on a vector **u** could be represented typically by

$$\mathbf{Tu} = (T_{ij}\mathbf{g}^i \otimes \mathbf{g}^j)\mathbf{u} = T_{ij}\mathbf{g}^i(\mathbf{g}^j \cdot \mathbf{u}) = T_{ij}\mathbf{g}^i u^j = T_{ij}u^j\mathbf{g}^i = \mathbf{v},$$

where **v** has covariant components

$$v_i = T_{ij}u^j.$$

The Kronecker symbol δ^i_j has corresponding matrix elements given by the 3×3 identity matrix I. It may be interpreted as a second-order mixed tensor, where whichever of the covariant or contravariant components occurs first is immaterial, since if we substitute $T = I$ in either of the transformation rules (1.81) or (1.83) we obtain $\overline{T} = I$ in view of eqn (1.60). Thus δ^i_j is a mixed tensor which has the same components on any co-ordinate system. The corresponding linear operator is just the identity operator $\mathbf{I}$, which for any vector $\mathbf{u}$ satisfies

$$\mathbf{Iu} = (\delta^i_j \mathbf{g}_i \otimes \mathbf{g}^j)\mathbf{u} = \delta^i_j \mathbf{g}_i u^j = \mathbf{g}_i u^i = \mathbf{u}.$$

The following representations of $\mathbf{I}$ may then be deduced:

$$\mathbf{I} = g_{ij}\mathbf{g}^i \otimes \mathbf{g}^j = g^{ij}\mathbf{g}_i \otimes \mathbf{g}_j = \mathbf{g}_j \otimes \mathbf{g}^j = \mathbf{g}^j \otimes \mathbf{g}_j. \tag{1.87}$$

Thus g_{ij}, g^{ij}, and δ^i_j are associated tensors.

Covariant, contravariant, and mixed tensors of higher order than two may be defined in terms of transformation rules following the pattern in eqns (1.76), (1.78), (1.80), and (1.82), though it may not be convenient to express these rules in matrix terms. For example, covariant and contravariant third-order tensors U_{ijk} and U^{ijk} respectively must follow the transformation rules:

$$\overline{U}_{ijk} = \frac{\partial x^l}{\partial \overline{x}^i}\frac{\partial x^m}{\partial \overline{x}^j}\frac{\partial x^n}{\partial \overline{x}^k}U_{lmn}, \quad \overline{U}^{ijk} = \frac{\partial \overline{x}^i}{\partial x^l}\frac{\partial \overline{x}^j}{\partial x^m}\frac{\partial \overline{x}^k}{\partial x^n}U^{lmn}. \tag{1.88}$$

The alternating symbol e_{ijk} defined by

$$e_{ijk} = e^{ijk} = \begin{cases} 1 & \text{if } (i, j, k) \text{ is an even permutation of } (1, 2, 3) \\ -1 & \text{if } (i, j, k) \text{ is an odd permutation of } (1, 2, 3) \\ 0 & \text{otherwise} \end{cases} \tag{1.89}$$

is not a (generalized) third-order tensor. Applying the left-hand transformation of eqns (1.88) gives, using the properties of determinants and eqns (1.61) and (1.67),

$$\frac{\partial x^l}{\partial \overline{x}^i}\frac{\partial x^m}{\partial \overline{x}^j}\frac{\partial x^n}{\partial \overline{x}^k}e_{lmn} = (\det B)e_{ijk} = J^{-1}e_{ijk} = \sqrt{\frac{\overline{g}}{g}}e_{ijk}. \tag{1.90}$$

Similarly we obtain

$$\frac{\partial \overline{x}^i}{\partial x^l}\frac{\partial \overline{x}^j}{\partial x^m}\frac{\partial \overline{x}^k}{\partial x^n}e^{lmn} = (\det A)e^{ijk} = Je^{ijk} = \sqrt{\frac{g}{\overline{g}}}e^{ijk}. \tag{1.91}$$

It follows that third-order covariant and contravariant tensors respectively are defined by

$$\varepsilon_{ijk} = \sqrt{g}e_{ijk} \tag{1.92}$$

and

$$\varepsilon^{ijk} = \frac{1}{\sqrt{g}}e^{ijk}. \tag{1.93}$$

Applying the appropriate transformation law to ε_{ijk} now gives $\sqrt{\overline{g}}\, e_{lmn} = \overline{\varepsilon}_{lmn}$, as required, and similarly for ε^{ijk}. These tensors, known as the *alternating tensors*, are required, for example, when forming correct vector expressions in curvilinear co-ordinate systems.

In particular, the vector product of two vectors **u** and **v** is given by

$$\mathbf{u} \times \mathbf{v} = \varepsilon^{ijk} u_j v_k \mathbf{g}_i = \varepsilon_{ijk} u^j v^k \mathbf{g}^i, \tag{1.94}$$

with summation over i, j, k. The component forms of the scalar triple product of vectors **u**, **v**, **w** are

$$\mathbf{u} \cdot (\mathbf{v} \times \mathbf{w}) = \varepsilon^{ijk} u_i v_j w_k = \varepsilon_{ijk} u^i v^j w^k. \tag{1.95}$$

The alternating symbols themselves may be called *relative* (rather than *absolute) tensors*, which means that when the tensor transformation law is applied as in eqns (1.90) and (1.91) a power of J (the *weight* of the relative tensor) appears on the right-hand side. Thus according to (1.90) e_{ijk} is a relative tensor of weight -1, while according to eqn (1.91) e^{ijk} (although it takes exactly the same values as e_{ijk}) is a relative tensor of weight 1.

1.6 Christoffel symbols and covariant differentiation

In curvilinear co-ordinates the base vectors will generally vary in magnitude and direction from one point to another, and this causes special problems for the differentiation of vector and tensor fields. In general, differentiation of covariant base vectors eqn (1.4) with respect to x^j satisfies

$$\frac{\partial \mathbf{g}_i}{\partial x^j} = \frac{\partial^2 \mathbf{r}}{\partial x^j \partial x^i} = \frac{\partial^2 \mathbf{r}}{\partial x^i \partial x^j} = \frac{\partial \mathbf{g}_j}{\partial x^i}. \tag{1.96}$$

Expressing the resulting vector (for a particular choice of i and j) as a linear combination of base vectors gives

$$\frac{\partial \mathbf{g}_i}{\partial x^j} = [ij, k]\mathbf{g}^k = \Gamma^k_{ij} \mathbf{g}_k, \tag{1.97}$$

with summation over k. The coefficients $[ij, k]$, Γ^k_{ij} in eqn (1.97) are called *Christoffel symbols of the first and second kinds*, respectively. Taking appropriate scalar products on eqn (1.97) gives

$$[ij, k] = \frac{\partial \mathbf{g}_i}{\partial x^j} \cdot \mathbf{g}_k \tag{1.98}$$

and

$$\Gamma^k_{ij} = \frac{\partial \mathbf{g}_i}{\partial x^j} \cdot \mathbf{g}^k. \tag{1.99}$$

Both $[ij, k]$ and Γ^k_{ij} are symmetric in i and j by eqn (1.96). We also have, by eqn (1.51),

$$\Gamma^k_{ij} = \frac{\partial \mathbf{g}_i}{\partial x^j} \cdot (g^{kl} \mathbf{g}_l) = g^{kl}[ij, l] \tag{1.100}$$

with summation over l. Similarly,

$$[ij, k] = g_{kl} \Gamma^l_{ij}. \tag{1.101}$$

Evaluating the scalar products in eqns (1.98) and (1.99) on background cartesians gives the formulas

$$[ij,k] = \frac{\partial^2 y_l}{\partial x^i \partial x^j}\frac{\partial y_l}{\partial x^k}, \quad \Gamma^k_{ij} = \frac{\partial^2 y_l}{\partial x^i \partial x^j}\frac{\partial x^k}{\partial y_l}. \tag{1.102}$$

Exercise 7. Using eqns (1.100), (1.102), with (1.33) and (1.34), verify the formula

$$\Gamma^k_{ij} = K^k \frac{\partial^2 x}{\partial x^i \partial x^j} + L^k \frac{\partial^2 y}{\partial x^i \partial x^j} + M^k \frac{\partial^2 z}{\partial x^i \partial x^j}, \tag{1.103}$$

where (in the obvious notation)

$$K^1 = \frac{(G_1 x_\xi + G_4 x_\eta + G_5 x_\varsigma)}{g}; \quad K^2 = \frac{(G_4 x_\xi + G_2 x_\eta + G_6 x_\varsigma)}{g};$$

$$K^3 = \frac{(G_5 x_\xi + G_6 x_\eta + G_3 x_\varsigma)}{g}$$

$$L^1 = \frac{(G_1 y_\xi + G_4 y_\eta + G_5 y_\varsigma)}{g}; \quad L^2 = \frac{(G_4 y_\xi + G_2 y_\eta + G_6 y_\varsigma)}{g};$$

$$L^3 = \frac{(G_5 y_\xi + G_6 y_\eta + G_3 y_\varsigma)}{g}$$

$$M^1 = \frac{(G_1 z_\xi + G_4 z_\eta + G_5 z_\varsigma)}{g}; \quad M^2 = \frac{(G_4 z_\xi + G_2 z_\eta + G_6 z_\varsigma)}{g};$$

$$M^3 = \frac{(G_5 z_\xi + G_6 z_\eta + G_3 z_\varsigma)}{g}.$$

Expressions for the derivatives of contravariant base vectors $\mathbf{g}^i$ may be obtained by differentiating eqn (1.6) with respect to x^k, which gives

$$\mathbf{g}^i \cdot \frac{\partial \mathbf{g}_j}{\partial x^k} + \frac{\partial \mathbf{g}^i}{\partial x^k} \cdot \mathbf{g}_j = 0.$$

Hence

$$\frac{\partial \mathbf{g}^i}{\partial x^k} \cdot \mathbf{g}_j = -\mathbf{g}^i \cdot \frac{\partial \mathbf{g}_j}{\partial x^k} = -\mathbf{g}^i \cdot \Gamma^l_{jk}\mathbf{g}_l = -\delta^i_l \Gamma^l_{jk} = -\Gamma^i_{jk}. \tag{1.104}$$

Comparison with eqn (1.99) now shows that

$$\frac{\partial \mathbf{g}^i}{\partial x^j} = -\Gamma^i_{jk}\mathbf{g}^k. \tag{1.105}$$

The metric tensor g_{ij} can also be differentiated:

$$\begin{aligned}\frac{\partial g_{ij}}{\partial x^k} &= \frac{\partial}{\partial x^k}(\mathbf{g}_i \cdot \mathbf{g}_j) = \mathbf{g}_i \cdot \frac{\partial \mathbf{g}_j}{\partial x^k} + \frac{\partial \mathbf{g}_i}{\partial x^k} \cdot \mathbf{g}_j = \mathbf{g}_i \cdot [jk,l]\mathbf{g}^l + [ik,l]\mathbf{g}^l \cdot \mathbf{g}_j \\ &= [jk,l]\delta^l_i + [ik,l]\delta^l_j = [jk,i] + [ik,j]. \end{aligned} \tag{1.106}$$

Exercise 8. By differentiating both sides of eqn (1.28), and using eqns (1.101) and (1.106), show that

$$\frac{\partial g^{lm}}{\partial x^k} = -g^{jl}\Gamma^m_{jk} - g^{jm}\Gamma^l_{jk}. \tag{1.107}$$

By simply substituting the result (1.106) for $\partial g_{ij}/\partial x^k$ into the right-hand side of the following equation, it is easy to verify the important result

$$[ij,k] = \frac{1}{2}\left(\frac{\partial g_{jk}}{\partial x^i} + \frac{\partial g_{ik}}{\partial x^j} - \frac{\partial g_{ij}}{\partial x^k}\right). \tag{1.108}$$

Neither $[ij,k]$ nor Γ^k_{ij} is a third-order tensor. In a system of cartesian co-ordinates, with constant base vectors, all components of the Christoffel symbols are zero, and tensor components which are all zero would remain zero under a transformation to a different co-ordinate system. In fact the transformation rule for Γ^k_{ij} under transformation to co-ordinates $\overline{x}^i$ may be derived as follows, using eqns (1.62), (1.68), and (1.96):

$$\begin{aligned}
\overline{\Gamma}^k_{ij} &= \frac{\partial \overline{\mathbf{g}}_i}{\partial \overline{x}^j}\cdot \overline{\mathbf{g}}^k = \frac{\partial}{\partial \overline{x}^j}\left(\frac{\partial x^m}{\partial \overline{x}^i}\mathbf{g}_m\right)\cdot \frac{\partial \overline{x}^k}{\partial x^n}\mathbf{g}^n \\
&= \frac{\partial^2 x^m}{\partial \overline{x}^i \partial \overline{x}^j}\frac{\partial \overline{x}^k}{\partial x^n}\mathbf{g}_m\cdot\mathbf{g}^n + \frac{\partial x^m}{\partial \overline{x}^i}\frac{\partial \overline{x}^k}{\partial x^n}\left(\frac{\partial \mathbf{g}_m}{\partial x^l}\frac{\partial x^l}{\partial \overline{x}^j}\right)\cdot\mathbf{g}^n \\
&= \frac{\partial^2 x^m}{\partial \overline{x}^i \partial \overline{x}^j}\frac{\partial \overline{x}^k}{\partial x^n}\delta^n_m + \frac{\partial x^m}{\partial \overline{x}^i}\frac{\partial \overline{x}^k}{\partial x^n}\frac{\partial x^l}{\partial \overline{x}^j}\Gamma^n_{ml}
\end{aligned}$$

from eqn (1.99). Thus we have

$$\overline{\Gamma}^k_{ij} = \frac{\partial^2 x^m}{\partial \overline{x}^i \partial \overline{x}^j}\frac{\partial \overline{x}^k}{\partial x^m} + \frac{\partial \overline{x}^k}{\partial x^n}\frac{\partial x^m}{\partial \overline{x}^i}\frac{\partial x^l}{\partial \overline{x}^j}\Gamma^n_{ml}. \tag{1.109}$$

This equation does not follow the transformation rule for a mixed third-order tensor because of the presence of the first term on the right side.

A useful special case occurs when we let the new co-ordinates $\overline{x}^i$ coincide with the background rectangular cartesian co-ordinates y_1, y_2, y_3. The components of the Christoffel symbol associated with the new co-ordinates are then identically zero, and eqn (1.109) becomes

$$0 = \frac{\partial^2 x^m}{\partial y_i \partial y_j}\frac{\partial y_k}{\partial x^m} + \frac{\partial y_k}{\partial x^n}\frac{\partial x^m}{\partial y_i}\frac{\partial x^l}{\partial y_j}\Gamma^n_{ml}.$$

Multiplying through by $\partial x^p/\partial y_k$ (implying summation over k), using a chain rule again, and re-arranging, we obtain

$$\frac{\partial^2 x^p}{\partial y_i \partial y_j} = -\frac{\partial x^m}{\partial y_i}\frac{\partial x^l}{\partial y_j}\Gamma^p_{ml}. \tag{1.110}$$

Contraction on i and j (putting $j = i$, implying summation) gives the formulas

$$\frac{\partial^2 x^p}{\partial y_i \partial y_i} = \nabla^2 x^p = -\frac{\partial x^m}{\partial y_i}\frac{\partial x^l}{\partial y_i}\Gamma^p_{ml} = -g^{ml}\Gamma^p_{ml} = -g^{ml}\frac{\partial^2 y_n}{\partial x^m \partial x^l}\frac{\partial x^p}{\partial y_n}, \tag{1.111}$$

where ∇^2 is the Laplacian operator

$$\frac{\partial^2}{\partial y_i \partial y_i} = \frac{\partial^2}{\partial x^2} + \frac{\partial^2}{\partial y^2} + \frac{\partial^2}{\partial z^2}.$$

Equation (1.110) can be written, using eqn (1.102), as

$$\frac{\partial^2 x^p}{\partial y_i \partial y_j} = -\frac{\partial x^m}{\partial y_i}\frac{\partial x^l}{\partial y_j}\frac{\partial x^p}{\partial y_k}\frac{\partial^2 y_k}{\partial x^m \partial x^l}. \tag{1.112}$$

Another formula which will be found useful later may be derived directly from eqn (1.112), i.e.

$$\frac{\partial^2 x^p}{\partial y_i \partial y_j}\frac{\partial y_k}{\partial x^p} = -\frac{\partial x^m}{\partial y_i}\frac{\partial x^l}{\partial y_j}\frac{\partial^2 y_k}{\partial x^m \partial x^l} \tag{1.113}$$

Exercise 9. Making use of eqn (1.17), deduce that

$$(\nabla^2 x^p)\frac{\partial y_k}{\partial x^p} = -g^{ml}\frac{\partial^2 y_k}{\partial x^m \partial x^l}. \tag{1.114}$$

Exercise 10. Derive eqn (1.113) more directly by taking the partial derivative with respect to y_j of the Chain Rule

$$\frac{\partial x^p}{\partial y_i}\frac{\partial y_k}{\partial x^p} = \delta_{ik}.$$

The transformation rule for $[ij, k]$, by comparison with eqn (1.109), may be shown to be

$$\overline{[ij, k]} = \frac{\partial^2 x^m}{\partial \overline{x}^i \partial \overline{x}^j}\frac{\partial x^p}{\partial \overline{x}^k} g_{mp} + \frac{\partial x^m}{\partial \overline{x}^i}\frac{\partial x^n}{\partial \overline{x}^j}\frac{\partial x^p}{\partial \overline{x}^k}[mn, p]. \tag{1.115}$$

From eqns (1.100) and (1.108) we obtain

$$\Gamma^k_{ij} = \frac{1}{2}g^{kl}\left(\frac{\partial g_{jl}}{\partial x^i} + \frac{\partial g_{il}}{\partial x^j} - \frac{\partial g_{ij}}{\partial x^l}\right). \tag{1.116}$$

Contraction on i and k then gives, exploiting the symmetry of g_{ij} and g^{ij},

$$\Gamma^i_{ij} = \frac{1}{2}g^{il}\frac{\partial g_{il}}{\partial x^j}. \tag{1.117}$$

Now if we regard the determinant g of (g_{ij}) formally as a function of nine elements g_{ij} (replacing g_{12} with $\frac{1}{2}(g_{12} + g_{21})$, etc.), we have

$$\frac{\partial g}{\partial g_{il}} = G^{il} = g g^{il},$$

where (G^{ij}) is the matrix of co-factors of (g_{ij}) given in eqns (1.33) and (1.34). So another chain rule gives

$$\frac{\partial g}{\partial x^j} = \frac{\partial g}{\partial g_{il}}\frac{\partial g_{il}}{\partial x^j} = g g^{il}\frac{\partial g_{il}}{\partial x^j},$$

leading to the useful expressions

$$\Gamma^i_{ij} = \frac{1}{2}\frac{1}{g}\frac{\partial g}{\partial x^j} = \frac{1}{2}\frac{\partial}{\partial x^j}(\ln g) = \frac{1}{\sqrt{g}}\frac{\partial}{\partial x^j}(\sqrt{g}). \tag{1.118}$$

Differentiating a vector field $\mathbf{u}$ with respect to x^j gives

$$\frac{\partial \mathbf{u}}{\partial x^j} = \frac{\partial}{\partial x^j}(u^i \mathbf{g}_i) = \frac{\partial u^i}{\partial x^j}\mathbf{g}_i + u^i \frac{\partial \mathbf{g}_i}{\partial x^j} = \frac{\partial u^i}{\partial x^j}\mathbf{g}_i + u^i \Gamma^k_{ij}\mathbf{g}_k = \left(\frac{\partial u^i}{\partial x^j} + \Gamma^i_{kj}u^k\right)\mathbf{g}_i,$$

with some re-arrangement of indices. Thus $\partial \mathbf{u}/\partial x^j$ (itself a vector field) is given by

$$\frac{\partial \mathbf{u}}{\partial x^j} = u^i_{,j}\mathbf{g}_i, \tag{1.119}$$

where

$$u^i_{,j} = \frac{\partial u^i}{\partial x^j} + \Gamma^i_{kj}u^k \tag{1.120}$$

is called the *covariant derivative* of the contravariant vector u^i.

A similar calculation gives

$$\frac{\partial \mathbf{u}}{\partial x^j} = u_{i,j}\mathbf{g}^i, \tag{1.121}$$

where the covariant derivative of the covariant vector u_i is given by

$$u_{i,j} = \frac{\partial u_i}{\partial x^j} - \Gamma^k_{ij}u_k. \tag{1.122}$$

Exercise 11. Using the definitions (1.120) and (1.122) and the transformation rules (1.70), (1.72), and (1.109), show that $u^i_{,j}$ and $u_{i,j}$ satisfy the transformation rules for mixed and covariant second-order tensors, respectively.

These tensors are associated, since the equations

$$\frac{\partial \mathbf{u}}{\partial x^j} = u^i_{,j}\mathbf{g}_i = u_{i,j}\mathbf{g}^i = u_{i,j}g^{ik}\mathbf{g}_k$$

imply, comparing coefficients, that

$$u^i_{,j} = g^{ik}u_{k,j}, \tag{1.123}$$

after some re-arrangement of indices; or, more simply, since

$$u^i_{,j}\mathbf{g}_i \otimes \mathbf{g}^j = u_{i,j}\mathbf{g}^i \otimes \mathbf{g}^j.$$

Clearly we also have

$$u^i_{,j} = \frac{\partial \mathbf{u}}{\partial x^j}\cdot \mathbf{g}^i \quad \text{and } u_{i,j} = \frac{\partial \mathbf{u}}{\partial x^j}\cdot \mathbf{g}_i. \tag{1.124}$$

Covariant differentiation can also be applied to tensor fields. With a second-order tensor $\mathbf{T}$ as given in eqn (1.86), we give the following example:

$$\begin{aligned}\frac{\partial \mathbf{T}}{\partial x^k} &= \frac{\partial}{\partial x^k}(T_{ij}\mathbf{g}^i \otimes \mathbf{g}^j) = \frac{\partial T_{ij}}{\partial x^k}\mathbf{g}^i \otimes \mathbf{g}^j + T_{ij}\frac{\partial \mathbf{g}^i}{\partial x^k} \otimes \mathbf{g}^j + T_{ij}\mathbf{g}^i \otimes \frac{\partial \mathbf{g}^j}{\partial x^k}\\ &= \frac{\partial T_{ij}}{\partial x^k}\mathbf{g}^i \otimes \mathbf{g}^j - T_{ij}\Gamma^i_{kl}\mathbf{g}^l \otimes \mathbf{g}^j - T_{ij}\mathbf{g}^i \otimes \Gamma^j_{kl}\mathbf{g}^l\\ &= \left(\frac{\partial T_{ij}}{\partial x^k} - \Gamma^l_{ki}T_{lj} - \Gamma^l_{kj}T_{il}\right)\mathbf{g}^i \otimes \mathbf{g}^j,\end{aligned}$$

after some rearrangement of indices, with the help of eqn (1.105). Thus

$$\frac{\partial \mathbf{T}}{\partial x^k} = T_{ij,k}\mathbf{g}^i \otimes \mathbf{g}^j, \tag{1.125}$$

where

$$T_{ij,k} = \frac{\partial T_{ij}}{\partial x^k} - \Gamma^l_{ik}T_{lj} - \Gamma^l_{jk}T_{il} \tag{1.126}$$

is a covariant tensor of order three. For example, if we put $T_{ij} = g_{ij}$, it follows, using eqns (1.16) and (1.102) and substituting into (1.126), that

$$g_{ij,k} = 0 \tag{1.127}$$

for all i, j, k. This result follows naturally from the tensor properties of the covariant derivative and the fact that in cartesian co-ordinate systems covariant derivatives reduce to straightforward partial derivatives. Since g_{ij} takes constant values in a cartesian system, the partial derivatives of these values are all zero, and these will transform to zero under tensor transformation to any other co-ordinate system. It can be shown similarly that

$$g^{ij}_{,k} = 0 \tag{1.128}$$

for all i, j, k, where the covariant derivative of general contravariant components T^{ij} is given by

$$T^{ij}_{,k} = \frac{\partial T^{ij}}{\partial x^k} + \Gamma^i_{lk}T^{lj} + \Gamma^j_{lk}T^{il}. \tag{1.129}$$

Covariant derivatives of third-order tensors may also be defined, but it will suffice here to mention the alternating tensor, which could be written as

$$\varepsilon^{ijk}\mathbf{g}_i \otimes \mathbf{g}_j \otimes \mathbf{g}_k = \varepsilon_{ijk}\mathbf{g}^i \otimes \mathbf{g}^j \otimes \mathbf{g}^k.$$

Since both covariant and contravariant components reduce to the array of constants (1.89) in a cartesian system, a similar argument to that used above for g^{ij} shows that the covariant derivatives must vanish, i.e.

$$\varepsilon^{ijk}_{,l} = 0 \tag{1.130}$$

and

$$\varepsilon_{ijk,l} = 0 \tag{1.131}$$

for all i, j, k, l.

It may be shown that the product rule for differentiation is valid for covariant differentiation; for example,

$$(T^{ij}u_k)_{,l} = T^{ij}u_{k,l} + T^{ij}_{,l}u_k.$$

1.7 Div, grad, and curl

The *divergence* of a vector field **u**, where

$$\mathbf{u} = U_1\mathbf{i}_1 + U_2\mathbf{i}_2 + U_3\mathbf{i}_3, \tag{1.132}$$

referred to background cartesian co-ordinates $\{y_i\}$, is the scalar defined by div $\mathbf{u} = \partial U_i/\partial y_i$, otherwise denoted by $\nabla \cdot \mathbf{u}$, with summation over i. In general curvilinear co-ordinates this transforms to the sum (summation over i) of covariant derivatives

$$\nabla \cdot \mathbf{u} = u^i_{,i}. \tag{1.133}$$

We may also deduce from eqn (1.119) that

$$\nabla \cdot \mathbf{u} = \mathbf{g}^i \cdot \frac{\partial \mathbf{u}}{\partial x^i}. \tag{1.134}$$

By eqns (1.120) and (1.118), we have

$$\begin{aligned}\nabla \cdot \mathbf{u} &= \frac{\partial u^i}{\partial x^i} + \Gamma^i_{ki} u^k = \frac{\partial u^i}{\partial x^i} + \Gamma^i_{ik} u^k \\ &= \frac{\partial u^i}{\partial x^i} + \frac{1}{\sqrt{g}} \frac{\partial}{\partial x^k} (\sqrt{g}) u^k = \frac{\partial u^i}{\partial x^i} + \frac{1}{\sqrt{g}} \frac{\partial}{\partial x^i} (\sqrt{g}) u^i,\end{aligned}$$

which gives the useful formula

$$\nabla \cdot \mathbf{u} = \frac{1}{\sqrt{g}} \frac{\partial}{\partial x^i} (\sqrt{g} u^i), \tag{1.135}$$

with summation over i. This is an expression for the divergence *in conservative form*. In general, conservative form is preferred for operator expressions when numerically solving partial differential equations (in particular, transport equations in fluid flow problems) because numerical accuracy is enhanced. More examples are given below.

A vector identity which recurs frequently in the following is:

$$\frac{\partial}{\partial x^1} (\mathbf{g}_2 \times \mathbf{g}_3) + \frac{\partial}{\partial x^2} (\mathbf{g}_3 \times \mathbf{g}_1)^- + \frac{\partial}{\partial x^3} (\mathbf{g}_1 \times \mathbf{g}_2) = \mathbf{0}. \tag{1.136}$$

Exercise 12. By writing each $\mathbf{g}_i$ as the appropriate $\partial \mathbf{r}/\partial x^i$, performing the differentiations using product rules, and finally exploiting the skew-symmetry of the vector product, verify eqn (1.136).

We write eqn (1.136) as

$$\sum_{i=1}^{3} \frac{\partial}{\partial x^i} (\mathbf{g}_j \times \mathbf{g}_k) = \mathbf{0}, \tag{1.137}$$

where for each i it is assumed that j and k are such that i, j, k are in cyclic order 1, 2, 3.

Now from eqns (1.135), (1.48) and (1.32) we have the two conservative forms

$$\nabla \cdot \mathbf{u} = \frac{1}{\sqrt{g}} \frac{\partial}{\partial x^i} (\sqrt{g} \mathbf{g}^i \cdot \mathbf{u}) = \frac{1}{\sqrt{g}} \sum_{i=1}^{3} \frac{\partial}{\partial x^i} \{(\mathbf{g}_j \times \mathbf{g}_k) \cdot \mathbf{u}\} \tag{1.138}$$

and the *non-conservative form*

$$\nabla \cdot \mathbf{u} = \frac{1}{\sqrt{g}} \sum_{i=1}^{3} (\mathbf{g}_j \times \mathbf{g}_k) \cdot \frac{\partial \mathbf{u}}{\partial x^i}, \tag{1.139}$$

using eqn (1.137).

The gradient operator was defined in eqns (1.12) and (1.13). We can also write

$$\nabla\varphi = \frac{1}{\sqrt{g}}\sum_{i=1}^{3}(\mathbf{g}_j \times \mathbf{g}_k)\frac{\partial\varphi}{\partial x^i} \tag{1.140}$$

in non-conservative form, using eqn (1.32), or, by eqn (1.137),

$$\nabla\varphi = \frac{1}{\sqrt{g}}\sum_{i=1}^{3}\frac{\partial}{\partial x^i}\{(\mathbf{g}_j \times \mathbf{g}_k)\varphi\} = \frac{1}{\sqrt{g}}\frac{\partial}{\partial x^i}(\sqrt{g}\mathbf{g}^i\varphi) \tag{1.141}$$

in conservative form, where again i, j, k when they appear together are always in cyclic order 1, 2, 3.

The *curl* of the vector with cartesian components in eqn (1.132) is the vector

$$curl\mathbf{u} = \begin{vmatrix} \mathbf{i}_1 & \mathbf{i}_2 & \mathbf{i}_3 \\ \partial/\partial y_1 & \partial/\partial y_2 & \partial/\partial y_3 \\ U_1 & U_2 & U_3 \end{vmatrix},$$

otherwise denoted by $\nabla \times \mathbf{u}$; this expression is equivalent to $e_{ijk}\frac{\partial U_k}{\partial y_j}\mathbf{i}_i$ with summation over i, j, k, which, making use of (1.93), generalizes to

$$\nabla \times \mathbf{u} = \varepsilon^{ijk}u_{k,j}\mathbf{g}_i = \frac{1}{\sqrt{g}}e^{ijk}u_{k,j}\mathbf{g}_i \tag{1.142}$$

in curvilinear co-ordinates. Since $u_{k,j} = \mathbf{g}_k \cdot \partial\mathbf{u}/x^j$, we have, writing out eqn (1.142) in full,

$$\nabla \times \mathbf{u} = \frac{1}{\sqrt{g}}(u_{2,3}\mathbf{g}_1 - u_{3,2}\mathbf{g}_1 + u_{3,1}\mathbf{g}_2 - u_{1,3}\mathbf{g}_2 + u_{1,2}\mathbf{g}_3 - u_{2,1}\mathbf{g}_3).$$

The expression in the brackets is

$$\left[\left(\mathbf{g}_2 \cdot \frac{\partial\mathbf{u}}{\partial x^3}\right)\mathbf{g}_1 - \left(\mathbf{g}_1 \cdot \frac{\partial\mathbf{u}}{\partial x^3}\right)\mathbf{g}_2 + \left(\mathbf{g}_3 \cdot \frac{\partial\mathbf{u}}{\partial x^1}\right)\mathbf{g}_2 - \left(\mathbf{g}_2 \cdot \frac{\partial\mathbf{u}}{\partial x^1}\right)\mathbf{g}_3 + \left(\mathbf{g}_1 \cdot \frac{\partial\mathbf{u}}{\partial x^2}\right)\mathbf{g}_3 - \left(\mathbf{g}_3 \cdot \frac{\partial\mathbf{u}}{\partial x^2}\right)\mathbf{g}_1\right]$$

after some re-arrangement, so $\nabla \times \mathbf{u}$ can be written as

$$\nabla \times \mathbf{u} = \frac{1}{\sqrt{g}}\sum_{i=1}^{3}\left\{(\mathbf{g}_j \times \mathbf{g}_k) \times \frac{\partial\mathbf{u}}{\partial x^i}\right\}, \tag{1.143}$$

after using well-known identities for vector triple products. Here again j and k are constrained, given any value of i, such that i, j, k are always in cyclic order 1, 2, 3. Equation (1.143) is a non-conservative form for $\nabla \times \mathbf{u}$. However, by eqn (1.137) we immediately have the conservative forms

$$\nabla \times \mathbf{u} = \frac{1}{\sqrt{g}}\sum_{i=1}^{3}\frac{\partial}{\partial x^i}\{(\mathbf{g}_j \times \mathbf{g}_k) \times \mathbf{u}\} \tag{1.144}$$

and, using eqn (1.32),

$$\nabla \times \mathbf{u} = \frac{1}{\sqrt{g}} \frac{\partial}{\partial x^i} (\sqrt{g} \mathbf{g}^i \times \mathbf{u}). \tag{1.145}$$

Note that, again by eqn (1.32), eqn (1.143) can be written as

$$\nabla \times \mathbf{u} = \mathbf{g}^i \times \frac{\partial \mathbf{u}}{\partial x^i} \tag{1.146}$$

with summation over i, which may be directly compared with eqn (1.134).

To obtain an expression for the *Laplacian* $\nabla^2 \varphi$ of a scalar field φ, where $\nabla^2 \varphi = \nabla \cdot (\nabla \varphi)$, using eqns (1.133) or (1.135), the contravariant component of $\nabla \varphi$ is needed. This is just

$$(\nabla \varphi)^i = g^{ij} \frac{\partial \varphi}{\partial x^j},$$

where the effect of the g^{ij} term is to 'raise the index' of the covariant vector $\partial \varphi / \partial x^j$. Then eqn (1.135) gives

$$\nabla^2 \varphi = \frac{1}{\sqrt{g}} \frac{\partial}{\partial x^i} \left(\sqrt{g} g^{ij} \frac{\partial \varphi}{\partial x^j} \right). \tag{1.147}$$

Alternatively, we have, using the expressions for div and grad in eqns (1.13) and (1.134),

$$\nabla^2 \varphi = \mathbf{g}^i \cdot \frac{\partial}{\partial x^i} \left(\mathbf{g}^j \frac{\partial \varphi}{\partial x^j} \right) \tag{1.148}$$

with summation over both i and j. Hence

$$\nabla^2 \varphi = \mathbf{g}^i \cdot \mathbf{g}^j \frac{\partial^2 \varphi}{\partial x^i \partial x^j} + \mathbf{g}^i \cdot \frac{\partial \mathbf{g}^j}{\partial x^i} \frac{\partial \varphi}{\partial x^j}. \tag{1.149}$$

But since, from eqn (1.148),

$$\nabla^2 x^k = \mathbf{g}^i \cdot \frac{\partial}{\partial x^i} \left(\mathbf{g}^j \frac{\partial x^k}{\partial x^j} \right) = \mathbf{g}^i \cdot \frac{\partial}{\partial x^i} (\mathbf{g}^j \delta^k_j) = \mathbf{g}^i \cdot \frac{\partial \mathbf{g}^k}{\partial x^i},$$

the identity (1.149) may be written in the form

$$\nabla^2 \varphi = g^{ij} \frac{\partial^2 \varphi}{\partial x^i \partial x^j} + (\nabla^2 x^j) \frac{\partial \varphi}{\partial x^j}. \tag{1.150}$$

Substituting $\varphi = x^k$ in eqn (1.147) gives another formula

$$\nabla^2 x^k = \frac{1}{\sqrt{g}} \frac{\partial}{\partial x^i} (\sqrt{g} g^{ij} \delta^k_j) = \frac{1}{\sqrt{g}} \frac{\partial}{\partial x^i} (\sqrt{g} g^{ik}). \tag{1.151}$$

Thus, eqn (1.147) also gives

$$\begin{aligned} \nabla^2 \varphi &= \frac{1}{\sqrt{g}} \frac{\partial}{\partial x^i} \left[\frac{\partial}{\partial x^j} (\sqrt{g} g^{ij} \varphi) - \varphi \frac{\partial}{\partial x^j} (\sqrt{g} g^{ij}) \right] \\ &= \frac{1}{\sqrt{g}} \frac{\partial^2}{\partial x^i \partial x^j} (\sqrt{g} g^{ij} \varphi) - \frac{1}{\sqrt{g}} \frac{\partial}{\partial x^i} (\varphi \sqrt{g} (\nabla^2 x^i)). \end{aligned} \tag{1.152}$$

We have seen in eqn (1.125) that, for a second-order tensor $\mathbf{T}$, $\partial\mathbf{T}/\partial x^k$ can be regarded as a linear operator, acting on vectors in E^3 to give vectors in E^3. When it acts on the contravariant base vector $\mathbf{g}^k$, the resulting vector is called the *divergence* of $\mathbf{T}$, and we write

$$\nabla \cdot \mathbf{T} = \left(\frac{\partial \mathbf{T}}{\partial x^k}\right)\mathbf{g}^k \tag{1.153}$$

with summation over k. In other words,

$$\nabla \cdot \mathbf{T} = (T^{ij}_{\ ,k}\mathbf{g}_i \otimes \mathbf{g}_j)\mathbf{g}^k = T^{ij}_{\ ,k}\mathbf{g}_i(\mathbf{g}_j \cdot \mathbf{g}^k) = T^{ij}_{\ ,k}\mathbf{g}_i\delta^k_j = T^{ij}_{\ ,j}\mathbf{g}_i, \tag{1.154}$$

expressed in terms of the covariant derivatives of the contravariant components of $\mathbf{T}$.

Exercise 13. Verify the formulas

$$\nabla \cdot \mathbf{T} = \frac{1}{\sqrt{g}}\frac{\partial}{\partial x^j}(\sqrt{g}T^{ij}\mathbf{g}_i) = \left(\frac{1}{\sqrt{g}}\frac{\partial}{\partial x^j}(\sqrt{g}T^{ij}) + T^{kj}\Gamma^i_{kj}\right)\mathbf{g}_i. \tag{1.155}$$

Exercise 14. Show that if p is a scalar field and $\mathbf{I}$ is the unit second-order tensor (defined in eqn (1.87)), then

$$\nabla \cdot (p\mathbf{I}) = \nabla p. \tag{1.156}$$

1.8 Summary of formulas in two dimensions

For two-dimensional situations in which field variables depend only on the rectangular cartesian co-ordinates x and y but not z, it is straightforward to establish the reduced form of the above results. We give a summary here of some of the main results for convenience.

With $y_1 = x$, $y_2 = y$, and curvilinear co-ordinates with $x^1 = \xi$, $x^2 = \eta$ (and occasionally finding it useful to put $y_3 = x^3 = z = \varsigma$), we have base vectors

$$\mathbf{g}_1 = \mathbf{i}x_\xi + \mathbf{j}y_\xi, \quad \mathbf{g}_2 = \mathbf{i}x_\eta + \mathbf{j}y_\eta, \quad \mathbf{g}_3 = \mathbf{k}, \tag{1.157}$$

where suffixes denote partial differentiation, e.g. $x_\xi = \partial x/\partial\xi$. The components of the covariant metric tensor are given by

$$\begin{pmatrix} g_{11} & g_{12} & g_{13} \\ g_{12} & g_{22} & g_{23} \\ g_{13} & g_{23} & g_{33} \end{pmatrix} = \begin{pmatrix} x_\xi^2 + y_\xi^2 & x_\xi x_\eta + y_\xi y_\eta & 0 \\ x_\xi x_\eta + y_\xi y_\eta & x_\eta^2 + y_\eta^2 & 0 \\ 0 & 0 & 1 \end{pmatrix}, \tag{1.158}$$

with determinant

$$g = g_{11}g_{22} - (g_{12})^2. \tag{1.159}$$

Moreover

$$\sqrt{g} = \mathbf{g}_1 \times \mathbf{g}_2 \cdot \mathbf{g}_3 = \begin{vmatrix} x_\xi & x_\eta & 0 \\ y_\xi & y_\eta & 0 \\ 0 & 0 & 1 \end{vmatrix} = (x_\xi y_\eta - x_\eta y_\xi). \tag{1.160}$$

The contravariant base vectors are

$$\begin{aligned}
\mathbf{g}^1 &= \frac{1}{\sqrt{g}}\mathbf{g}_2 \times \mathbf{k} = \frac{1}{\sqrt{g}}(\mathbf{i}y_\eta - \mathbf{j}x_\eta), \\
\mathbf{g}^2 &= \frac{1}{\sqrt{g}}\mathbf{k} \times \mathbf{g}_1 = \frac{1}{\sqrt{g}}(-\mathbf{i}y_\xi + \mathbf{j}x_\xi), \\
\mathbf{g}^3 &= \mathbf{k},
\end{aligned} \tag{1.161}$$

and in comparison with eqn (1.38) we have

$$\xi_x = y_\eta/\sqrt{g}, \quad \xi_y = -x_\eta/\sqrt{g}, \quad \eta_x = -y_\xi/\sqrt{g}, \quad \eta_y = x_\xi/\sqrt{g}. \tag{1.162}$$

The components of the contravariant metric tensor are given by

$$\begin{pmatrix} g^{11} & g^{12} & g^{13} \\ g^{12} & g^{22} & g^{23} \\ g^{13} & g^{23} & g^{33} \end{pmatrix} = \begin{pmatrix} g_{22}/g & -g_{12}/g & 0 \\ -g_{12}/g & g_{11}/g & 0 \\ 0 & 0 & 1 \end{pmatrix}. \tag{1.163}$$

Note that the two-dimensional version of eqn (1.136) is

$$\frac{\partial}{\partial \xi}(\sqrt{g}\mathbf{g}^1) + \frac{\partial}{\partial \eta}(\sqrt{g}\mathbf{g}^2) = \frac{\partial}{\partial \xi}(\mathbf{g}^2 \times \mathbf{k}) + \frac{\partial}{\partial \eta}(\mathbf{k} \times \mathbf{g}^1) = \mathbf{0}. \tag{1.164}$$

From eqn (1.141) we have the two-dimensional form for $\nabla\varphi$:

$$\begin{aligned}
\nabla\varphi &= \frac{1}{\sqrt{g}}\left\{\frac{\partial}{\partial \xi}(\sqrt{g}\mathbf{g}^1\varphi) + \frac{\partial}{\partial \eta}(\sqrt{g}\mathbf{g}^2\varphi)\right\} \\
&= \frac{1}{\sqrt{g}}\left\{\frac{\partial}{\partial \xi}[(\mathbf{i}y_\eta - \mathbf{j}x_\eta)\varphi]\right\} + \frac{1}{\sqrt{g}}\left\{\frac{\partial}{\partial \eta}[-\mathbf{i}y_\xi + \mathbf{j}x_\xi]\varphi\right\} \\
&= \mathbf{i}\frac{1}{\sqrt{g}}\left\{\frac{\partial}{\partial \xi}(y_\eta\varphi) - \frac{\partial}{\partial \eta}(y_\xi\varphi)\right\} + \mathbf{j}\frac{1}{\sqrt{g}}\left\{-\frac{\partial}{\partial \xi}(x_\eta\varphi) + \frac{\partial}{\partial \eta}(x_\xi\varphi)\right\}
\end{aligned} \tag{1.165}$$

in conservative form. Thus the cartesian components of $\nabla\varphi$ are given by

$$\frac{\partial \varphi}{\partial x} = \frac{1}{\sqrt{g}}\left\{\frac{\partial}{\partial \xi}(y_\eta\varphi) - \frac{\partial}{\partial \eta}(y_\xi\varphi)\right\}$$

and

$$\frac{\partial \varphi}{\partial y} = \frac{1}{\sqrt{g}}\left\{-\frac{\partial}{\partial \xi}(x_\eta\varphi) + \frac{\partial}{\partial \eta}(x_\xi\varphi)\right\} \tag{1.166}$$

in conservative form. By further differentiation, or directly from eqn (1.13), we obtain the non-conservative forms

$$\frac{\partial \varphi}{\partial x} = \frac{1}{\sqrt{g}}\left(y_\eta\frac{\partial \varphi}{\partial \xi} - y_\xi\frac{\partial \varphi}{\partial \eta}\right)$$

and

$$\frac{\partial \varphi}{\partial y} = \frac{1}{\sqrt{g}}\left(-x_\eta\frac{\partial \varphi}{\partial \xi} + x_\xi\frac{\partial \varphi}{\partial \eta}\right). \tag{1.167}$$

From eqns (1.138) and (1.161) we deduce a conservative form for the divergence $\nabla \cdot \mathbf{u}$:

$$\begin{aligned}\nabla \cdot \mathbf{u} &= \frac{1}{\sqrt{g}}\left\{\frac{\partial}{\partial \xi}(\sqrt{g}\mathbf{g}^1 \cdot \mathbf{u}) + \frac{\partial}{\partial \eta}(\sqrt{g}\mathbf{g}^2 \cdot \mathbf{u})\right\} \\ &= \frac{1}{\sqrt{g}}\left\{\frac{\partial}{\partial \xi}(y_\eta U_1 - x_\eta U_2) + \frac{\partial}{\partial \eta}(-y_\xi U_1 + x_\xi U_2)\right\}, \end{aligned} \tag{1.168}$$

where $\mathbf{u} = U_1\mathbf{i} + U_2\mathbf{j}$. Again by further differentiation, or directly from eqn (1.134), we deduce the non-conservative form:

$$\nabla \cdot \mathbf{u} = \frac{1}{\sqrt{g}}\left(y_\eta \frac{\partial U_1}{\partial \xi} - x_\eta \frac{\partial U_2}{\partial \xi} - y_\xi \frac{\partial U_1}{\partial \eta} + x_\xi \frac{\partial U_2}{\partial \eta}\right). \tag{1.169}$$

A similar treatment of eqns (1.145) and (1.146) yields a conservative form for $\nabla \times \mathbf{u}$:

$$\nabla \times \mathbf{u} = \mathbf{k}\frac{1}{\sqrt{g}}\left\{\frac{\partial}{\partial \xi}(x_\eta U_1 + y_\eta U_2) - \frac{\partial}{\partial \eta}(x_\xi U_1 + y_\xi U_2)\right\} \tag{1.170}$$

and the non-conservative form

$$\nabla \times \mathbf{u} = \mathbf{k}\frac{1}{\sqrt{g}}\left(x_\eta \frac{\partial U_1}{\partial \xi} + y_\eta \frac{\partial U_2}{\partial \xi} - x_\xi \frac{\partial U_1}{\partial \eta} - y_\xi \frac{\partial U_2}{\partial \eta}\right). \tag{1.171}$$

Making use of both eqns (1.166) and (1.168), we obtain the two-dimensional Laplacian $\nabla^2\varphi = \nabla \cdot (\nabla\varphi)$ in conservative form:

$$\begin{aligned}\nabla^2\varphi = &\frac{1}{\sqrt{g}}\frac{\partial}{\partial \xi}\left[\frac{y_\eta}{\sqrt{g}}\left(\frac{\partial}{\partial \xi}[y_\eta\varphi] - \frac{\partial}{\partial \eta}[y_\xi\varphi]\right) - \frac{x_\eta}{\sqrt{g}}\left(-\frac{\partial}{\partial \xi}[x_\eta\varphi] + \frac{\partial}{\partial \eta}[x_\xi\varphi]\right)\right] \\ &+\frac{1}{\sqrt{g}}\frac{\partial}{\partial \eta}\left[-\frac{y_\xi}{\sqrt{g}}\left(\frac{\partial}{\partial \xi}[y_\eta\varphi] - \frac{\partial}{\partial \eta}[y_\xi\varphi]\right) + \frac{x_\xi}{\sqrt{g}}\left(-\frac{\partial}{\partial \xi}[x_\eta\varphi] + \frac{\partial}{\partial \eta}[x_\xi\varphi]\right)\right].\end{aligned} \tag{1.172}$$

Note that eqn (1.152) also gives, making use of eqn (1.163), the form

$$\begin{aligned}\nabla^2\varphi = \frac{1}{\sqrt{g}}\Bigg\{&\frac{\partial^2}{\partial \xi^2}\left(\frac{g_{22}}{\sqrt{g}}\varphi\right) - 2\frac{\partial^2}{\partial \xi \partial \eta}\left(\frac{g_{12}}{\sqrt{g}}\varphi\right) + \frac{\partial^2}{\partial \eta^2}\left(\frac{g_{11}}{\sqrt{g}}\varphi\right) \\ &-\frac{\partial}{\partial \xi}[(\sqrt{g}\nabla^2\xi)\varphi] - \frac{\partial}{\partial \eta}[(\sqrt{g}\nabla^2\eta)\varphi]\Bigg\}.\end{aligned} \tag{1.173}$$

Exercise 15. Using eqns (1.111), (1.162), and (1.163), show that

$$\nabla^2\xi = \frac{g_{22}}{g^{3/2}}(x_\eta y_{\xi\xi} - y_\eta x_{\xi\xi}) - 2\frac{g_{12}}{g^{3/2}}(x_\eta y_{\xi\eta} - y_\eta x_{\xi\eta}) + \frac{g_{11}}{g^{3/2}}(x_\eta y_{\eta\eta} - y_\eta x_{\eta\eta}), \tag{1.174}$$

$$\nabla^2\eta = \frac{g_{22}}{g^{3/2}}(y_\xi x_{\xi\xi} - x_\xi y_{\xi\xi}) - 2\frac{g_{12}}{g^{3/2}}(y_\xi x_{\xi\eta} - x_\xi y_{\xi\eta}) + \frac{g_{11}}{g^{3/2}}(y_\xi x_{\eta\eta} - x_\xi y_{\eta\eta}). \tag{1.175}$$

1.9 The Riemann-Christoffel tensor

The covariant third-order tensor $u_{i,jk}$ formed from the covariant components u_i of a vector by using eqn (1.126) to obtain the covariant derivatives of the covariant second-order tensor $u_{i,j}$ given by eqn (1.122) is found to be

$$u_{i,jk} = \frac{\partial^2 u_i}{\partial x^k \partial x^j} - \Gamma^l_{ij}\frac{\partial u_l}{\partial x^k} - \Gamma^l_{jk}\frac{\partial u_i}{\partial x^l} - \Gamma^l_{ik}\frac{\partial u_l}{\partial x^j} - \frac{\partial \Gamma^l_{ij}}{\partial x^k}u_l + \Gamma^l_{jk}\Gamma^m_{il}u_m + \Gamma^l_{ik}\Gamma^m_{lj}u_m. \tag{1.176}$$

We can investigate the commutativity of successive covariant differentiations by subtracting from this expression a similar one with j and k interchanged. This gives

$$u_{i,jk} - u_{i,kj} = \left(\frac{\partial \Gamma^l_{ik}}{\partial x^j} - \frac{\partial \Gamma^l_{ij}}{\partial x^k} + \Gamma^m_{ik}\Gamma^l_{mj} - \Gamma^m_{ij}\Gamma^l_{mk}\right)u_l = R^l_{.ijk}u_l, \tag{1.177}$$

where

$$R^l_{.ijk} = \frac{\partial \Gamma^l_{ik}}{\partial x^j} - \frac{\partial \Gamma^l_{ij}}{\partial x^k} + \Gamma^m_{ik}\Gamma^l_{mj} - \Gamma^m_{ij}\Gamma^l_{mk}. \tag{1.178}$$

The left-hand side of eqn (1.177) represents a covariant third-order tensor, and it follows that, for the right-hand side also to represent a tensor, $R^l_{.ijk}$ must be a mixed fourth-order tensor. It is called the *Riemann-Christoffel tensor*. In fact, since our background space is Euclidean and the Christoffel symbols all vanish when we take a rectangular cartesian set of co-ordinates, all the components of $R^l_{.ijk}$ will also be zero in cartesian co-ordinates, and, moreover, will always transform to zero under tensor transformation rules for any other choice of co-ordinates. We may also prove that

$$R^l_{.ijk} = 0 \tag{1.179}$$

in E^3 for all i, j, k, l, by substituting directly for Γ^l_{ik} from eqn (1.102) (which also assumes the existence of a background cartesian system) into eqn (1.178).

It will not be necessary here to consider non-Euclidean three-dimensional spaces, in which non-vanishing Riemann-Christoffel tensors may exist, but in Chapter 3 we shall need to consider a two-dimensional version of eqn (1.178) on an arbitrary curved surface within a three-dimensional Euclidean space. So we now set out some general results for three-dimensional non-Euclidean Riemann-Christoffel tensors (as defined by eqn (1.178)) to serve as a basis for comparison with the Riemann-Christoffel tensor defined on a curved two-dimensional surface, to appear later. First we note that the covariant associated tensor R_{ijkl}, called the *curvature tensor*, given by

$$R_{ijkl} = g_{im}R^m_{..jkl} \tag{1.180}$$

may be defined. Using eqns (1.100), (1.108), the symmetry of g_{ij} and g^{ij}, and the relation $g_{im}g^{jm} = \delta^j_i$, it may be shown that

$$R_{ijkl} = \frac{1}{2}\left(\frac{\partial^2 g_{il}}{\partial x^j \partial x^k} + \frac{\partial^2 g_{jk}}{\partial x^i \partial x^l} - \frac{\partial^2 g_{ik}}{\partial x^j \partial x^l} - \frac{\partial^2 g_{jl}}{\partial x^i \partial x^k}\right)$$
$$+ g^{mn}([jk, m][il, n] - [jl, m][ik, n]). \tag{1.181}$$

It follows that R_{ijkl} is skew-symmetric in the first two indices as well as the last two, that is

$$R_{ijkl} = -R_{ijlk}$$

and

$$R_{ijkl} = -R_{jikl}.$$

Furthermore

$$R_{ijkl} = R_{klij}.$$

Hence R_{ijkl} in any curved three-dimensional space has only six independent components, namely R_{1212}, R_{2323}, R_{1313}, R_{1213}, R_{1223}, and R_{2313}.

In a Euclidean space, by eqns (1.179) and (1.180) all components of R_{ijkl} are zero, and so, according to eqn (1.177), second-order covariant derivatives of arbitrary covariant vectors u_i satisfy

$$u_{i,jk} = u_{i,kj}.$$

It can be shown that a similar commutative property applies to second-order covariant derivatives of covariant or contravariant tensors of any order in a Euclidean space.

1.10 Orthogonal curvilinear co-ordinates

A curvilinear co-ordinate system $\{x^i\} = \{\xi, \eta, \varsigma\}$ is orthogonal if the covariant base vectors at any point are mutually orthogonal. It follows that the contravariant base vectors are parallel to their respective covariant base vectors and also mutually orthogonal. So we have

$$g_{12} = g_{23} = g_{13} = 0 \tag{1.182}$$

and

$$g^{12} = g^{23} = g^{13} = 0.$$

An example mentioned above is a spherical polar co-ordinate system with metric tensor given by eqn (1.19).

It is convenient to put

$$\sqrt{g_{11}} = h_1, \quad \sqrt{g_{22}} = h_2, \quad \sqrt{g_{33}} = h_3, \tag{1.183}$$

with

$$\sqrt{g^{11}} = 1/h_1, \quad \sqrt{g^{22}} = 1/h_2, \quad \sqrt{g^{33}} = 1/h_3,$$

where h_1, h_2, h_3 are called *scale factors*. Then $g = g_{11}\, g_{22}\, g_{33}$ and

$$\sqrt{g} = h_1 h_2 h_3. \tag{1.184}$$

Using eqns (1.108) and (1.100) to evaluate Christoffel symbols, we get the useful expressions

$$[ii, i] = h_i \frac{\partial h_i}{\partial x^i}, \quad \Gamma^i_{ii} = \frac{1}{h_i} \frac{\partial h_i}{\partial x^i}, \quad \text{(no summation over } i\text{)} \tag{1.185}$$

$$[ii, j] = -h_i \frac{\partial h_i}{\partial x^j},$$

$$\Gamma^j_{ii} = -\frac{h_i}{(h_j)^2} \frac{\partial h_i}{\partial x^j}, \quad (i, j \text{ different, no summation over } i) \tag{1.186}$$

$$[ij, i] = [ji, i] = h_i \frac{\partial h_i}{\partial x^j},$$

$$\Gamma^i_{ij} = \Gamma^i_{ji} = \frac{1}{h_i} \frac{\partial h_i}{\partial x^j}, \quad (i, j \text{ different, no summation over } i) \tag{1.187}$$

$$[ij, k] = 0, \quad \Gamma^k_{ij} = 0, \quad (i, j, k \text{ all different}). \tag{1.188}$$

Exercise 16. Use the identity $\nabla^2 x^m = -g^{ij}\Gamma^m_{ij}$ (see eqn (1.111)) to prove the identities

$$\nabla^2 \xi = \frac{1}{\sqrt{g}} \frac{\partial}{\partial \xi}\left(\frac{h_2 h_3}{h_1}\right), \quad \nabla^2 \eta = \frac{1}{\sqrt{g}} \frac{\partial}{\partial \eta}\left(\frac{h_3 h_1}{h_2}\right), \quad \nabla^2 \varsigma = \frac{1}{\sqrt{g}} \frac{\partial}{\partial \varsigma}\left(\frac{h_1 h_2}{h_3}\right). \tag{1.189}$$

Note that these formulas can be substituted directly into eqn (1.150) to obtain an expression for the Laplacian $\nabla^2 \phi$ of a general scalar field ϕ in an orthogonal curvilinear co-ordinate system.

Exercise 17. Use eqn (1.189) together with eqn (1.114) to show that there is a relationship between the orthogonal curvilinear co-ordinates and cartesian co-ordinates y_k, $k = 1, 2, 3$, given by

$$\frac{\partial}{\partial \xi}\left(\frac{h_2 h_3}{h_1} \frac{\partial y_k}{\partial \xi}\right) + \frac{\partial}{\partial \eta}\left(\frac{h_3 h_1}{h_2} \frac{\partial y_k}{\partial \eta}\right) + \frac{\partial}{\partial \varsigma}\left(\frac{h_1 h_2}{h_3} \frac{\partial y_k}{\partial \varsigma}\right) = 0, \quad k = 1, 2, 3. \tag{1.190}$$

More identities may be obtained from the vanishing of the six independent components of the curvature tensor given by eqn (1.181). These are the six *Lamé Equations:*

$$\frac{\partial}{\partial x^j}\left(\frac{1}{h_j} \frac{\partial h_k}{\partial x^j}\right) + \frac{\partial}{\partial x^k}\left(\frac{1}{h_k} \frac{\partial h_j}{\partial x^k}\right) + \frac{1}{(h_i)^2} \frac{\partial h_j}{\partial x^i} \frac{\partial h_k}{\partial x^i} = 0,$$

$$\frac{1}{h_j} \frac{\partial h_i}{\partial x^j} \frac{\partial h_j}{\partial x^k} + \frac{1}{h_k} \frac{\partial h_i}{\partial x^k} \frac{\partial h_k}{\partial x^j} - \frac{\partial^2 h_i}{\partial x^j \partial x^k} = 0, \tag{1.191}$$

where in each equation i, j, k must all be different and taken in the cyclic order 1, 2, 3. So the three scale parameters must satisfy six compatibility equations.

1.11 Tangential and normal derivatives – an introduction

The rates of change of scalar functions in directions tangential to co-ordinate curves and normal to co-ordinate surfaces are often needed in grid-generation work in connection with the formulation of boundary conditions. These derivatives may be obtained by taking the scalar product of the gradient vector of the given function with a unit vector in the required direction.

Given a set of curvilinear co-ordinates ξ, η, ς in E^3, consider the ξ co-ordinate curve (on which η and ς are constant) at some point P in space. Tangential to the curve is the covariant base vector $\mathbf{g}_1$, and a unit vector in this direction is given by $\mathbf{g}_1/|\mathbf{g}_1|$. Thus the *tangential derivative* of the scalar function φ at the point P in the ξ-direction is given by

$$\left(\frac{\partial\varphi}{\partial T}\right)^{\xi} = \frac{\mathbf{g}_1}{|\mathbf{g}_1|}\cdot\nabla\varphi.$$

Reverting to the notation x^i for the curvilinear co-ordinates, we can write the tangential derivative in the x^i-direction as

$$\left(\frac{\partial\varphi}{\partial T}\right)^{x^i} = \frac{\mathbf{g}_i}{|\mathbf{g}_i|}\cdot\nabla\varphi = \frac{\mathbf{g}_i}{|\mathbf{g}_i|}\cdot\mathbf{g}^j\frac{\partial\varphi}{\partial x^j} = \frac{1}{|\mathbf{g}_i|}\delta_i^j\frac{\partial\varphi}{\partial x^j},$$

where we have made use of eqn (1.13), and summation is assumed over j but *not* i. So the tangential derivative is, by the usual properties of the Kronecker symbol,

$$\left(\frac{\partial\varphi}{\partial T}\right)^{x^i} = \frac{1}{|\mathbf{g}_i|}\frac{\partial\varphi}{\partial x^i} = \frac{1}{\sqrt{g_{ii}}}\frac{\partial\varphi}{\partial x^i} \tag{1.192}$$

with no summation over i.

The rate of change of φ at a point P in a direction normal to the x^1 co-ordinate surface (the surface on which x^1 is constant) passing through P may be obtained similarly when it is recalled that the contravariant vector $\mathbf{g}^1$ is normal to this surface. Thus we can write the *normal derivative* of φ to the x^1 surface as

$$\left(\frac{\partial\varphi}{\partial n}\right)^{x^1} = \frac{\mathbf{g}^1}{|\mathbf{g}^1|}\cdot\nabla\varphi,$$

so that the normal derivative to the x^i co-ordinate surface is given, using eqn (1.15), by

$$\left(\frac{\partial\varphi}{\partial n}\right)^{x^i} = \frac{\mathbf{g}^i}{|\mathbf{g}^i|}\cdot\mathbf{g}^j\frac{\partial\varphi}{\partial x^j} = \frac{1}{|\mathbf{g}^i|}g^{ij}\frac{\partial\varphi}{\partial x^j} = \frac{1}{\sqrt{g^{ii}}}g^{ij}\frac{\partial\varphi}{\partial x^j}, \tag{1.193}$$

where summation is assumed over j, but *not* over i, throughout.

2

Classical differential geometry of space-curves

2.1 Vector approach

We consider smooth curves in E^3 specified in terms of rectangular cartesian co-ordinates x, y, z (or y_1, y_2, y_3). Such curves are generated by three smooth functions of a single real parameter, say t:

$$x = x(t), \quad y = y(t), \quad z = z(t),$$

so that the position vector $\mathbf{r}$ of points on the curve relative to some origin O is given by

$$\mathbf{r} = \mathbf{r}(t) = x(t)\mathbf{i} + y(t)\mathbf{j} + z(t)\mathbf{k}. \tag{2.1}$$

A frequently-quoted example is the circular helix, which could be specified by

$$x = a\cos t, \quad y = a\sin t, \quad z = ct, \tag{2.2}$$

where a and c are constants. This curve lies on the surface of the circular cylinder $x^2 + y^2 = a^2$, and makes complete turns about the z-axis as t increases by 2π.

Distance along a curve is measured by the arc-length parameter s, where differentials of arc-length $\mathrm{d}s$ satisfy the equation

$$\mathrm{d}s^2 = \mathrm{d}x^2 + \mathrm{d}y^2 + \mathrm{d}z^2 = \mathrm{d}\mathbf{r} \cdot \mathrm{d}\mathbf{r}, \tag{2.3}$$

or, in terms of derivatives with respect to the parameter t,

$$\dot{s}^2 = \dot{x}^2 + \dot{y}^2 + \dot{z}^2 = \dot{\mathbf{r}} \cdot \dot{\mathbf{r}}, \tag{2.4}$$

where the dot denotes $\frac{\mathrm{d}}{\mathrm{d}t}$, or

$$\mathrm{d}s = \sqrt{\dot{x}^2 + \dot{y}^2 + \dot{z}^2}\,\mathrm{d}t. \tag{2.5}$$

An *intrinsic* definition of a space-curve is given when s is used as a parameter in eqn (2.1). For example, the circular helix (2.2) gives

$$\mathrm{d}s = \sqrt{(-a\sin t)^2 + (a\cos t)^2 + c^2}\,\mathrm{d}t = \sqrt{a^2 + c^2}\,\mathrm{d}t$$

from eqn (2.5), which may be integrated to give $t = (a^2+c^2)^{-\frac{1}{2}}s$ and the intrinsic form

$$x = a\cos\left(\frac{s}{\sqrt{a^2+c^2}}\right), \quad y = a\sin\left(\frac{s}{\sqrt{a^2+c^2}}\right), \quad z = \frac{cs}{\sqrt{a^2+c^2}}. \tag{2.6}$$

Thus the length of this curve in one complete turn about the z-axis is $2\pi\sqrt{a^2+c^2}$, the distance advanced in the direction of the z-axis during this turn being $2\pi c$.

Differentiating eqn (2.1) with respect to t at a point P on the curve gives a vector

$$\dot{\mathbf{r}} = \mathbf{i}\dot{x} + \mathbf{j}\dot{y} + \mathbf{k}\dot{z}, \tag{2.7}$$

which is tangential to the curve. The magnitude of this vector is $\dot{s} = \sqrt{\dot{x}^2+\dot{y}^2+\dot{z}^2}$, so that a unit tangent vector is

$$\mathbf{t} = \frac{\dot{\mathbf{r}}}{\dot{s}} = \frac{\mathrm{d}\mathbf{r}}{\mathrm{d}s} = \mathbf{i}\frac{\mathrm{d}x}{\mathrm{d}s} + \mathbf{j}\frac{\mathrm{d}y}{\mathrm{d}s} + \mathbf{k}\frac{\mathrm{d}z}{\mathrm{d}s}. \tag{2.8}$$

Differentiating the identity

$$\mathbf{t}\cdot\mathbf{t} = 1$$

with respect to s gives

$$\mathbf{t}\cdot\frac{\mathrm{d}\mathbf{t}}{\mathrm{d}s} = 0,$$

which shows that the vector $\mathrm{d}\mathbf{t}/\mathrm{d}s$ is perpendicular to $\mathbf{t}$. The direction of this vector at a point P, given by the unit vector $\mathbf{n}$, is the direction of the *principal normal* to the curve at that point, and we write

$$\frac{\mathrm{d}\mathbf{t}}{\mathrm{d}s} = \kappa\mathbf{n}, \tag{2.9}$$

where κ is called the *curvature* at that point. We usually assume that κ is a non-negative function of s, although there are circumstances where it would be convenient to let it take negative values in order to allow $\mathbf{n}$ to be a continuous vector function of s. If $\mathrm{d}\mathbf{t}/\mathrm{d}s = \mathbf{0}$, which would be the case everywhere when the curve is a straight line, then $\kappa = 0$, but $\mathbf{n}$ is not uniquely defined. The plane containing the tangent and the principal normal at a point on a curve is called the *osculating plane* at that point.

Introducing the shorthand notation $\frac{\mathrm{d}}{\mathrm{d}s}(\;) = (\;)'$, we have

$$\mathbf{t} = \mathbf{r}'$$

and

$$\mathbf{t}' = \mathbf{r}'' = \mathbf{i}x'' + \mathbf{j}y'' + \mathbf{k}z'' = \kappa\mathbf{n}. \tag{2.10}$$

Hence

$$\kappa = |\mathbf{r}''| = \sqrt{(x'')^2 + (y'')^2 + (z'')^2}. \tag{2.11}$$

The Chain Rule for differentiation gives

$$\dot{\mathbf{r}} = \mathbf{r}'\dot{s} \tag{2.12}$$

and

$$\ddot{\mathbf{r}} = \mathbf{r}''\dot{s}^2 + \mathbf{r}'\ddot{s}. \tag{2.13}$$

Hence we obtain

$$\begin{aligned}\kappa^2 = \mathbf{r}'' \cdot \mathbf{r}'' &= \frac{1}{\dot{s}^4}\left(\ddot{\mathbf{r}} - \dot{\mathbf{r}}\frac{\ddot{s}}{\dot{s}}\right) \cdot \left(\ddot{\mathbf{r}} - \dot{\mathbf{r}}\frac{\ddot{s}}{\dot{s}}\right) \\ &= \frac{1}{\dot{s}^4}\left(\ddot{\mathbf{r}} \cdot \ddot{\mathbf{r}} - 2\frac{\ddot{s}}{\dot{s}}\dot{\mathbf{r}} \cdot \ddot{\mathbf{r}} + \frac{\ddot{s}^2}{\dot{s}^2}\dot{\mathbf{r}} \cdot \dot{\mathbf{r}}\right).\end{aligned}$$

But using the identity $\dot{\mathbf{r}} \cdot \dot{\mathbf{r}} = \dot{s}^2$ together with its derivative $\dot{\mathbf{r}} \cdot \ddot{\mathbf{r}} = \dot{s}\ddot{s}$ reduces this expression for κ to

$$\kappa = \frac{1}{(\dot{\mathbf{r}} \cdot \dot{\mathbf{r}})^{\frac{3}{2}}}\sqrt{(\dot{\mathbf{r}} \cdot \dot{\mathbf{r}})(\ddot{\mathbf{r}} \cdot \ddot{\mathbf{r}}) - (\dot{\mathbf{r}} \cdot \ddot{\mathbf{r}})^2}. \tag{2.14}$$

By the Lagrange identity (1.40) this is equivalent to

$$\kappa = \frac{1}{(\dot{\mathbf{r}} \cdot \dot{\mathbf{r}})^{\frac{3}{2}}}\sqrt{(\dot{\mathbf{r}} \times \ddot{\mathbf{r}}) \cdot (\dot{\mathbf{r}} \times \ddot{\mathbf{r}})} = \frac{|\dot{\mathbf{r}} \times \ddot{\mathbf{r}}|}{|\dot{\mathbf{r}}|^3}. \tag{2.15}$$

For curves in two dimensions (the Oxy plane) eqn (2.15) reduces to the well-known formula

$$\kappa = \frac{|\dot{x}\ddot{y} - \dot{y}\ddot{x}|}{(\dot{x}^2 + \dot{y}^2)^{\frac{3}{2}}}, \tag{2.16}$$

and if x is used as a parameter in place of t, this becomes

$$\kappa = \frac{|\mathrm{d}^2y/\mathrm{d}x^2|}{[1 + (\mathrm{d}y/\mathrm{d}x)^2]^{\frac{3}{2}}}. \tag{2.17}$$

The *radius of curvature* ρ at a given point is given by

$$\rho = \frac{1}{\kappa}. \tag{2.18}$$

Exercise 1. Show that for the twisted curve given by the parametric form $x = \ln\cos t$, $y = \ln\sin t$, $z = \sqrt{2}t$ $(0 < t < \pi/2)$, we have $\kappa = \frac{1}{\sqrt{2}}\sin 2t$.

2.2 The Serret-Frenet equations

Given a unit tangent vector $\mathbf{t}$ and a unit principal normal $\mathbf{n}$ at a point on a curve in E^3, we can define a third unit vector $\mathbf{b}$, called the *unit binormal vector*, orthogonal to both of them, such that

$$\mathbf{b} = \mathbf{t} \times \mathbf{n}. \tag{2.19}$$

The system of vectors $(\mathbf{t}, \mathbf{n}, \mathbf{b})$ then forms a right-handed set of unit vectors, which forms the *moving trihedron* as s varies, i.e. as we move along the curve. By definition, we also have $\mathbf{t} = \mathbf{n} \times \mathbf{b}$ and $\mathbf{n} = \mathbf{b} \times \mathbf{t}$. The planes at a point of the curve containing the directions $\mathbf{n}$ and $\mathbf{b}$, and the directions $\mathbf{b}$ and $\mathbf{t}$, are called the *normal plane* and the *rectifying plane*, respectively, (see Fig. 2.1) at that point. Differentiating eqn (2.19)

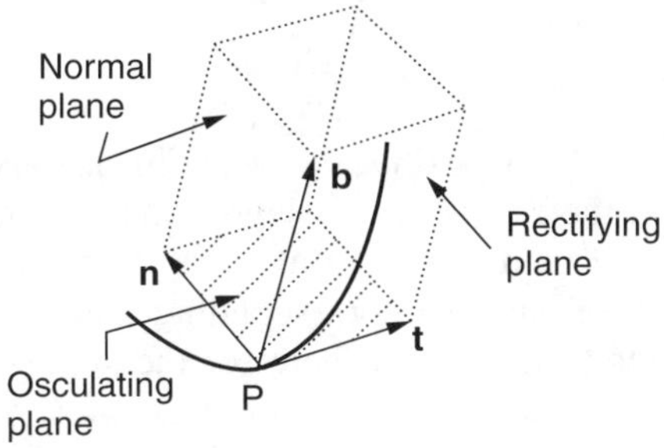

Fig. 2.1 Osculating, normal, and rectifying planes at a point P on a curve.

with respect to s gives

$$\frac{\mathrm{d}\mathbf{b}}{\mathrm{d}s} = \frac{\mathrm{d}\mathbf{t}}{\mathrm{d}s} \times \mathbf{n} + \mathbf{t} \times \frac{\mathrm{d}\mathbf{n}}{\mathrm{d}s} = \kappa\mathbf{n} \times \mathbf{n} + \mathbf{t} \times \frac{\mathrm{d}\mathbf{n}}{\mathrm{d}s} = \mathbf{t} \times \frac{\mathrm{d}\mathbf{n}}{\mathrm{d}s}, \tag{2.20}$$

since $\mathbf{n} \times \mathbf{n} = 0$. We deduce that $\mathrm{d}\mathbf{b}/\mathrm{d}s$ is a vector orthogonal to $\mathbf{t}$. Moreover, since $\mathbf{b}$ is defined as a unit vector, with $\mathbf{b} \cdot \mathbf{b} = 1$, it follows that $\mathbf{b} \cdot \mathrm{d}\mathbf{b}/\mathrm{d}s = 0$, so that $\mathrm{d}\mathbf{b}/\mathrm{d}s$ is also orthogonal to $\mathbf{b}$.

Hence $\mathrm{d}\mathbf{b}/\mathrm{d}s$ can only be parallel to $\mathbf{n}$, and we can write

$$\frac{\mathrm{d}\mathbf{b}}{\mathrm{d}s} = -\tau\mathbf{n},$$

where the scalar τ is called the *torsion* of the curve at a point. Not all writers follow the convention of including a negative sign in this equation; the significance of the sign can be shown by considering the special case of the circular helix given by eqn (2.6). Here we obtain

$$\mathbf{t} = \mathbf{r}' = \frac{1}{\sqrt{a^2+c^2}}\left\{-a\mathbf{i}\sin\frac{s}{\sqrt{a^2+c^2}} + a\mathbf{j}\cos\frac{s}{\sqrt{a^2+c^2}} + c\mathbf{k}\right\},$$

$$\mathbf{t}' = \mathbf{r}'' = \kappa\mathbf{n} = \frac{1}{a^2+c^2}\left\{-a\mathbf{i}\cos\frac{s}{\sqrt{a^2+c^2}} - a\mathbf{j}\sin\frac{s}{\sqrt{a^2+c^2}}\right\},$$

so that

$$\kappa = \frac{a}{a^2+c^2} \tag{2.21}$$

and

$$\mathbf{n} = -\mathbf{i}\cos\frac{s}{\sqrt{a^2+c^2}} - \mathbf{j}\sin\frac{s}{\sqrt{a^2+c^2}}.$$

Hence

$$\mathbf{b} = \mathbf{t} \times \mathbf{n} = \frac{1}{\sqrt{a^2+c^2}}\left\{c\mathbf{i}\sin\frac{s}{\sqrt{a^2+c^2}} - c\mathbf{j}\cos\frac{s}{\sqrt{a^2+c^2}} + a\mathbf{k}\right\}$$

and

$$\mathbf{b}' = \frac{c}{a^2+c^2}\left\{\mathbf{i}\cos\frac{s}{\sqrt{a^2+c^2}} + \mathbf{j}\sin\frac{s}{\sqrt{a^2+c^2}}\right\} = -\frac{c}{a^2+c^2}\mathbf{n}.$$

Thus

$$\tau = \frac{c}{a^2 + c^2}, \tag{2.22}$$

a constant, in this case. Now if c is positive, so is τ. This is the case for a 'right-handed' helix, which twists around the z-axis in the same sense as a right-handed screwdriver advancing in the direction of increasing z. But negative values of c give a left-handed helix with the opposite sense, and the corresponding value of the torsion is negative by eqn (2.22). Thus the convention used here for the sign of τ associates positive τ with right-handed twisting about some axis and negative τ with left-handed twisting. Right or left-handed twisting of a space-curve is also associated with a particular sense of rotation of the osculating plane with increasing s. If the osculating plane does not rotate, then $\mathbf{b}$ is a constant vector, the torsion is zero at every point, and the curve must lie in a plane.

To evaluate $\mathrm{d}\mathbf{n}/\mathrm{d}s$, we can differentiate $\mathbf{n} = \mathbf{b} \times \mathbf{t}$ to obtain

$$\frac{\mathrm{d}\mathbf{n}}{\mathrm{d}s} = \mathbf{b} \times \frac{\mathrm{d}\mathbf{t}}{\mathrm{d}s} + \frac{\mathrm{d}\mathbf{b}}{\mathrm{d}s} \times \mathbf{t} = \kappa \mathbf{b} \times \mathbf{n} - \tau \mathbf{n} \times \mathbf{t} = -\kappa \mathbf{t} + \tau \mathbf{b}.$$

Summarising, we have the *Serret-Frenet* formulas:

$$\begin{aligned} \frac{\mathrm{d}\mathbf{t}}{\mathrm{d}s} &= \qquad \kappa \mathbf{n} \\ \frac{\mathrm{d}\mathbf{n}}{\mathrm{d}s} &= -\kappa \mathbf{t} \qquad + \tau \mathbf{b} \\ \frac{\mathrm{d}\mathbf{b}}{\mathrm{d}s} &= \qquad -\tau \mathbf{n}, \end{aligned} \tag{2.23}$$

where the skew-symmetric nature of the array of scalars on the right-hand side serves as an aide-memoire.

If at a point in space initial values of $\mathbf{t}$ and $\mathbf{n}$ (and hence of $\mathbf{b}$ also, since $\mathbf{b} = \mathbf{t} \times \mathbf{n}$) are given, and if functions $\kappa(s)$ and $\tau(s)$ of arc-length parameter s are also specified, then in principle eqns (2.23) can be integrated to determine $\mathbf{t}$, $\mathbf{n}$, and $\mathbf{b}$ as functions of s, and thus determine the space-curve in its entirety. The fundamental theorem of space-curves states that specification of the functions $\kappa(s)$ and $\tau(s)$ determines a space-curve uniquely apart from its precise position in space. A proof is not difficult, though we do not include one here.

A useful formula for torsion may be derived as follows. We begin with the identities $\mathbf{t} = \mathbf{r}'$, $\mathbf{t}' = \mathbf{r}''$, and $\mathbf{t}'' = \mathbf{r}'''$, where the prime again stands for $\mathrm{d}/\mathrm{d}s$. Now, regarding everything as a function of s, we have

$$\mathbf{t}'' = (\mathbf{t}')' = (\kappa \mathbf{n})' = \kappa \mathbf{n}' + \kappa' \mathbf{n} = \kappa(-\kappa \mathbf{t} + \tau \mathbf{b}) + \kappa' \mathbf{n} = -\kappa^2 \mathbf{t} + \kappa' \mathbf{n} + \kappa \tau \mathbf{b}.$$

Hence

$$\mathbf{r}'' \times \mathbf{r}''' = \mathbf{t}' \times \mathbf{t}'' = (\kappa \mathbf{n}) \times (-\kappa^2 \mathbf{t} + \kappa' \mathbf{n} + \kappa \tau \mathbf{b}) = \kappa^3 \mathbf{b} + \kappa^2 \tau \mathbf{t}.$$

Finally

$$\mathbf{r}' \cdot (\mathbf{r}'' \times \mathbf{r}''') = \mathbf{t} \cdot (\kappa^3 \mathbf{b} + \kappa^2 \tau \mathbf{t}) = \kappa^2 \tau,$$

so that

$$\tau = \frac{1}{\kappa^2} \mathbf{r}' \cdot (\mathbf{r}'' \times \mathbf{r}''') = \rho^2 \mathbf{r}' \cdot (\mathbf{r}'' \times \mathbf{r}''') = \frac{\mathbf{r}' \cdot (\mathbf{r}'' \times \mathbf{r}''')}{(\mathbf{r}'' \cdot \mathbf{r}'')}, \tag{2.24}$$

which can also be expressed as

$$\tau = \rho^2 \begin{vmatrix} x' & y' & z' \\ x'' & y'' & z'' \\ x''' & y''' & z''' \end{vmatrix}. \tag{2.25}$$

Exercise 2. If a curve is given in terms of a parameter t, show, making use of eqns (2.12), (2.13), and a similar equation for $\dddot{\mathbf{r}}$, that

$$\tau = \rho^2 \frac{\dot{\mathbf{r}} \cdot (\ddot{\mathbf{r}} \times \dddot{\mathbf{r}})}{\dot{s}^6} = \rho^2 \frac{\dot{\mathbf{r}} \cdot (\ddot{\mathbf{r}} \times \dddot{\mathbf{r}})}{(\dot{\mathbf{r}} \cdot \dot{\mathbf{r}})^3}, \tag{2.26}$$

where ρ may be obtained from eqns (2.15) and (2.18).

Exercise 3. For the curve given by the parametric form

$$x = a(3t - t^3), \quad y = 3at^2, \quad z = a(3t + t^3),$$

where a is a constant, show that $\kappa = \tau = \frac{1}{3}a^{-1}(1 + t^2)^{-2}$.

2.3 Generalized co-ordinate approach

It may be instructive to derive the Serret-Frenet formulas using generalized co-ordinates. In the process we introduce the concept of intrinsic differentiation. Given a set of curvilinear co-ordinates x^1, x^2, x^3, a space-curve may be specified in terms of a parameter t and the functions

$$x^i = x^i(t), \quad i = 1, 2, 3.$$

Arc-length is given in terms of the basic metric

$$\mathrm{d}s^2 = g_{ij}\,\mathrm{d}x^i\,\mathrm{d}x^j, \tag{2.27}$$

or in terms of the parameter t as

$$s = \int_{t_0}^{t} \sqrt{g_{ij} \frac{\mathrm{d}x^i}{\mathrm{d}t} \frac{\mathrm{d}x^j}{\mathrm{d}t}}\,\mathrm{d}t, \tag{2.28}$$

measured from some point on the curve where $t = t_0$. Equation (2.27) can be rewritten as

$$g_{ij} \frac{\mathrm{d}x^i}{\mathrm{d}s} \frac{\mathrm{d}x^j}{\mathrm{d}s} = 1, \tag{2.29}$$

which according to eqn (1.55) may be interpreted as implying that the vector whose contravariant components are $\mathrm{d}x^i/\mathrm{d}s$, $i = 1, 2, 3$, has magnitude unity. Of course, this is just the unit tangent vector $\mathbf{t}$ to the curve; here we write

$$t^i = \frac{\mathrm{d}x^i}{\mathrm{d}s}. \tag{2.30}$$

For any vector field $\mathbf{u}$ we have by the Chain Rule

$$\frac{\mathrm{d}\mathbf{u}}{\mathrm{d}s} = \frac{\partial \mathbf{u}}{\partial x^j} \frac{\mathrm{d}x^j}{\mathrm{d}s} = u^i_{,j} \frac{\mathrm{d}x^j}{\mathrm{d}s} \mathbf{g}_i = u_{i,j} \frac{\mathrm{d}x^j}{\mathrm{d}s} \mathbf{g}^i$$

in terms of the covariant derivatives, by eqns (1.119) and (1.121). Since $\mathbf{du}/\mathrm{d}s$ is a vector quantity, we can define the *intrinsic derivatives* of the contravariant and covariant components of **u** by:

$$\frac{\delta u^i}{\delta s} = u^i_{,j}\frac{\mathrm{d}x^j}{\mathrm{d}s} = u^i_{,j}t^j, \qquad \frac{\delta u_i}{\delta s} = u_{i,j}\frac{\mathrm{d}x^j}{\mathrm{d}s} = u_{i,j}t^j, \tag{2.31}$$

so that

$$\frac{\mathbf{du}}{\mathrm{d}s} = \frac{\delta u^i}{\delta s}\mathbf{g}_i = \frac{\delta u_i}{\delta s}\mathbf{g}^i.$$

Another equation which will be useful later is

$$\frac{\delta u^i}{\delta s} = u^i_{,j}\frac{\mathrm{d}x^j}{\mathrm{d}s} = \left(\frac{\partial u^i}{\partial x^j} + \Gamma^i_{kj}u^k\right)\frac{\mathrm{d}x^j}{\mathrm{d}s} = \frac{\mathrm{d}u^i}{\mathrm{d}s} + \Gamma^i_{kj}u^k\frac{\mathrm{d}x^j}{\mathrm{d}s}. \tag{2.32}$$

Similarly, intrinsic derivatives of second-order tensors can be defined in terms of covariant derivatives. For example, the intrinsic derivatives of contravariant second-order tensor components T^{ij} are

$$\frac{\delta T^{ij}}{\delta s} = T^{ij}_{,k}\frac{\mathrm{d}x^k}{\mathrm{d}s}, \tag{2.33}$$

where the covariant derivatives of T^{ij} are given by eqn (1.129). Thus it follows, from eqn (1.128) that

$$\frac{\delta g^{ij}}{\delta s} = 0.$$

Now rewriting eqn (2.29) as $g^{ij}t_it_j = 1$ and taking intrinsic derivatives (with the usual product rule) gives

$$\frac{\delta g^{ij}}{\delta s}t_it_j + g^{ij}\frac{\delta t_i}{\delta s}t_j + g^{ij}t_i\frac{\delta t_j}{\delta s} = 0 + 2g^{ij}t_i\frac{\delta t_j}{\delta s} = 0,$$

using the symmetry of g^{ij}. Thus

$$g^{ij}t_i\frac{\delta t_j}{\delta s} = 0. \tag{2.34}$$

From eqn (1.54) it follows that $\delta t_j/\delta s$ represent the covariant components of a vector orthogonal to **t**, and we can put

$$\frac{\delta t_i}{\delta s} = \kappa n_i, \tag{2.35}$$

where n_i are the covariant components of the unit vector **n** which satisfies

$$g^{ij}n_in_j = 1 \tag{2.36}$$

and

$$g^{ij}t_in_j = 0. \tag{2.37}$$

We shall assume here that κ is non-negative.

Further intrinsic differentiation gives

$$g^{ij} n_i \frac{\delta n_j}{\delta s} = 0, \tag{2.38}$$

which implies that $\delta n_j/\delta s$ is a vector orthogonal to n_i; moreover,

$$\begin{aligned} g^{ij} t_i \frac{\delta n_j}{\delta s} + g^{ij} \frac{\delta t_i}{\delta s} n_j &= g^{ij} t_i \frac{\delta n_j}{\delta s} + g^{ij} (\kappa n_i) n_j = g^{ij} t_i \frac{\delta n_j}{\delta s} + \kappa \\ &= g^{ij} t_i \frac{\delta n_j}{\delta s} + \kappa g^{ij} t_i t_j = g^{ij} t_i \left(\frac{\delta n_j}{\delta s} + \kappa t_j \right) = 0. \end{aligned}$$

We deduce that $\left(\frac{\delta n_j}{\delta s} + \kappa t_j\right)$ are covariant components of a vector orthogonal to **t**, and define

$$\frac{\delta n_i}{\delta s} + \kappa t_i = \tau b_i, \tag{2.39}$$

where **b** is a unit vector satisfying

$$g^{ij} b_i b_j = 1 \tag{2.40}$$

and

$$g^{ij} t_i b_j = 0, \tag{2.41}$$

but with sense yet to be defined. Now **b** is orthogonal to **n** as well as **t**, since from eqn (2.39)

$$g^{ij} n_i b_j = \frac{1}{\tau} \left(g^{ij} n_i \frac{\delta n_i}{\delta s} + \kappa g^{ij} n_i t_j \right) = 0, \tag{2.42}$$

using eqns (2.37) and (2.38), assuming that the scalar τ is non-zero. So we can choose the sense of **b** such that {**t**, **n**, **b**} form a right-handed set of unit vectors.

Any vector **u** can be expanded on a rectangular cartesian basis as

$$\mathbf{u} = (\mathbf{u} \cdot \mathbf{i})\mathbf{i} + (\mathbf{u} \cdot \mathbf{j})\mathbf{j} + (\mathbf{u} \cdot \mathbf{k})\mathbf{k}$$

and, in the same way,

$$\mathbf{u} = (\mathbf{u} \cdot \mathbf{t})\mathbf{t} + (\mathbf{u} \cdot \mathbf{n})\mathbf{n} + (\mathbf{u} \cdot \mathbf{b})\mathbf{b}.$$

Applying this to the vector $\delta \mathbf{b}/\delta s$, taking covariant components with respect to the given curvilinear co-ordinates, we get

$$\begin{aligned} \frac{\delta b_i}{\delta s} &= \left(g^{jk} t_j \frac{\delta b_k}{\delta s} \right) t_i + \left(g^{jk} n_j \frac{\delta b_k}{\delta s} \right) n_i + \left(g^{jk} b_j \frac{\delta b_k}{\delta s} \right) b_i \\ &= - \left(g^{jk} \frac{\delta t_j}{\delta s} b_k \right) t_i - \left(g^{jk} \frac{\delta n_j}{\delta s} b_k \right) n_i + 0, \end{aligned}$$

after differentiating eqns (2.40), (2.41), and (2.42). Hence from eqns (2.35) and (2.39)

$$\frac{\delta b_i}{\delta s} = -\kappa (g^{jk} n_j b_k) t_i - g^{jk} (-\kappa t_j + \tau b_j) b_k n_i = -\tau n_i,$$

again making use of eqns (2.40), (2.41), and (2.42).

Summarizing, we have the Serret-Frenet formulas in covariant form:

$$\begin{aligned}\frac{\delta t_i}{\delta s} &= \qquad \kappa n_i \\ \frac{\delta n_i}{\delta s} &= -\kappa t_i \qquad + \tau b_i \\ \frac{\delta b_i}{\delta s} &= \qquad -\tau n_i.\end{aligned} \tag{2.43}$$

Alternatively, having established that $\{\mathbf{t}, \mathbf{n}, \mathbf{b}\}$ form an orthonormal set, we can write the vector product $\mathbf{b} = \mathbf{t} \times \mathbf{n}$ in component form as

$$b^i = \varepsilon^{ijk} t_j n_k,$$

and, making use of eqn (1.130), we have

$$\begin{aligned}\frac{\delta b^i}{\delta s} &= \varepsilon^{ijk} t_j \frac{\delta n_k}{\delta s} + \varepsilon^{ijk} \frac{\delta t_j}{\delta s} n_k = \varepsilon^{ijk} t_j(-\kappa t_k + \tau b_k) + \kappa \varepsilon^{ijk} n_j n_k \\ &= \tau \varepsilon^{ijk} t_j b_k,\end{aligned} \tag{2.44}$$

the other terms vanishing due to the symmetry of $t_j t_k$ and $n_j n_k$ and the skew-symmetry of ε^{ijk} in the summed indices j, k. The right-hand side now gives the scalar τ multiplied by the contravariant component of the vector product $\mathbf{t} \times \mathbf{b}$, which because of the choice of the sense of $\mathbf{b}$ is equal to $-\mathbf{n}$, with contravariant component n^i.

Hence

$$\frac{\delta b^i}{\delta s} = -\tau n^i,$$

with associated covariant components

$$\frac{\delta b_i}{\delta s} = -\tau n_i$$

as expected.

2.4 Metric tensor of a space-curve

In this section we parametrise a space-curve C directly by a curvilinear co-ordinate ξ, so that

$$\mathbf{r} = (x(\xi), y(\xi), z(\xi)) \tag{2.45}$$

on the curve. In the context of grid generation, space-curves appear as boundaries of surfaces and as edges of three-dimensional blocks, and it is convenient to map a given finite length of space-curve onto an interval of the ξ-axis, say $0 \leqslant \xi \leqslant 1$. A uniformly spaced set of points in the ξ-interval will then map to a set of points along the curve.

Arc-length along the curve is then defined by

$$(\mathrm{d}s)^2 = \mathrm{d}\mathbf{r} \cdot \mathrm{d}\mathbf{r} = \frac{\mathrm{d}\mathbf{r}}{\mathrm{d}\xi} \cdot \frac{\mathrm{d}\mathbf{r}}{\mathrm{d}\xi}(\mathrm{d}\xi)^2 = g_{11}(\mathrm{d}\xi)^2,$$

where

$$g_{11} = \frac{d\mathbf{r}}{d\xi} \cdot \frac{d\mathbf{r}}{d\xi} = \mathbf{g}_1 \cdot \mathbf{g}_1 \tag{2.46}$$

is, by comparison with eqns (1.18), the solitary component of the metric tensor of the curve, and $\mathbf{g}_1$ is tangential to C at any point.

Hence

$$\frac{ds}{d\xi} = \sqrt{g_{11}} \tag{2.47}$$

and the total length of C, if ξ varies monotonically from 0 to 1, is

$$L = \int_0^1 \sqrt{g_{11}}\, d\xi. \tag{2.48}$$

A unit tangent vector to the curve is

$$\mathbf{t} = \frac{d\mathbf{r}}{ds} = \frac{d\mathbf{r}}{d\xi}\frac{d\xi}{ds} = \frac{1}{\sqrt{g_{11}}}\frac{d\mathbf{r}}{d\xi}. \tag{2.49}$$

We also have

$$\frac{d}{d\xi}\left(\frac{d\mathbf{r}}{d\xi}\right) = \frac{d}{d\xi}(\sqrt{g_{11}}\mathbf{t}) = \frac{d}{d\xi}(\sqrt{g_{11}})\mathbf{t} + \sqrt{g_{11}}\frac{d\mathbf{t}}{d\xi}$$

and, since we already know that $d\mathbf{t}/ds = \kappa\mathbf{n}$, it follows that

$$\frac{d^2\mathbf{r}}{d\xi^2} = \frac{1}{2}(g_{11})^{-\frac{1}{2}}\frac{dg_{11}}{d\xi}\mathbf{t} + g_{11}\kappa\mathbf{n} = \frac{1}{2}\frac{1}{g_{11}}\frac{dg_{11}}{d\xi}\frac{d\mathbf{r}}{d\xi} + \kappa g_{11}\mathbf{n}, \tag{2.50}$$

which is the *curve identity.*

Taking the scalar product of both sides with $\mathbf{n}$ gives the expression for curvature

$$\kappa = \frac{1}{g_{11}}\frac{d^2\mathbf{r}}{d\xi^2} \cdot \mathbf{n}. \tag{2.51}$$

The Jacobian matrix of the transformation $\xi \to (x, y, z)$ may be defined as the 3×1 column vector

$$\mathcal{J} = \begin{pmatrix} x_\xi \\ y_\xi \\ z_\xi \end{pmatrix}, \tag{2.52}$$

where suffixes denote derivatives with respect to ξ. Clearly we have

$$g_{11} = (x_\xi)^2 + (y_\xi)^2 + (z_\xi)^2 = \mathcal{J}^T\mathcal{J}. \tag{2.53}$$

Suppose now that the curve is parametrized by a different parameter χ, $0 \leqslant \chi \leqslant 1$, where $\chi = \chi(\xi)$, for some function $\chi(\xi)$ satisfying $\chi(0) = 0$, $\chi(1) = 1$. A set of points on the curve could be generated by dividing the χ interval into n equal divisions, with $\chi_i = i/n$, $i = 0, 1, 2, \ldots, n$, and locating the points on C corresponding to these values of χ. The distribution of points may, however, not be suitable for various reasons to serve as a grid. A more satisfactory distribution might be obtained by varying the choice of parameter values χ_i according to the choice of the function $\chi(\xi)$, so that a *uniformly* distributed set $\{\xi_i\}$ of values in the ξ interval will map onto a *non-uniformly* distributed set of values $\{\chi_i\}$ in the χ interval.

Here we put

$$(\mathrm{d}s)^2 = \frac{\mathrm{d}\mathbf{r}}{\mathrm{d}\chi} \cdot \frac{\mathrm{d}\mathbf{r}}{\mathrm{d}\chi} (\mathrm{d}\chi)^2 = \tilde{g}_{11} (\mathrm{d}\chi)^2,$$

so that $\mathrm{d}s/\mathrm{d}\chi = \sqrt{\tilde{g}_{11}}$. Clearly we have

$$g_{11} = \frac{\mathrm{d}\mathbf{r}}{\mathrm{d}\xi} \cdot \frac{\mathrm{d}\mathbf{r}}{\mathrm{d}\xi} = \frac{\mathrm{d}\mathbf{r}}{\mathrm{d}\chi} \cdot \frac{\mathrm{d}\mathbf{r}}{\mathrm{d}\chi} \left(\frac{\mathrm{d}\chi}{\mathrm{d}\xi}\right)^2 = \tilde{g}_{11} \left(\frac{\mathrm{d}\chi}{\mathrm{d}\xi}\right)^2. \tag{2.54}$$

If we wanted grid points on C to be evenly distributed in the sense that the length of the curve between neighbouring points was always the same, we would require that

$$\int_{\chi_i}^{\chi_{i+1}} \sqrt{\tilde{g}_{11}}\, \mathrm{d}\chi = \mathit{const.} = L/n, \tag{2.55}$$

where L is the total length of the curve. This equation is an example of an *equidistribution principle*; these principles in general are prescriptions for controlling the density of grid points.

Equation (2.55) suggests that the mapping $\xi \to \chi$ should satisfy the differential equation

$$\frac{\mathrm{d}\chi}{\mathrm{d}\xi} = \frac{L}{\sqrt{\tilde{g}_{11}}}, \tag{2.56}$$

or, in approximate form,

$$\frac{\chi_{i+1} - \chi_i}{\xi_{i+1} - \xi_i} = \frac{L}{\sqrt{\tilde{g}_{11}}}, \tag{2.57}$$

with $\sqrt{\tilde{g}_{11}}$ evaluated, say, at the mid-point $\chi_{i+\frac{1}{2}}$ of the corresponding χ interval. Here the ξ_i values are evenly distributed, so that $(\xi_{i+1} - \xi_i) = 1/n$. Thus the spacing of points on the χ interval is proportional to $1/\sqrt{\tilde{g}_{11}}$.

Equation (2.57) is satisfactory for numerical work except for cases where $\tilde{g}_{11}$ becomes too small, or zero. A better choice might be based on

$$\frac{\mathrm{d}\chi}{\mathrm{d}\xi} = \frac{c}{\sqrt{1 + \tilde{g}_{11}}} \tag{2.58}$$

instead of eqn (2.56), for some constant c, or

$$\frac{\mathrm{d}\chi}{\mathrm{d}\xi} = \frac{c}{\sqrt{1 + \alpha^2 \tilde{g}_{11}}}, \tag{2.59}$$

where extra flexibility in controlling grid density is provided by the parameter α.

The optimal spacing of grid points may also be influenced by the curvature κ of C. To obtain higher grid-point density in regions of high curvature, the equation

$$\frac{\mathrm{d}\chi}{\mathrm{d}\xi} = \frac{c}{(1 + \beta^2 |\kappa|)\sqrt{1 + \alpha^2 \tilde{g}_{11}}} \tag{2.60}$$

may be used.

Thus the grid generation procedure may be represented in general by

$$\frac{\mathrm{d}\chi}{\mathrm{d}\xi} = c\varphi(\chi), \quad \chi(0) = 0, \tag{2.61}$$

where $\varphi(\chi)$ is some *weight function* satisfying

$$\int_0^1 \frac{1}{\varphi(\chi)} \mathrm{d}\chi = c.$$

Note that the procedure involves the *physical space* E^3 of the space-curve, the one-dimensional *computational space* of the ξ parameter, and the intermediate *parametric space* of χ.

3

Differential geometry of surfaces in E^3

3.1 Equations of surfaces

The classical differential geometry of curves and surfaces is presented in many textbooks, of which we quote several in the bibliography. Here we are concerned with surfaces embedded in three-dimensional Euclidean space, which may be represented by a variety of mathematical expressions in terms of rectangular cartesian co-ordinates x, y, z.

A function of two variables $f(x, y)$ gives a surface

$$z = f(x, y). \tag{3.1}$$

A function of three variables $F(x, y, z)$ gives a surface

$$F(x, y, z) = 0. \tag{3.2}$$

Two real parameters u and v may be used to give a surface $\mathbf{r} = \mathbf{r}(u, v)$, or, in terms of three functions of two variables,

$$x = x(u, v), \quad y = y(u, v), \quad z = z(u, v). \tag{3.3}$$

The parameters u and v here may be referred to as *surface* (or *Gaussian*) *co-ordinates*.

We assume that all functions are continuously differentiable unless otherwise stated. In each of the above cases the functions may be defined only for a restricted range of the independent variables. For example, the hemispherical surface obtained by slicing the unit sphere $x^2 + y^2 + z^2 = 1$ in half through the plane Oxy and discarding the part with $z < 0$ may be expressed as

$$z = f(x, y) = +\sqrt{1 - x^2 - y^2} \quad \text{with } x^2 + y^2 \leqslant 1$$

or

$$F(x, y, z) = x^2 + y^2 + z^2 - 1 = 0 \quad \text{with } 0 \leqslant z \leqslant 1 \quad \text{and } x^2 + y^2 \leqslant 1$$

or, in terms of spherical polar co-ordinates θ, φ as parameters (Fig. 3.1),

$$x = \sin\theta\cos\varphi, \quad y = \sin\theta\sin\varphi, \quad z = \cos\theta, \quad \text{with } 0 \leqslant \theta \leqslant \pi/2, \quad 0 \leqslant \varphi \leqslant 2\pi. \tag{3.4}$$

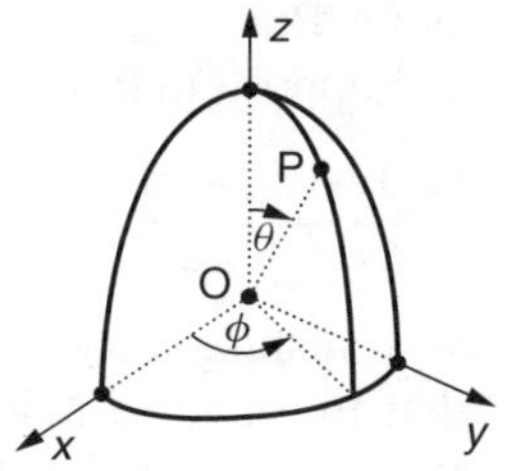

Fig. 3.1 Surface of sphere with spherical polar co-ordinates.

Standard relations between the functions in (3.1) and (3.2), in the case where they represent the same surface, may be obtained as follows. Suppose that at a point P on the surface with co-ordinates (x_0, y_0, z_0) the partial derivative $\partial F/\partial z \neq 0$. Then the Implicit Function Theorem of calculus implies that in some neighbourhood of P eqn (3.2) can be solved for z as a function of x and y, giving eqn (3.1). Thus in this neighbourhood we have, for all x and y,

$$F(x, y, f(x, y)) = 0. \tag{3.5}$$

Now differentiating partially with respect to x gives

$$\frac{\partial F}{\partial x} + \frac{\partial F}{\partial z}\frac{\partial f}{\partial x} = 0,$$

from which we derive

$$\frac{\partial f}{\partial x} = -\frac{\partial F}{\partial x}\left(\frac{\partial F}{\partial z}\right)^{-1}, \tag{3.6}$$

and similarly

$$\frac{\partial f}{\partial y} = -\frac{\partial F}{\partial y}\left(\frac{\partial F}{\partial z}\right)^{-1}. \tag{3.7}$$

Of course, eqn (3.1) is just a special case of (3.3), in which $x = u$, $y = v$, and $z = f(u, v)$.

Given a representation (3.2) of a surface, the tangent plane to the surface at a point P with co-ordinates (x_0, y_0, z_0) is the best linear approximation to the surface at P. Consideration of the first-order increment formula $\delta F = F_x\delta x + F_y\delta y + F_z\delta z = 0$ from P to a neighbouring point on the surface, where the partial derivatives of F at P are denoted by F_x, F_y, F_z, shows that the equation of the tangent plane is

$$F_x(x - x_0) + F_y(y - y_0) + F_z(z - z_0) = 0.$$

It follows that the vector of partial derivatives (F_x, F_y, F_z), i.e. the gradient vector ∇F, is normal to the tangent plane, and hence normal to the surface, at P. A point P on the surface is said to be *non-singular* if $\nabla F \neq \mathbf{0}$ there; thus it is possible to specify the normal to the surface at a non-singular point.

Another approach to non-singular points comes from considering the matrix M of partial derivatives of the representation (3.3):

$$\mathrm{M} = \begin{pmatrix} \partial x/\partial u & \partial y/\partial u & \partial z/\partial u \\ \partial x/\partial v & \partial y/\partial v & \partial z/\partial v \end{pmatrix}. \tag{3.8}$$

If this matrix has rank 2 at a point $P(x_0, y_0, z_0)$, at least one of the second-order sub-determinants is non-zero there. Suppose that this is

$$\frac{\partial x}{\partial u}\frac{\partial y}{\partial v} - \frac{\partial y}{\partial u}\frac{\partial x}{\partial v} \neq 0.$$

This is the condition for the non-vanishing of the Jacobian of the transformation $(u, v) \to (x, y)$, which implies that the inverse mapping $u = u(x, y)$, $v = v(x, y)$ exists in some neighbourhood of the projected point (x_0, y_0) in the plane Oxy. Hence it is possible to write

$$z = z(u, v) = z(u(x, y), v(x, y)) = f(x, y),$$

as in eqn (3.1), which is equivalent to

$$F(x, y, z) = f(x, y) - z = 0$$

It follows that F has non-zero gradient $(\partial f/\partial x, \partial f/\partial y, -1)$ and thus P is non-singular according to our previous definition.

Singular points arise when the rank of M is 0 or 1. When $rank\text{M} = 1$ at all points, the surface reduces to a curve. For example, the equations

$$x = u + v, \quad y = (u + v)^2, \quad z = (u + v)^3$$

represent a curve.

It is possible for one sub-determinant of M to be zero, as in the case of the circular cylinder of radius a

$$x = a\cos u, \quad y = a\sin u, \quad z = v \tag{3.9}$$

and for two, as in the case of the planar surface

$$x = u, \quad y = v, \quad z = 1,$$

but $rank\text{M} = 2$ at all points in both cases.

Singular points may arise out of the nature of the surface, but may now also appear due to a chosen parametrization (3.3) in the situation where $rank\text{M} < 2$. For example, for the spherical surface (3.4) we have

$$\text{M} = \begin{pmatrix} \cos\theta\cos\varphi & \cos\theta\sin\varphi & -\sin\theta \\ -\sin\theta\sin\varphi & \sin\theta\cos\varphi & 0 \end{pmatrix},$$

with $rank\text{M} = 2$, except for the singular point at the 'North Pole' of the sphere (Fig. 3.1), where $\theta = 0$ and $rank\text{M} = 1$.

A surface with an obvious singular point is the right circular cone (Fig. 3.2) with semi-vertical angle α, which can be represented in terms of the parameters u, φ, by the parametric equations

$$x = u\sin\alpha\cos\varphi, \quad y = u\sin\alpha\sin\varphi, \quad z = u\cos\alpha. \tag{3.10}$$

Here

$$\text{M} = \begin{pmatrix} \sin\alpha\cos\varphi & \sin\alpha\sin\varphi & \cos\alpha \\ -u\sin\alpha\sin\varphi & u\sin\alpha\cos\varphi & 0 \end{pmatrix},$$

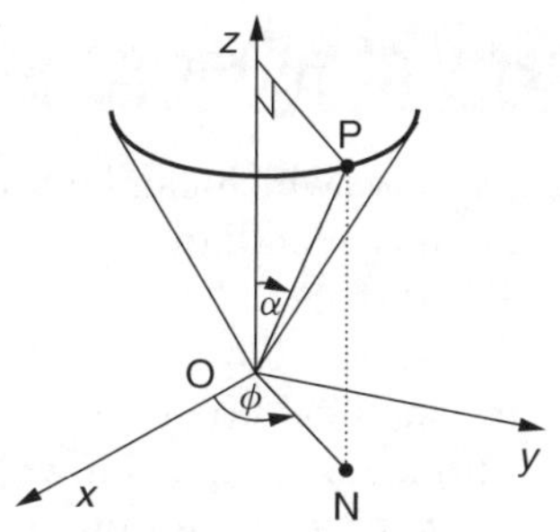

Fig. 3.2 Right circular cone with semi-angle α and vertex O.

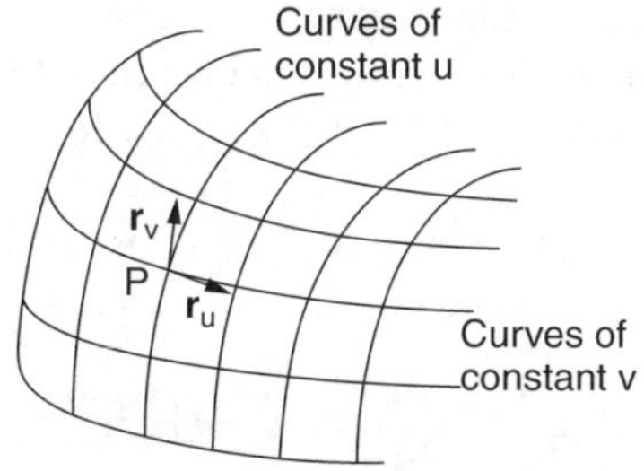

Fig. 3.3 Surface with curvilinear co-ordinates u, v, and base vectors $\mathbf{r}_u$, $\mathbf{r}_v$.

with *rank*M $= 2$, except at the vertex of the cone, where $u = 0$ and *rank*M $= 1$. The vertex will clearly be a singular point whatever the parametrization of the surface.

In general, when a surface is parametrized as in eqn (3.3), we may assume that the surface can be covered by a grid consisting of two families of co-ordinate curves; on the members of one family u varies but v is constant, while on the other v varies and u is constant (Fig. 3.3). Through a point P of the surface there will generally pass one member of each family, and tangent vectors in the directions of these curves will be given by $\partial\mathbf{r}/\partial u$ and $\partial\mathbf{r}/\partial v$, which we shall write here as $\mathbf{r}_u$ and $\mathbf{r}_v$ respectively.

Now the rows of M as defined in eqn (3.8) are just the cartesian components of $\mathbf{r}_u$ and $\mathbf{r}_v$, and the condition *rank*M $= 2$ is equivalent to the condition that $\mathbf{r}_u$ and $\mathbf{r}_v$ should be non-zero and independent, i.e.

$$\mathbf{r}_u \times \mathbf{r}_v \neq \mathbf{0}. \tag{3.11}$$

This condition holds good under a change of parametrization from (u, v) to $(\overline{u}, \overline{v})$, given some relation $u = u(\overline{u}, \overline{v})$, $v = v(\overline{u}, \overline{v})$, provided that the Jacobian of the transformation is non-zero, that is,

$$\frac{\partial u}{\partial \overline{u}}\frac{\partial v}{\partial \overline{v}} - \frac{\partial u}{\partial \overline{v}}\frac{\partial v}{\partial \overline{u}} \neq 0.$$

For, using Chain Rules, we have

$$\mathbf{r}_{\overline{u}} \times \mathbf{r}_{\overline{v}} = \left(\mathbf{r}_u\frac{\partial u}{\partial \overline{u}} + \mathbf{r}_v\frac{\partial v}{\partial \overline{u}}\right) \times \left(\mathbf{r}_u\frac{\partial u}{\partial \overline{v}} + \mathbf{r}_v\frac{\partial v}{\partial \overline{v}}\right) = \left(\frac{\partial u}{\partial \overline{u}}\frac{\partial v}{\partial \overline{v}} - \frac{\partial u}{\partial \overline{v}}\frac{\partial v}{\partial \overline{u}}\right)(\mathbf{r}_u \times \mathbf{r}_v).$$

3.2 Intrinsic geometry of surfaces

Intrinsic properties of surfaces are those which can be formulated without reference to the space (in the present case E^3) in which they are embedded, in particular without reference to vectors normal to the surface. These properties depend essentially on a certain quadratic differential form, the so-called *first fundamental form*, which contains all the basic information about the metrical properties of the surface.

Here we assume that a given surface S is parametrized as in eqn (3.3), with parameters u^α, $\alpha = 1, 2$ (or, where this would make for clarity, u, v). It is again convenient to use superscript indices, as the u^α will serve as curvilinear co-ordinates for the surface, but the convention now is that *Greek indices can take only the values 1 or 2.* Moreover, repeated Greek indices in an expression normally imply summation over the values 1 and 2, according to the corresponding summation convention. We shall also retain the background rectangular cartesian co-ordinates y_1, y_2, y_3 (or x, y, z), and a set of curvilinear co-ordinates x^1, x^2, x^3 (or ξ, η, ς) for the space E^3. On S we have $y_i = y_i(u^1, u^2)$, $i = 1, 2, 3$.

At a point P on S two co-ordinate curves u^α = constant intersect, and vectors in the tangent plane at P, tangential to these curves, are

$$\mathbf{a}_\alpha = \frac{\partial \mathbf{r}}{\partial u^\alpha}, \quad \alpha = 1, 2. \tag{3.12}$$

With reference to the background space we have

$$\mathbf{a}_\alpha = \frac{\partial \mathbf{r}}{\partial y_i}\frac{\partial y_i}{\partial u^\alpha} = \mathbf{i}_i \frac{\partial y_i}{\partial u^\alpha} = \frac{\partial \mathbf{r}}{\partial x^i}\frac{\partial x^i}{\partial u^\alpha} = \mathbf{g}_i \frac{\partial x^i}{\partial u^\alpha}, \tag{3.13}$$

where $\mathbf{i}_i$ and $\mathbf{g}_i$ are the unit basis vectors and covariant base-vectors of the background three-dimensional cartesian and curvilinear systems, respectively. Here the index i is taken to be summed from 1 to 3.

A curve C on the surface S can be given in terms of a parameter t by

$$u^\alpha = u^\alpha(t), \quad \alpha = 1, 2. \tag{3.14}$$

An infinitesimal line-segment of C may be represented by

$$\mathbf{dr} = \frac{\partial \mathbf{r}}{\partial u^\alpha} \mathrm{d}u^\alpha = \mathbf{a}_\alpha \, \mathrm{d}u^\alpha, \tag{3.15}$$

with α summed over the values 1 and 2. Hence differentials of arc-length ds are given by

$$\mathrm{d}s^2 = \mathbf{dr} \cdot \mathbf{dr} = (\mathbf{a}_\alpha \, \mathrm{d}u^\alpha) \cdot (\mathbf{a}_\beta \, \mathrm{d}u^\beta) = \mathbf{a}_\alpha \cdot \mathbf{a}_\beta \, \mathrm{d}u^\alpha \, \mathrm{d}u^\beta.$$

This gives the first fundamental form referred to above:

$$\mathrm{d}s^2 = a_{\alpha\beta} \, \mathrm{d}u^\alpha \, \mathrm{d}u^\beta, \tag{3.16}$$

with

$$a_{\alpha\beta} = \mathbf{a}_\alpha \cdot \mathbf{a}_\beta = \frac{\partial y_i}{\partial u^\alpha}\frac{\partial y_i}{\partial u^\beta} = \left(\mathbf{g}_i \frac{\partial x^i}{\partial u^\alpha}\right) \cdot \left(\mathbf{g}_j \frac{\partial x^j}{\partial u^\beta}\right) = g_{ij} \frac{\partial x^i}{\partial u^\alpha}\frac{\partial x^j}{\partial u^\beta}, \tag{3.17}$$

where g_{ij} is the covariant metric tensor of the background curvilinear system; $a_{\alpha\beta}$ is called the *covariant metric tensor* of the surface. The justification for calling it a covariant tensor is that it has second-order covariant tensor properties *with respect to transformations of surface co-ordinates*, in the sense that a transformation from (u^1, u^2) to a different pair $(\overline{u}^1, \overline{u}^2)$ produces a new set of values

$$\overline{a}_{\alpha\beta} = \frac{\partial y_i}{\partial \overline{u}^\alpha}\frac{\partial y_i}{\partial \overline{u}^\beta} = \frac{\partial u^\gamma}{\partial \overline{u}^\alpha}\frac{\partial y_i}{\partial u^\gamma}\frac{\partial u^\delta}{\partial \overline{u}^\beta}\frac{\partial y_i}{\partial u^\delta} = \frac{\partial u^\gamma}{\partial \overline{u}^\alpha}\frac{\partial u^\delta}{\partial \overline{u}^\beta}a_{\gamma\delta} \tag{3.18}$$

by eqn (3.17). Thus $a_{\alpha\beta}$ transforms according to eqn (1.76), except that the indices γ, δ are summed only over the values 1 and 2.

The first fundamental form is also commonly denoted in terms of u, v as

$$\mathrm{d}s^2 = E(\mathrm{d}u)^2 + 2F\,\mathrm{d}u\,\mathrm{d}v + G(\mathrm{d}v)^2, \tag{3.19}$$

where clearly

$$E = a_{11}, \quad F = a_{12} = a_{21}, \quad G = a_{22}. \tag{3.20}$$

We also have the formula

$$\frac{\mathrm{d}s}{\mathrm{d}t} = \sqrt{a_{\alpha\beta}\frac{\mathrm{d}u^\alpha}{\mathrm{d}t}\frac{\mathrm{d}u^\beta}{\mathrm{d}t}}, \tag{3.21}$$

giving arc-length

$$s = \int_{t_0}^{t} \sqrt{a_{\alpha\beta}\dot{u}^\alpha\dot{u}^\beta}\,\mathrm{d}t \tag{3.22}$$

measured from some point on the curve with $t = t_0$, with derivatives with respect to t denoted by a dot.

Exercise 1. For the surface $z = f(x, y)$ parametrized with $u^1 = x$, $u^2 = y$, show that

$$a_{11} = 1 + \left(\frac{\partial f}{\partial x}\right)^2, \quad a_{12} = \left(\frac{\partial f}{\partial x}\right)\left(\frac{\partial f}{\partial y}\right), \quad a_{22} = 1 + \left(\frac{\partial f}{\partial y}\right)^2. \tag{3.23}$$

Exercise 2. For the surface $\mathbf{r} = \mathbf{r}(u, v) = (u + v^2, u^2 + v, uv)$, show that

$$a_{11} = 1 + 4u^2 + v^2, \quad a_{12} = 2u + 2v + uv, \quad a_{22} = 1 + u^2 + 4v^2.$$

The quadratic form (3.16) must be positive definite, and this holds if and only if

$$a_{11} > 0 \quad \text{and } a = a_{11}a_{22} - (a_{12})^2 > 0, \tag{3.24}$$

where a is the second-order determinant of the 2×2 matrix $(a_{\alpha\beta})$.

If inequalities (3.24) are both satisfied, then we necessarily also have $a_{22} > 0$. Arc-length along the u^1 and u^2 co-ordinate curves is given by $\sqrt{a_{11}}\,\mathrm{d}u^1$ and $\sqrt{a_{22}}\,\mathrm{d}u^2$ respectively. The second inequality in eqn (3.24) corresponds to the inequality

$$|\mathbf{a}_1 \times \mathbf{a}_2|^2 = |\mathbf{a}_1|^2|\mathbf{a}_2|^2 - (\mathbf{a}_1 \cdot \mathbf{a}_2)^2 > 0,$$

which is guaranteed if the point P under consideration is non-singular. It follows, incidentally, that

$$|\mathbf{a}_1 \times \mathbf{a}_2| = \sqrt{a}. \tag{3.25}$$

Exercise 3. Show (by taking determinants of both sides of eqn (3.18)) that the property of positive definiteness of $a_{\alpha\beta}$ is preserved under co-ordinate transformations provided the Jacobian of the transformation is non-zero.

Vectors in the tangent plane to S at a point P of the surface are linear combinations of the tangent vectors $\mathbf{a}_1$ and $\mathbf{a}_2$, and may be called *surface vectors*. The vectors $\mathbf{a}_\alpha$ thus serve as covariant base vectors for the tangent plane, and a surface vector $\mathbf{A}$ with $\mathbf{A} = A^\alpha \mathbf{a}_\alpha$ has contravariant components A^α, $\alpha = 1, 2$. An equation analogous to eqn (1.55) now gives the length of a surface vector $\mathbf{A}$:

$$|\mathbf{A}| = \sqrt{a_{\alpha\beta} A^\alpha A^\beta}. \tag{3.26}$$

We can also define surface contravariant base vectors $\mathbf{a}^\alpha$, $\alpha = 1, 2$, such that

$$\mathbf{a}^\alpha \cdot \mathbf{a}_\beta = \delta^\alpha_\beta, \tag{3.27}$$

and the surface contravariant metric tensor

$$a^{\alpha\beta} = \mathbf{a}^\alpha \cdot \mathbf{a}^\beta. \tag{3.28}$$

The symmetric 2×2 matrix array corresponding to (3.28) is the inverse of that corresponding to eqn (3.17), that is,

$$a_{\alpha\beta} a^{\alpha\gamma} = \delta^\gamma_\beta. \tag{3.29}$$

Thus we have, explicitly,

$$a^{11} = a_{22}/a, \quad a^{12} = a^{21} = -a_{12}/a, \quad a^{22} = a_{11}/a. \tag{3.30}$$

Exercise 4. Show that $\mathbf{a}^\alpha = a^{\alpha\beta}\mathbf{a}_\beta$ with

$$\mathbf{a}^1 = \frac{1}{a}(a_{22}\mathbf{a}_1 - a_{12}\mathbf{a}_2) \quad \text{and } \mathbf{a}^2 = \frac{1}{a}(-a_{12}\mathbf{a}_1 + a_{11}\mathbf{a}_2). \tag{3.31}$$

Exercise 5. A surface of revolution may be generated in E^3 by rotating the curve in the cartesian plane Oxz given in parametric form by $x = f(u)$, $z = g(u)$ about the axis Oz. This gives the parametric form

$$x = f(u)\cos v, \quad y = f(u)\sin v, \quad z = g(u), \tag{3.32}$$

where u and v may be used as surface co-ordinates. Show that the covariant surface base vectors, with $u = u^1$ and $v = u^2$, are

$$\mathbf{a}_1 = (f'(u)\cos v, f'(u)\sin v, g'(u)), \quad \mathbf{a}_2 = (-f(u)\sin v, f(u)\cos v, 0)$$

in background cartesian co-ordinates and that the covariant metric tensor has components

$$a_{\alpha\beta} = \begin{pmatrix} f'^2 + g'^2 & 0 \\ 0 & f^2 \end{pmatrix}, \tag{3.33}$$

which are functions of u but not v, while the contravariant metric tensor is

$$a^{\alpha\beta} = \begin{pmatrix} (f'^2 + g'^2)^{-1} & 0 \\ 0 & f^{-2} \end{pmatrix}. \tag{3.34}$$

A surface vector $\mathbf{A}$ has covariant and contravariant components with respect to the surface base vectors given, respectively, by

$$A_\alpha = \mathbf{A} \cdot \mathbf{a}_\alpha, \qquad A^\alpha = \mathbf{A} \cdot \mathbf{a}^\alpha. \tag{3.35}$$

Since eqn (3.16) can be written

$$1 = a_{\alpha\beta} \frac{\mathrm{d}u^\alpha}{\mathrm{d}s} \frac{\mathrm{d}u^\beta}{\mathrm{d}s}, \tag{3.36}$$

it follows by comparison with eqn (3.26) that $\mathrm{d}u^\alpha/\mathrm{d}s = \lambda^\alpha$ represents the contravariant components of a unit surface vector. Viewed from E^3 this vector $\boldsymbol{\lambda}$ has cartesian components

$$\lambda_i = \frac{\mathrm{d}y_i}{\mathrm{d}s} = \frac{\partial y_i}{\partial u^\alpha} \frac{\mathrm{d}u^\alpha}{\mathrm{d}s} = \frac{\partial y_i}{\partial u^\alpha} \lambda^\alpha \tag{3.37}$$

(which may be regarded as a set of direction cosines) and background contravariant curvilinear components

$$\lambda^i = \frac{\mathrm{d}x^i}{\mathrm{d}s} = \frac{\partial x^i}{\partial u^\alpha} \frac{\mathrm{d}u^\alpha}{\mathrm{d}s} = \frac{\partial x^i}{\partial u^\alpha} \lambda^\alpha. \tag{3.38}$$

The angle θ between directions specified by unit surface vectors $\boldsymbol{\lambda}$ and $\boldsymbol{\mu}$ each satisfying $a_{\alpha\beta}\lambda^\alpha\lambda^\beta = 1$ and $a_{\alpha\beta}\mu^\alpha\mu^\beta = 1$ is given by

$$\cos\theta = \boldsymbol{\lambda} \cdot \boldsymbol{\mu} = \lambda_i \mu_i$$

in cartesian components, or, by eqn (3.37),

$$\cos\theta = \frac{\partial y_i}{\partial u^\alpha} \frac{\partial y_i}{\partial u^\beta} \lambda^\alpha \mu^\beta = a_{\alpha\beta}\lambda^\alpha\mu^\beta \tag{3.39}$$

using eqn (3.17). This result may be compared with the general equations for a scalar product in eqn (1.54).

Unit surface vectors $\boldsymbol{\lambda}$, $\boldsymbol{\mu}$ tangential to the u^1 and u^2 co-ordinate curves at a point must have contravariant components given by, respectively,

$$\lambda^1 = \frac{1}{\sqrt{a_{11}}}, \quad \lambda^2 = 0 \quad \text{and } \mu^1 = 0, \quad \mu^2 = \frac{1}{\sqrt{a_{22}}}. \tag{3.40}$$

According to eqn (3.39) the angle θ between the co-ordinate curves is given by

$$\cos\theta = a_{12}\lambda^1\mu^2 = \frac{a_{12}}{\sqrt{a_{11}a_{22}}}. \tag{3.41}$$

The co-ordinate curves form an orthogonal network if $a_{12} = F = 0$ everywhere.

The element of surface area $\mathrm{d}\sigma$ given by the parallelogram with sides formed by the line-segments $\mathbf{a}_1\,\mathrm{d}u^1$, $\mathbf{a}_2\,\mathrm{d}u^2$ tangential to the co-ordinate curves at a point is

$$\mathrm{d}\sigma = \sqrt{a}\,\mathrm{d}u^1\,\mathrm{d}u^2, \tag{3.42}$$

analogously to eqn (1.44).

It is sometimes convenient to have at our disposal the two-dimensional alternating symbol $e^{\alpha\beta} = e_{\alpha\beta}$ satisfying

$$e^{11} = e_{11} = e^{22} = e_{22} = 0, \quad e^{12} = e_{12} = 1, \quad e^{21} = e_{21} = -1. \tag{3.43}$$

Under transformations from surface co-ordinates (u^1, u^2) to $(\overline{u}^1, \overline{u}^2)$ we then have

$$\frac{\partial \overline{u}^\gamma}{\partial u^\alpha}\frac{\partial \overline{u}^\delta}{\partial u^\beta}e^{\alpha\beta} = \frac{\partial \overline{u}^\gamma}{\partial u^1}\frac{\partial \overline{u}^\delta}{\partial u^2} - \frac{\partial \overline{u}^\gamma}{\partial u^2}\frac{\partial \overline{u}^\delta}{\partial u^1} = Je^{\gamma\delta},$$

and, similarly,

$$\frac{\partial u^\alpha}{\partial \overline{u}^\gamma}\frac{\partial u^\beta}{\partial \overline{u}^\delta}e_{\alpha\beta} = J^{-1}e_{\gamma\delta},$$

where J is the Jacobian of the transformation:

$$J = \begin{vmatrix} \dfrac{\partial \overline{u}^1}{\partial u^1} & \dfrac{\partial \overline{u}^1}{\partial u^2} \\ \dfrac{\partial \overline{u}^2}{\partial u^1} & \dfrac{\partial \overline{u}^2}{\partial u^2} \end{vmatrix}.$$

Thus $e^{\alpha\beta}$ and $e_{\alpha\beta}$ transform like relative tensors. Derivations similar to those resulting in the definitions (1.92) and (1.93) show that absolute (surface) tensors are given by $\varepsilon^{\alpha\beta}$ and $\varepsilon_{\alpha\beta}$, where

$$\varepsilon^{\alpha\beta} = \frac{1}{\sqrt{a}}e^{\alpha\beta}, \quad \varepsilon_{\alpha\beta} = \sqrt{a}e_{\alpha\beta}. \tag{3.44}$$

Note that

$$\varepsilon_{\alpha\beta}\varepsilon^{\alpha\gamma} = e_{\alpha\beta}e^{\alpha\gamma} = \delta^\gamma_\beta. \tag{3.45}$$

Exercise 6. Show that

$$\varepsilon^{\alpha\beta}\varepsilon^{\gamma\delta}a_{\alpha\gamma}a_{\beta\delta} = \varepsilon_{\alpha\beta}\varepsilon_{\gamma\delta}a^{\alpha\gamma}a^{\beta\delta} = 2.$$

We can now say, for example, that, given two unit surface vectors $\boldsymbol{\lambda}$ and $\boldsymbol{\mu}$, a consistent tensor expression is given by $\varepsilon_{\alpha\beta}\lambda^\alpha\mu^\beta$. Comparing this with eqn (1.94), we see that it represents the component of the vector product of $\boldsymbol{\lambda}$ and $\boldsymbol{\mu}$ in the direction of a unit vector normal to the surface, a direction which remains invariant under changes of surface co-ordinates. Thus, if θ is the angle between the directions $\boldsymbol{\lambda}, \boldsymbol{\mu}$ (in the sense of a positive right-handed screw rotation from $\boldsymbol{\lambda}$ to $\boldsymbol{\mu}$), we have

$$\sin\theta = \varepsilon_{\alpha\beta}\lambda^\alpha\mu^\beta. \tag{3.46}$$

The condition for the orthogonality of two surface directions $\boldsymbol{\lambda}, \boldsymbol{\mu}$ may now be written either, using eqn (3.39), as

$$a_{\alpha\beta}\lambda^\alpha\mu^\beta = 0, \tag{3.47}$$

or, from eqn (3.46), as

$$\varepsilon_{\alpha\beta}\lambda^\alpha\mu^\beta = \pm 1. \tag{3.48}$$

In this section it may be seen how the metrical properties of a surface, i.e. the measurement of lengths of curves on the surface, angles between intersecting curves, and areas, may be derived with reference to the first fundamental form of the surface, and in particular to the covariant metric tensor $a_{\alpha\beta}$. Since $a_{\alpha\beta}$ itself was derived by using the properties of the enveloping Euclidean space, we may say that the metrical properties of the surface are *induced* by the Euclidean metric of E^3. However, once formulas for length, angle, and area have been established in terms of $a_{\alpha\beta}$, measurement may be carried out on the surface without regard to the embedding space.

When two surfaces (or parts of two surfaces) are such that they admit surface co-ordinate systems which give identical first fundamental forms, then the intrinsic geometry of the surfaces is the same. The existence of these surface co-ordinates implies the existence of a mapping between points on the surfaces such that corresponding curves have the same length or intersect at the same angle. This means that there is no difference locally between the surfaces as far as measurement of lengths, angles, and areas is concerned, no matter how different the surfaces may appear from the standpoint of the enveloping three-dimensional space. Such surfaces are called *isometric*.

A simple example is the isometry between the circular cylinder (3.9) and a portion of a plane. The first fundamental form of the cylinder is easily seen to be

$$a^2(\mathrm{d}\varphi)^2 + (\mathrm{d}z)^2$$

which is equivalent to $(\mathrm{d}x)^2 + (\mathrm{d}y)^2$ for the plane with cartesian co-ordinates under the transformation $a\varphi \to x$ and $z \to y$.

3.3 Surface covariant differentiation

Surface Christoffel symbols of first and second kinds can be defined. It is important to note at the outset, however, that there is no immediate surface equivalent of eqn (1.97), since in general $\partial \mathbf{a}_\alpha/\partial u^\beta$ is not a surface vector and cannot be expressed in terms of $\mathbf{a}_1$ and $\mathbf{a}_2$. But we can effectively use the equivalents of eqns (1.98) and (1.99) to define the surface Christoffel symbols, and then the same procedure that led to eqn (1.108) gives

$$[\alpha\beta, \gamma] = \frac{\partial \mathbf{a}_\alpha}{\partial u^\beta} \cdot \mathbf{a}_\gamma = \frac{\partial^2 \mathbf{r}}{\partial u^\alpha \partial u^\beta} \cdot \frac{\partial \mathbf{r}}{\partial u^\gamma} = \frac{1}{2}\left(\frac{\partial a_{\alpha\gamma}}{\partial u^\beta} + \frac{\partial a_{\beta\gamma}}{\partial u^\alpha} - \frac{\partial a_{\alpha\beta}}{\partial u^\gamma}\right) \tag{3.49}$$

and

$$\Gamma^\gamma_{\alpha\beta} = \frac{\partial \mathbf{a}_\alpha}{\partial u^\beta} \cdot \mathbf{a}^\gamma = a^{\gamma\delta}[\alpha\beta, \delta], \tag{3.50}$$

where the Greek indices can take only the values 1 and 2.

Explicitly, using eqns (3.30), we can write, putting $u = u^1$, $v = u^2$,

$$[11,1] = \frac{1}{2}\frac{\partial a_{11}}{\partial u}, \quad [11,2] = \frac{\partial a_{12}}{\partial u} - \frac{1}{2}\frac{\partial a_{11}}{\partial v}, \quad [12,1] = [21,1] = \frac{1}{2}\frac{\partial a_{11}}{\partial v},$$
$$[12,2] = [21,2] = \frac{1}{2}\frac{\partial a_{22}}{\partial u}, \quad [22,1] = \frac{\partial a_{12}}{\partial v} - \frac{1}{2}\frac{\partial a_{22}}{\partial u}, \quad [22,2] = \frac{1}{2}\frac{\partial a_{22}}{\partial v} \tag{3.51}$$

and

$$\Gamma^1_{11} = \frac{1}{2a}\left[a_{22}\frac{\partial a_{11}}{\partial u} + a_{12}\left(\frac{\partial a_{11}}{\partial v} - 2\frac{\partial a_{12}}{\partial u}\right)\right]$$

$$\Gamma^1_{12} = \Gamma^1_{21} = \frac{1}{2a}\left[a_{22}\frac{\partial a_{11}}{\partial v} - a_{12}\frac{\partial a_{22}}{\partial u}\right]$$

$$\Gamma^1_{22} = \frac{1}{2a}\left[a_{22}\left(2\frac{\partial a_{12}}{\partial v} - \frac{\partial a_{22}}{\partial u}\right) - a_{12}\frac{\partial a_{22}}{\partial v}\right] \tag{3.52}$$
$$\Gamma^2_{22} = \frac{1}{2a}\left[a_{11}\frac{\partial a_{22}}{\partial v} + a_{12}\left(\frac{\partial a_{22}}{\partial u} - 2\frac{\partial a_{12}}{\partial v}\right)\right]$$
$$\Gamma^2_{11} = \frac{1}{2a}\left[a_{11}\left(2\frac{\partial a_{12}}{\partial u} - \frac{\partial a_{11}}{\partial v}\right) - a_{12}\frac{\partial a_{11}}{\partial u}\right]$$
$$\Gamma^2_{12} = \Gamma^2_{21} = \frac{1}{2a}\left[a_{11}\frac{\partial a_{22}}{\partial u} - a_{12}\frac{\partial a_{11}}{\partial v}\right].$$

There should be no difficulty in practice in distinguishing between the surface Christoffel symbols and their three-dimensional version in eqn (1.99). The use of either Greek or Roman indices will indicate the context.

Exercise 7. For the surface of revolution given by eqn (3.32), show that the Christoffel symbols are

$$[11, 1] = f'f'' + g'g'', \quad [12, 2] = [21, 2] = ff'$$
$$[22, 1] = -ff', \quad [11, 2] = [12, 1] = [21, 1] = [22, 2] = 0 \tag{3.53}$$

and

$$\Gamma^1_{11} = (f'^2 + g'^2)^{-1}(f'f'' + g'g''), \quad \Gamma^2_{12} = \Gamma^2_{21} = f^{-1}f'$$
$$\Gamma^1_{22} = -(f'^2 + g'^2)^{-1}ff', \quad \Gamma^1_{12} = \Gamma^1_{21} = \Gamma^2_{11} = \Gamma^2_{22} = 0. \tag{3.54}$$

Exercise 8. For the spherical surface defined by eqn (3.4), with $u = \theta$, $v = \phi$, show that the only non-vanishing Christoffel symbols are

$$[12, 2] = [21, 2] = -[22, 1] = -\Gamma^1_{22} = \sin\theta\cos\theta, \quad \Gamma^2_{12} = \Gamma^2_{21} = \cot\theta.$$

It is straightforward to obtain the results analogous to eqns (1.106) and (1.107), namely

$$\frac{\partial a_{\alpha\beta}}{\partial u^\gamma} = [\gamma\alpha, \beta] + [\gamma\beta, \alpha] = a_{\delta\beta}\Gamma^\delta_{\gamma\alpha} + a_{\delta\alpha}\Gamma^\delta_{\gamma\beta} \tag{3.55}$$

and

$$\frac{\partial a^{\alpha\beta}}{\partial u^\gamma} = -a^{\delta\alpha}\Gamma^\beta_{\delta\gamma} - a^{\delta\beta}\Gamma^\alpha_{\delta\gamma}. \tag{3.56}$$

Surface covariant differentiation and intrinsic surface derivatives along a surface curve can now be defined. Following eqns (1.120), (1.122), and (2.31), a surface vector **A**, for example, has surface covariant derivatives

$$A^\alpha_{,\beta} = \frac{\partial \mathbf{A}}{\partial u^\beta}\cdot\mathbf{a}^\alpha = \frac{\partial A^\alpha}{\partial u^\beta} + \Gamma^\alpha_{\gamma\beta}A^\gamma, \quad A_{\alpha,\beta} = \frac{\partial \mathbf{A}}{\partial u^\beta}\cdot\mathbf{a}_\alpha = \frac{\partial A_\alpha}{\partial u^\beta} - \Gamma^\gamma_{\alpha\beta}A_\gamma \tag{3.57}$$

and intrinsic derivatives

$$\frac{\delta A^\alpha}{\delta s} = A^\alpha_{,\beta}\frac{du^\beta}{ds} = A^\alpha_{,\beta}\lambda^\beta, \quad \frac{\delta A_\alpha}{\delta s} = A_{\alpha,\beta}\frac{du^\beta}{ds} = A_{\alpha,\beta}\lambda^\beta \tag{3.58}$$

along a surface curve with unit tangent vector $\boldsymbol{\lambda}$.

Moreover, we can have surface covariant and contravariant second-order tensors with covariant derivatives given by formulas corresponding to eqns (1.126) and (1.129), but with Greek rather than Roman indices. In particular, for the contravariant metric tensor $a^{\alpha\beta}$, we can deduce that its surface covariant derivatives are

$$\begin{aligned} a^{\alpha\beta}_{,\gamma} &= \frac{\partial a^{\alpha\beta}}{\partial u^\gamma} + \Gamma^\alpha_{\delta\gamma} a^{\delta\beta} + \Gamma^\beta_{\delta\gamma} a^{\alpha\delta} = \frac{\partial(\mathbf{a}^\alpha \cdot \mathbf{a}^\beta)}{\partial u^\gamma} + \Gamma^\alpha_{\delta\gamma} a^{\delta\beta} + \Gamma^\beta_{\delta\gamma} a^{\alpha\delta} \\ &= \mathbf{a}^\alpha \cdot \frac{\partial \mathbf{a}^\beta}{\partial u^\gamma} + \frac{\partial \mathbf{a}^\alpha}{\partial u^\gamma} \cdot \mathbf{a}^\beta + \Gamma^\alpha_{\delta\gamma} a^{\delta\beta} + \Gamma^\beta_{\delta\gamma} a^{\alpha\delta} \\ &= -\Gamma^\beta_{\delta\gamma} a^{\alpha\delta} - \Gamma^\alpha_{\delta\gamma} a^{\delta\beta} + \Gamma^\alpha_{\delta\gamma} a^{\delta\beta} + \Gamma^\beta_{\delta\gamma} a^{\alpha\delta} = 0, \end{aligned} \tag{3.59}$$

where we have made use of the surface equivalent of eqn (1.105). Similarly we can show that

$$a_{\alpha\beta,\gamma} = 0 \tag{3.60}$$

for all α, β, γ ranging over the values 1, 2.

These results are equivalent to those for the metric tensor for general curvilinear co-ordinates in Chapter 1. Note, however, that the supplementary argument given there based on the existence of a background rectangular cartesian system in which all the covariant derivatives are necessarily zero no longer applies, since a general two-dimensional surface does not admit a cartesian system (unless it is planar).

The surface equation analogous to eqn (1.118) is

$$\Gamma^\alpha_{\alpha\beta} = \frac{1}{\sqrt{a}} \frac{\partial}{\partial u^\beta}(\sqrt{a}), \tag{3.61}$$

and this may be used in the process of proving that the covariant derivatives of $\varepsilon^{\alpha\beta}$ and $\varepsilon_{\alpha\beta}$ satisfy

$$\varepsilon^{\alpha\beta}_{,\gamma} = \varepsilon_{\alpha\beta,\gamma} = 0. \tag{3.62}$$

Investigation of the commutativity of repeated covariant differentiation of an arbitrary surface covariant vector A_α leads to the equation

$$A_{\alpha,\beta\gamma} - A_{\alpha,\gamma\beta} = R^\delta_{.\alpha\beta\gamma} A_\delta, \tag{3.63}$$

as in eqn (1.177), where the mixed *surface Riemann-Christoffel tensor* is given by

$$R^\delta_{.\alpha\beta\gamma} = \frac{\partial \Gamma^\delta_{\alpha\gamma}}{\partial u^\beta} - \frac{\partial \Gamma^\delta_{\alpha\beta}}{\partial u^\gamma} + \Gamma^\mu_{\alpha\gamma}\Gamma^\delta_{\mu\beta} - \Gamma^\mu_{\alpha\beta}\Gamma^\delta_{\mu\gamma} \tag{3.64}$$

similarly to eqn (1.178). Moreover, we can define the covariant fourth-order *surface curvature tensor*

$$R_{\alpha\beta\gamma\delta} = a_{\alpha\mu} R^\mu_{.\beta\gamma\delta}. \tag{3.65}$$

Note that these fourth-order surface tensors do not in general vanish identically, unlike their three-dimensional counterparts in a Euclidean space, as considered in Chapter 1. They have the same skew-symmetric properties discussed there, however, and it follows that the only possible non-vanishing components of $R_{\alpha\beta\gamma\delta}$ are

$$R_{1212} = R_{2121} = -R_{1221} = -R_{2112}.$$

3.4 Geodesic curves

Another aspect of the intrinsic geometry of surfaces arises from the problem of determining the curve of minimum length which joins two given points on the surface. For example, when the surface is a plane in E^3, the shortest distance between two points will be a straight line. If the general problem is approached through the usual calculus of variations, differential equations are obtained which any solution must satisfy. Curves which satisfy these equations are called *geodesics*, although in fact not all solutions are necessarily curves of minimum length.

According to eqn (3.22) the length of a curve C on a surface with co-ordinates u^1, u^2 and covariant metric tensor $a_{\alpha\beta}$ is given by

$$L = \int_{t_1}^{t_2} \sqrt{a_{\alpha\beta}\dot{u}^\alpha \dot{u}^\beta}\, \mathrm{d}t = \int_{t_1}^{t_2} f(u^1, u^2, \dot{u}^1, \dot{u}^2)\, \mathrm{d}t, \tag{3.66}$$

where t is used to parametrize the curve, each $a_{\alpha\beta}$ in general is a function of u^1 and u^2, and we assume that t takes the fixed values t_1, t_2 at the given end-points. Moreover, f is formally regarded as a function of four independent variables. Neighbouring curves, having first-order variations δu^α and $\delta \dot{u}^\alpha$ in the values of u^α and $\dot{u}^\alpha$ at corresponding values of t, but still having the same end-points, have a first-order variation in length

$$\delta L = \int_{t_1}^{t_2} \delta f\, \mathrm{d}t = \int_{t_1}^{t_2} \left(\frac{\partial f}{\partial u^\alpha}\delta u^\alpha + \frac{\partial f}{\partial \dot{u}^\alpha}\delta \dot{u}^\alpha \right) \mathrm{d}t$$

with summation over α. Integration by parts on the last term gives

$$\delta L = \left[\frac{\partial f}{\partial \dot{u}^\alpha}\delta u^\alpha \right]_{t_1}^{t_2} + \int_{t_1}^{t_2} \left\{ \frac{\partial f}{\partial u^\alpha} - \frac{\mathrm{d}}{\mathrm{d}t}\left(\frac{\partial f}{\partial \dot{u}^\alpha} \right) \right\} \delta u^\alpha\, \mathrm{d}t,$$

and the integrated part vanishes because of the fixed end-point requirement.

If L is to be a minimum for C, the first-order variation δL must be zero for arbitrary variations δu^α. Thus

$$\int_{t_1}^{t_2} \left\{ \frac{\partial f}{\partial u^\alpha} - \frac{\mathrm{d}}{\mathrm{d}t}\left(\frac{\partial f}{\partial \dot{u}^\alpha} \right) \right\} \delta u^\alpha\, \mathrm{d}t = 0$$

for arbitrary (first-order) variations $\delta u^1, \delta u^2$. A standard argument of the calculus of variations then leads to the conclusion that the two differential equations

$$\frac{\partial f}{\partial u^\alpha} - \frac{\mathrm{d}}{\mathrm{d}t}\left(\frac{\partial f}{\partial \dot{u}^\alpha} \right) = 0, \quad \alpha = 1, 2, \tag{3.67}$$

must hold everywhere on C. Curves for which these equations hold are geodesics. Equations (3.67) are the *Euler-Lagrange equations* (or just the *Euler equations*) for the variational problem $\delta L = 0$ with L given by (3.66).

Now since $f = \sqrt{a_{\alpha\beta}(u^1, u^2)\dot{u}^\alpha \dot{u}^\beta}$, we have

$$\frac{\partial f}{\partial u^\gamma} = \frac{1}{2f}\frac{\partial a_{\alpha\beta}}{\partial u^\gamma}\dot{u}^\alpha \dot{u}^\beta \quad \text{and} \quad \frac{\partial f}{\partial \dot{u}^\gamma} = \frac{1}{f}a_{\gamma\beta}\dot{u}^\beta.$$

Hence, with γ in place of α, eqn (3.67) becomes

$$\frac{1}{2f}\frac{\partial a_{\alpha\beta}}{\partial u^\gamma}\dot{u}^\alpha\dot{u}^\beta - \frac{\mathrm{d}}{\mathrm{d}t}\left(\frac{1}{f}a_{\gamma\beta}\dot{u}^\beta\right) = 0, \quad \gamma = 1, 2. \tag{3.68}$$

A simpler form of these equations results if we express the parametric form of the solution curve in terms of arc-length s rather than t. Along a solution curve we now have $f = 1$, a constant, by eqn (3.36). Thus eqn (3.68) becomes

$$\frac{\mathrm{d}}{\mathrm{d}s}\left(a_{\gamma\beta}\frac{\mathrm{d}u^\beta}{\mathrm{d}s}\right) - \frac{1}{2}\frac{\partial a_{\alpha\beta}}{\partial u^\gamma}\frac{\mathrm{d}u^\alpha}{\mathrm{d}s}\frac{\mathrm{d}u^\beta}{\mathrm{d}s} = 0, \tag{3.69}$$

that is,

$$a_{\gamma\beta}\frac{\mathrm{d}^2u^\beta}{\mathrm{d}s^2} + \frac{\partial a_{\gamma\beta}}{\partial u^\alpha}\frac{\mathrm{d}u^\alpha}{\mathrm{d}s}\frac{\mathrm{d}u^\beta}{\mathrm{d}s} - \frac{1}{2}\frac{\partial a_{\alpha\beta}}{\partial u^\gamma}\frac{\mathrm{d}u^\alpha}{\mathrm{d}s}\frac{\mathrm{d}u^\beta}{\mathrm{d}s} = 0,$$

which may be expressed as

$$a_{\gamma\beta}\frac{\mathrm{d}^2u^\beta}{\mathrm{d}s^2} + \tfrac{1}{2}\left(\frac{\partial a_{\gamma\beta}}{\partial u^\alpha} + \frac{\partial a_{\gamma\alpha}}{\partial u^\beta} - \frac{\partial a_{\beta\alpha}}{\partial u^\gamma}\right)\frac{\mathrm{d}u^\beta}{\mathrm{d}s}\frac{\mathrm{d}u^\alpha}{\mathrm{d}s} = 0,$$

exploiting the symmetry of $(\mathrm{d}u^\beta/\mathrm{d}s)(\mathrm{d}u^\alpha/\mathrm{d}s)$ with respect to β and α. In other words, using eqn (3.49),

$$a_{\gamma\beta}\frac{\mathrm{d}^2u^\beta}{\mathrm{d}s^2} + [\beta\alpha, \gamma]\frac{\mathrm{d}u^\beta}{\mathrm{d}s}\frac{\mathrm{d}u^\alpha}{\mathrm{d}s} = 0, \quad \gamma = 1, 2. \tag{3.70}$$

Re-writing α as δ, multiplying through by $a^{\alpha\gamma}$ (implying summation over γ), and using eqn (3.50), now gives

$$\frac{\mathrm{d}^2u^\alpha}{\mathrm{d}s^2} + a^{\alpha\gamma}[\beta\delta, \gamma]\frac{\mathrm{d}u^\beta}{\mathrm{d}s}\frac{\mathrm{d}u^\delta}{\mathrm{d}s} = \frac{\mathrm{d}^2u^\alpha}{\mathrm{d}s^2} + \Gamma^\alpha_{\beta\delta}\frac{\mathrm{d}u^\beta}{\mathrm{d}s}\frac{\mathrm{d}u^\delta}{\mathrm{d}s} = 0, \quad \alpha = 1, 2. \tag{3.71}$$

Equations (3.70) and (3.71) are second-order differential equations which define geodesic curves, curves whose length is stationary (rather than necessarily being a minimum) with respect to small variations given two fixed end-points. If we are given starting values of u^α and $\mathrm{d}u^\alpha/\mathrm{d}s$ at a point on a surface, we can in principle solve either of these pairs of equations and obtain a geodesic curve.

By the two-dimensional version of eqn (2.32), we can express eqn (3.71) in the simple form

$$\frac{\delta}{\delta s}\left(\frac{\mathrm{d}u^\alpha}{\mathrm{d}s}\right) = 0. \tag{3.72}$$

In practice eqns (3.71) may be quite complicated. For the surface of revolution as defined by eqn (3.32) they become, using eqn (3.54),

$$\frac{\mathrm{d}^2u}{\mathrm{d}s^2} + (f'^2 + g'^2)^{-1}(f'f'' + g'g'')\left(\frac{\mathrm{d}u}{\mathrm{d}s}\right)^2 - (f'^2 + g'^2)ff'\left(\frac{\mathrm{d}v}{\mathrm{d}s}\right)^2 = 0,$$

$$\frac{\mathrm{d}^2v}{\mathrm{d}s^2} + 2f^{-1}f'\left(\frac{\mathrm{d}u}{\mathrm{d}s}\right)\left(\frac{\mathrm{d}v}{\mathrm{d}s}\right) = 0. \tag{3.73}$$

The first equation shows that circular cross-sections (*parallels*) of the surface $u =$ constant can be geodesics only when $f'(u) = 0$, i.e. when the radius of the cross-section is stationary with respect to u. The second equation shows that meridians $v =$ constant are geodesics. More generally, if the second equation is multiplied through by f^2 it can then be integrated directly to give

$$f^2 \frac{\mathrm{d}v}{\mathrm{d}s} = h$$

where h is a constant of integration. An explicit form for a geodesic then follows, since by eqn (3.33)

$$f^4\,\mathrm{d}v^2 = h^2\,\mathrm{d}s^2 = h^2[(f'^2 + g'^2)\,\mathrm{d}u^2 + f^2\,\mathrm{d}v^2],$$

and then, re-arranging,

$$f^2(f^2 - h^2)\,\mathrm{d}v^2 = h^2(f'^2 + g'^2)\,\mathrm{d}u^2,$$

so that

$$v = C \pm \int \frac{h\sqrt{f'^2 + g'^2}}{f\sqrt{f^2 - h^2}}\,\mathrm{d}u \tag{3.74}$$

in terms of two constants of integration C, h. For example, in the case of a circular cylinder, with $f(u) = a$, and $g(u) = cu$, where a and c are constants, we see that the relationship between u and v must be linear. It follows that geodesics on a circular cylinder are helices.

If we put $u^1 = u$, $u^2 = v$, the two eqns (3.71) may be combined into one by assuming that the solution can be represented as a relation between the parameters u, v. Since, with $\mathrm{d}/\mathrm{d}s$ denoted by $(\ )'$,

$$\frac{\mathrm{d}v}{\mathrm{d}u} = \frac{v'}{u'} \quad \text{and} \quad \frac{\mathrm{d}^2 v}{\mathrm{d}u^2} = \frac{\mathrm{d}}{\mathrm{d}u}\left(\frac{v'}{u'}\right)\frac{1}{u'} = \frac{u'v'' - v'u''}{u'^3},$$

substituting for u'' and v'' from eqn (3.71) leads directly to

$$\frac{\mathrm{d}^2 v}{\mathrm{d}u^2} - \Gamma^1_{22}\left(\frac{\mathrm{d}v}{\mathrm{d}u}\right)^3 + (\Gamma^2_{22} - 2\Gamma^1_{12})\left(\frac{\mathrm{d}v}{\mathrm{d}u}\right)^2 + (2\Gamma^2_{12} - \Gamma^1_{11})\left(\frac{\mathrm{d}v}{\mathrm{d}u}\right) + \Gamma^2_{11} = 0. \tag{3.75}$$

For a planar surface on which we take surface co-ordinates to be rectangular cartesians $u = X$, $v = Y$, the Christoffel symbols are all zero, and eqn (3.75) reduces to

$$\frac{\mathrm{d}^2 Y}{\mathrm{d}X^2} = 0,$$

with straight line solutions $Y = mX + C$ as expected.

Another, non-intrinsic, feature of geodesics arises when we consider their curvature as viewed from E^3. For any curve on the surface with points $\mathbf{r} = \mathbf{r}(u^1, u^2)$, there are unit tangent vectors

$$\frac{\mathrm{d}\mathbf{r}}{\mathrm{d}s} = \frac{\partial \mathbf{r}}{\partial u^1}\frac{\mathrm{d}u^1}{\mathrm{d}s} + \frac{\partial \mathbf{r}}{\partial u^2}\frac{\mathrm{d}u^2}{\mathrm{d}s} = \mathbf{a}_1\frac{\mathrm{d}u^1}{\mathrm{d}s} + \mathbf{a}_2\frac{\mathrm{d}u^2}{\mathrm{d}s}. \tag{3.76}$$

Further differentiation gives

$$\frac{d^2\mathbf{r}}{ds^2} = \mathbf{a}_1\frac{d^2u^1}{ds^2} + \mathbf{a}_2\frac{d^2u^2}{ds^2} + \frac{\partial \mathbf{a}_1}{\partial u^1}\left(\frac{du^1}{ds}\right)^2$$
$$+\left(\frac{\partial \mathbf{a}_1}{\partial u^2} + \frac{\partial \mathbf{a}_2}{\partial u^1}\right)\left(\frac{du^1}{ds}\right)\left(\frac{du^2}{ds}\right) + \frac{\partial \mathbf{a}_2}{\partial u^2}\left(\frac{du^2}{ds}\right)^2$$
$$= \mathbf{a}_\beta\frac{d^2u^\beta}{ds^2} + \frac{\partial \mathbf{a}_\beta}{\partial u^\gamma}\left(\frac{du^\beta}{ds}\right)\left(\frac{du^\gamma}{ds}\right),$$

and hence, with d/ds denoted by ( )$'$,

$$\mathbf{r}''\cdot\mathbf{a}_\alpha = \mathbf{a}_\alpha\cdot\mathbf{a}_\beta u^{\beta\prime\prime} + \mathbf{a}_\alpha\cdot\frac{\partial \mathbf{a}_\beta}{\partial u^\gamma}u^{\beta\prime}u^{\gamma\prime} = a_{\alpha\beta}\frac{d^2u^\beta}{ds^2} + [\beta\gamma,\alpha]\frac{du^\beta}{ds}\frac{du^\gamma}{ds}, \qquad (3.77)$$

using eqns (3.17) and (3.49).

Comparison with eqn (3.70) now shows that for a geodesic

$$\mathbf{r}''\cdot\mathbf{a}_\alpha = 0, \quad \alpha = 1, 2, \qquad (3.78)$$

which implies that the direction of $\mathbf{r}''$ is orthogonal to the tangent plane to the surface at any point. Recalling the significance of $\mathbf{r}''$ from Chapter 2, we deduce that geodesics have the property that *the principal normal is always in the same direction as the normal to the surface*.

3.5 Surface Frenet equations and geodesic curvature

In Chapter 2 we looked at the curvature and torsion of a space-curve and derived the Serret-Frenet equations. Now we discuss a general curve on a surface on which u^1 and u^2 are surface co-ordinates. The curve is given by

$$u^\alpha = u^\alpha(s)$$

in terms of arc-length as parameter, and the unit surface contravariant tangent vector λ^α at any point satisfies

$$a_{\alpha\beta}\lambda^\alpha\lambda^\beta = 1 \qquad (3.79)$$

by eqn (3.36). We can apply intrinsic differentiation to this equation, bearing in mind that the $a_{\alpha\beta}$s are effectively constant with respect to intrinsic differentiation, because of eqns (2.33) and (3.60).

We obtain

$$a_{\alpha\beta}\lambda^\alpha\frac{\delta\lambda^\beta}{\delta s} + a_{\alpha\beta}\frac{\delta\lambda^\alpha}{\delta s}\lambda^\beta = 2a_{\alpha\beta}\lambda^\alpha\frac{\delta\lambda^\beta}{\delta s} = 0, \qquad (3.80)$$

using the symmetry of $a_{\alpha\beta}$. Thus by eqn (3.39) $\delta\lambda^\beta/\delta s$ are the contravariant components of a surface vector which is orthogonal to λ^α.

Now let a contravariant unit vector parallel to $\delta\lambda^\alpha/\delta s$ be ν^α. We choose the sense of ν^α such that

$$\varepsilon_{\alpha\beta}\lambda^\alpha\nu^\beta = 1, \qquad (3.81)$$

or, equivalently,

$$\nu^\alpha = \varepsilon^{\beta\alpha}\lambda_\beta, \tag{3.82}$$

in terms of the associated covariant components λ_β, where lowering the index is represented here by

$$\lambda_\alpha = a_{\alpha\beta}\lambda^\beta$$

rather than eqn (1.53). To verify eqn (3.82), note that, using eqn (3.45),

$$\varepsilon_{\alpha\beta}\lambda^\alpha\nu^\beta = \varepsilon_{\alpha\beta}\lambda^\alpha(\varepsilon^{\gamma\beta}\lambda_\gamma) = \delta^\gamma_\alpha\lambda^\alpha\lambda_\gamma = \lambda^\gamma\lambda_\gamma = a_{\beta\gamma}\lambda^\beta\lambda^\gamma = 1 \tag{3.83}$$

and

$$a_{\alpha\beta}\lambda^\alpha\nu^\beta = a_{\alpha\beta}\lambda^\alpha(\varepsilon^{\gamma\beta}\lambda_\gamma) = \varepsilon^{\gamma\beta}\lambda_\beta\lambda_\gamma = 0,$$

due to the symmetric and skew-symmetric natures of $\lambda_\beta\lambda_\gamma$ and $\varepsilon^{\gamma\beta}$, respectively.

Similarly we can verify that

$$\lambda^\alpha = -\varepsilon^{\beta\alpha}\nu_\beta. \tag{3.84}$$

We now write

$$\frac{\delta\lambda^\alpha}{\delta s} = \kappa_g\nu^\alpha, \tag{3.85}$$

where the scalar magnitude κ_g is called the *geodesic curvature* of the curve at the point in question.

Because of eqns (2.33) and (3.62), $\varepsilon^{\alpha\beta}$ also acts as a constant for the purposes of intrinsic differentiation. Hence eqn (3.82) yields

$$\frac{\delta\nu^\alpha}{\delta s} = \varepsilon^{\beta\alpha}\frac{\delta\lambda_\beta}{\delta s} = \kappa_g\varepsilon^{\beta\alpha}\nu_\beta,$$

using the associated covariant components of the surface vector quantities $\delta\boldsymbol{\lambda}/\delta s$, $\boldsymbol{\nu}$ in eqn (3.85). So, from eqn (3.84), we obtain

$$\frac{\delta\nu^\alpha}{\delta s} = -\kappa_g\lambda^\alpha. \tag{3.86}$$

Equations (3.85) and (3.86) constitute the *surface-Frenet equations* for curves on surfaces.

For an actual geodesic curve, eqn (3.72) shows that $\delta\lambda^\alpha/\delta s = 0$. This means that *the geodesic curvature of a geodesic is zero.*

The surface-Frenet equations may be expressed, using the surface form of eqn (2.32), as

$$\frac{\mathrm{d}\lambda^\alpha}{\mathrm{d}s} + \Gamma^\alpha_{\beta\gamma}\lambda^\beta\frac{\mathrm{d}u^\gamma}{\mathrm{d}s} = \frac{\mathrm{d}\lambda^\alpha}{\mathrm{d}s} + \Gamma^\alpha_{\beta\gamma}\lambda^\beta\lambda^\gamma = \kappa_g\nu^\alpha,$$
$$\frac{\mathrm{d}\nu^\alpha}{\mathrm{d}s} + \Gamma^\alpha_{\beta\gamma}\nu^\beta\frac{\mathrm{d}u^\gamma}{\mathrm{d}s} = \frac{\mathrm{d}\nu^\alpha}{\mathrm{d}s} + \Gamma^\alpha_{\beta\gamma}\nu^\beta\lambda^\gamma = -\kappa_g\lambda^\alpha. \tag{3.87}$$

An explicit formula for κ_g may be obtained by multiplying the first of these equations through by $\varepsilon_{\nu\alpha}\lambda^\nu$, implying summation over α and ν, and making use of eqn (3.83).

We obtain

$$\begin{aligned}\kappa_g &= \varepsilon_{\nu\alpha}\lambda^{\nu}\frac{\mathrm{d}\lambda^{\alpha}}{\mathrm{d}s} + \varepsilon_{\nu\alpha}\Gamma^{\alpha}_{\beta\gamma}\lambda^{\nu}\lambda^{\beta}\lambda^{\gamma}\\ &= \sqrt{a}\left(e_{\nu\alpha}\frac{\mathrm{d}u^{\nu}}{\mathrm{d}s}\frac{\mathrm{d}^2u^{\alpha}}{\mathrm{d}s^2} + e_{\nu\alpha}\Gamma^{\alpha}_{\beta\gamma}\frac{\mathrm{d}u^{\nu}}{\mathrm{d}s}\frac{\mathrm{d}u^{\beta}}{\mathrm{d}s}\frac{\mathrm{d}u^{\gamma}}{\mathrm{d}s}\right)\\ &= \sqrt{a}\left(\frac{\mathrm{d}u}{\mathrm{d}s}\frac{\mathrm{d}^2v}{\mathrm{d}s^2} - \frac{\mathrm{d}v}{\mathrm{d}s}\frac{\mathrm{d}^2u}{\mathrm{d}s^2} + \Gamma^2_{\beta\gamma}\frac{\mathrm{d}u}{\mathrm{d}s}\frac{\mathrm{d}u^{\beta}}{\mathrm{d}s}\frac{\mathrm{d}u^{\gamma}}{\mathrm{d}s} - \Gamma^1_{\beta\gamma}\frac{\mathrm{d}v}{\mathrm{d}s}\frac{\mathrm{d}u^{\beta}}{\mathrm{d}s}\frac{\mathrm{d}u^{\gamma}}{\mathrm{d}s}\right).\end{aligned}$$

Hence

$$\begin{aligned}\frac{\kappa_g}{\sqrt{a}} &= \frac{\mathrm{d}u}{\mathrm{d}s}\frac{\mathrm{d}^2v}{\mathrm{d}s^2} - \frac{\mathrm{d}v}{\mathrm{d}s}\frac{\mathrm{d}^2u}{\mathrm{d}s^2} + \Gamma^2_{11}\left(\frac{\mathrm{d}u}{\mathrm{d}s}\right)^3 - \Gamma^1_{22}\left(\frac{\mathrm{d}v}{\mathrm{d}s}\right)^3\\ &\quad + (2\Gamma^2_{12} - \Gamma^1_{11})\left(\frac{\mathrm{d}u}{\mathrm{d}s}\right)^2\left(\frac{\mathrm{d}v}{\mathrm{d}s}\right) - (2\Gamma^1_{12} - \Gamma^2_{22})\left(\frac{\mathrm{d}v}{\mathrm{d}s}\right)^2\left(\frac{\mathrm{d}u}{\mathrm{d}s}\right). \qquad (3.88)\end{aligned}$$

Example: We calculate the geodesic curvature of a 'circle of latitude' (a parallel) $\theta = \theta_0$ of a sphere of radius a with surface co-ordinates given by spherical polars θ, ϕ.

Take $u^1 = \theta$, $u^2 = \phi$, with the usual background cartesian co-ordinates defining the spherical surface in the parametric form $y_1 = x = a\sin\theta\cos\phi$, $y_2 = y = a\sin\theta\sin\phi$, $y_3 = z = a\cos\theta$. The circle of latitude has radius $r = a\sin\theta_0$, and arc-length s can be measured from some given meridian as $s = r\phi$. Hence this curve is $u^{\alpha} = u^{\alpha}(s)$, where

$$u^1 = \theta_0, u^2 = s/r = s/(a\sin\theta_0).$$

Now we have the unit tangent λ^{α}, with

$$\lambda^1 = \frac{\mathrm{d}u^1}{\mathrm{d}s} = 0, \quad \lambda^2 = \frac{\mathrm{d}u^2}{\mathrm{d}s} = \frac{1}{a\sin\theta_0}.$$

By eqn (3.17), $a_{11} = \left(\frac{\partial x}{\partial\theta}\right)^2 + \left(\frac{\partial y}{\partial\theta}\right)^2 + \left(\frac{\partial z}{\partial\theta}\right)^2 = a^2$, $a_{12} = a_{21} = 0$, $a_{22} = a^2\sin^2\theta$, similarly, and the only non-vanishing Christoffel symbols are, from eqn (3.49), $[22, 1] = [12, 2] = [21, 2] = a^2\sin\theta\cos\theta$, and $\Gamma^1_{22} = \Gamma^2_{12} = \Gamma^2_{21} = \sin\theta\cos\theta$. Hence the first of eqns (3.87) becomes, with $\alpha = 1, 2$,

$$0 + (\sin\theta_0\cos\theta_0)\left(\frac{1}{a\sin\theta_0}\right)^2 = \kappa_g\nu^1 \quad \text{and } 0 + 0 = \kappa_g\nu^2.$$

Thus $\nu^2 = 0$, and since $\boldsymbol{\nu}$ is a unit vector, we have $a_{\alpha\beta}\nu^{\alpha}\nu^{\beta} = a^2(\nu^1)(\nu^1) = 1$. Hence $\nu^1 = a^{-1}$, and

$$\kappa_g = a^{-1}\cot\theta_0.$$

Of course, viewed from E^3, a parallel is just a circle with curvature $\kappa = r^{-1} = a^{-1}\mathrm{cosec}\theta_0$. Note that the 'equator', the parallel with $\theta = \pi/2$, is a geodesic with geodesic curvature zero.

3.6 The second fundamental form

The basic geometry of surfaces in E^3 depends on two quadratic differential forms, the first of which generates the metrical properties considered above. The study of the shape of a surface S as viewed from the enveloping space gives rise to another quadratic differential form $b_{\alpha\beta}\,\mathrm{d}u^\alpha\,\mathrm{d}u^\beta$ in the differentials of surface co-ordinates u^1, u^2. We consider an arbitrary surface curve C on S passing through a point P at which the curve has tangent vector $\mathbf{t} = \mathrm{d}\mathbf{r}/\mathrm{d}s$ (in the direction in which the arc-length parameter s is increasing) and principal normal $\mathbf{n}$. Then

$$\frac{\mathrm{d}\mathbf{t}}{\mathrm{d}s} = \kappa\mathbf{n} = \boldsymbol{\kappa}, \tag{3.89}$$

where κ is the curvature of C at P and $\boldsymbol{\kappa}$ is the curvature vector.

If $\mathbf{N}$ is a unit normal to the surface at P, we can decompose $\boldsymbol{\kappa}$ into

$$\boldsymbol{\kappa} = \boldsymbol{\kappa}_N + \boldsymbol{\kappa}_g, \tag{3.90}$$

where $\boldsymbol{\kappa}_N$ is in the direction of $\mathbf{N}$, κ_N (called the *normal curvature* of C at P) is the component $\boldsymbol{\kappa}\cdot\mathbf{N}$, and $\boldsymbol{\kappa}_g$ is tangential to the surface at P. Since $\mathbf{N}\cdot\mathbf{t} = 0$, differentiation with respect to s as parameter gives

$$\mathbf{N}\cdot\frac{\mathrm{d}\mathbf{t}}{\mathrm{d}s} + \frac{\mathrm{d}\mathbf{N}}{\mathrm{d}s}\cdot\mathbf{t} = \mathbf{N}\cdot(\boldsymbol{\kappa}_N + \boldsymbol{\kappa}_g) + \frac{\mathrm{d}\mathbf{N}}{\mathrm{d}s}\cdot\frac{\mathrm{d}\mathbf{r}}{\mathrm{d}s} = \kappa_N + \frac{\mathrm{d}\mathbf{N}}{\mathrm{d}s}\cdot\frac{\mathrm{d}\mathbf{r}}{\mathrm{d}s} = 0,$$

which gives

$$\kappa_N = -\frac{\mathrm{d}\mathbf{N}}{\mathrm{d}s}\cdot\frac{\mathrm{d}\mathbf{r}}{\mathrm{d}s} = -\frac{\partial\mathbf{N}}{\partial u^\alpha}\frac{\mathrm{d}u^\alpha}{\mathrm{d}s}\cdot\mathbf{a}_\beta\frac{\mathrm{d}u^\beta}{\mathrm{d}s} = b_{\alpha\beta}\frac{\mathrm{d}u^\alpha}{\mathrm{d}s}\frac{\mathrm{d}u^\beta}{\mathrm{d}s} = \frac{b_{\alpha\beta}\,\mathrm{d}u^\alpha\,\mathrm{d}u^\beta}{\mathrm{d}s^2}, \tag{3.91}$$

where it is convenient to make $b_{\alpha\beta}$ explicitly symmetric, with

$$b_{\alpha\beta} = -\frac{1}{2}\left(\frac{\partial\mathbf{N}}{\partial u^\alpha}\cdot\mathbf{a}_\beta + \frac{\partial\mathbf{N}}{\partial u^\beta}\cdot\mathbf{a}_\alpha\right). \tag{3.92}$$

We can also write

$$\kappa_N = b_{\alpha\beta}\lambda^\alpha\lambda^\beta, \tag{3.93}$$

where λ^α represents a unit surface contravariant tangent vector to C.

Note that the sign of κ_N does not depend on the orientation of C (the direction in which s is measured), but it does depend on the direction in which the surface normal $\mathbf{N}$ is taken.

Since $\mathbf{N}$ is normal to the surface tangent vectors, differentiating $\mathbf{N}\cdot\mathbf{a}_\alpha = 0$ with respect to u^β gives

$$\frac{\partial\mathbf{N}}{\partial u^\beta}\cdot\mathbf{a}_\alpha = -\mathbf{N}\cdot\frac{\partial\mathbf{a}_\alpha}{\partial u^\beta} = -\mathbf{N}\cdot\frac{\partial^2\mathbf{r}}{\partial u^\alpha\partial u^\beta},$$

which we may note is in fact already symmetric in α and β. Thus an alternative expression for $b_{\alpha\beta}$ is

$$b_{\alpha\beta} = \mathbf{N}\cdot\frac{\partial^2\mathbf{r}}{\partial u^\alpha\partial u^\beta}. \tag{3.94}$$

The *second fundamental form* $b_{\alpha\beta}\,\mathrm{d}u^\alpha\,\mathrm{d}u^\beta$ may be written, putting $u^1 = u$, $u^2 = v$, as

$$L(\mathrm{d}u)^2 + 2M\,\mathrm{d}u\,\mathrm{d}v + N(\mathrm{d}v)^2,$$

where

$$b_{11} = L = \mathbf{N}\cdot\frac{\partial^2\mathbf{r}}{\partial u^2},\quad b_{12} = b_{21} = M = \mathbf{N}\cdot\frac{\partial^2\mathbf{r}}{\partial u\partial v},\quad b_{22} = N = \mathbf{N}\cdot\frac{\partial^2\mathbf{r}}{\partial v^2}. \tag{3.95}$$

If we define the direction of $\mathbf{N}$ in the usual right-handed sense with respect to the tangent vectors $\mathbf{a}_1, \mathbf{a}_2$, we can write

$$\mathbf{N} = \frac{\mathbf{a}_1\times\mathbf{a}_2}{|\mathbf{a}_1\times\mathbf{a}_2|} = \frac{1}{\sqrt{a}}\mathbf{a}_1\times\mathbf{a}_2, \tag{3.96}$$

by eqn (3.25). Hence eqns (3.95) can be written as scalar triple products

$$L = \frac{1}{\sqrt{a}}\left[\frac{\partial^2\mathbf{r}}{\partial u^2}, \frac{\partial\mathbf{r}}{\partial u}, \frac{\partial\mathbf{r}}{\partial v}\right],\quad M = \frac{1}{\sqrt{a}}\left[\frac{\partial^2\mathbf{r}}{\partial u\partial v}, \frac{\partial\mathbf{r}}{\partial u}, \frac{\partial\mathbf{r}}{\partial v}\right],\quad N = \frac{1}{\sqrt{a}}\left[\frac{\partial^2\mathbf{r}}{\partial v^2}, \frac{\partial\mathbf{r}}{\partial u}, \frac{\partial\mathbf{r}}{\partial v}\right] \tag{3.97}$$

or as determinants, for example, with cartesian co-ordinates x, y, z,

$$L = \frac{1}{\sqrt{a}}\begin{vmatrix} \dfrac{\partial^2 x}{\partial u^2} & \dfrac{\partial^2 y}{\partial u^2} & \dfrac{\partial^2 z}{\partial u^2} \\ \dfrac{\partial x}{\partial u} & \dfrac{\partial y}{\partial u} & \dfrac{\partial z}{\partial u} \\ \dfrac{\partial x}{\partial v} & \dfrac{\partial y}{\partial v} & \dfrac{\partial z}{\partial v} \end{vmatrix}, \tag{3.98}$$

and similarly for M and N.

Exercise 9. For the surface of revolution defined by eqn (3.32) show that

$$\begin{aligned} L &= (f'^2(u) + g'^2(u))^{-\frac{1}{2}}[f'(u)g''(u) - f''(u)g'(u)], \qquad M = 0, \\ N &= (f'^2(u) + g'^2(u))^{-\frac{1}{2}} f(u)g'(u). \end{aligned} \tag{3.99}$$

The second fundamental form is directly related to the distance, to second order in small quantities, between points on the surface in a neighbourhood of the surface point P and the tangent plane at P. The increment in position vector from P with co-ordinates (u, v) to a neighbouring point $\mathrm{Q}(u + \delta u, v + \delta v)$ is, by Taylor Series expansion,

$$\begin{aligned} \delta\mathbf{r} &= \mathbf{r}(u + \delta u, v + \delta v) - \mathbf{r}(u, v) \\ &= \mathbf{a}_1\delta u + \mathbf{a}_2\delta v + \frac{1}{2}\left\{\frac{\partial^2\mathbf{r}}{\partial u^2}(\delta u)^2 + 2\frac{\partial^2\mathbf{r}}{\partial u\partial v}\delta u\partial v + \frac{\partial^2\mathbf{r}}{\partial v^2}(\delta v)^2\right\} \end{aligned}$$

to second order in $\delta u, \delta v$. The distance between Q and the tangent plane through P is given by projecting $\delta\mathbf{r}$ in the direction of the unit normal $\mathbf{N}$ to the tangent plane, which gives

$$\mathbf{N}\cdot\boldsymbol{\delta}\mathbf{r} = \frac{1}{2}\mathbf{N}\cdot\left\{\frac{\partial^2\mathbf{r}}{\partial u^2}(\delta u)^2 + 2\frac{\partial^2\mathbf{r}}{\partial u\partial v}\delta u\delta v + \frac{\partial^2\mathbf{r}}{\partial v^2}(\delta v)^2\right\}$$
$$= \frac{1}{2}\left\{L(\delta u)^2 + 2M\delta u\delta v + N(\delta v)^2\right\} = \frac{1}{2}b_{\alpha\beta}\delta u^\alpha\delta u^\beta, \qquad (3.100)$$

since $\mathbf{N}$ is perpendicular to the surface tangent vectors $\mathbf{a}_1$ and $\mathbf{a}_2$. The sign of the resulting expression will be different for points on S lying on different sides of the tangent plane. So if the second fundamental form is positive definite or negative definite, which is the case if $\det(b_{\alpha\beta}) > 0$ at P, all points in the immediate neighbourhood of P will lie on the same side of the tangent plane. Such points P may be called *elliptic*. But if $\det(b_{\alpha\beta}) < 0$, there will be points in a neighbourhood of P lying on different sides of the tangent plane, and P may be called *hyperbolic*.The third possibility is that the second fundamental form is positive or negative semi-definite, which occurs when $\det(b_{\alpha\beta}) = 0$. This is the case, for example, for a circular cylinder, all points on the surface of which may be called *parabolic*.

Example: A surface which exhibits all three types of points is the *torus*, the surface (Fig. 3.4) formed by revolving the circle

$$(x-b)^2 + z^2 = a^2$$

in the Oxz plane, where $b > a$, about the z-axis (Fig. 3.5). This is a surface of revolution of the form eqn (3.32), with $f(u) = b + a\cos u$, and $g(u) = a\sin u$. Substituting into eqns (3.99), we obtain

$$L = a, \quad M = 0, \quad N = (b + a\cos u)\cos u.$$

Hence $\det(b_{\alpha\beta}) = LN - M^2 = a\cos u(b + a\cos u)$. Since $(b + a\cos u) > 0$ for all u, it follows that the sign of $\det(b_{\alpha\beta})$ is the same as the sign of $\cos u$. Hence there are

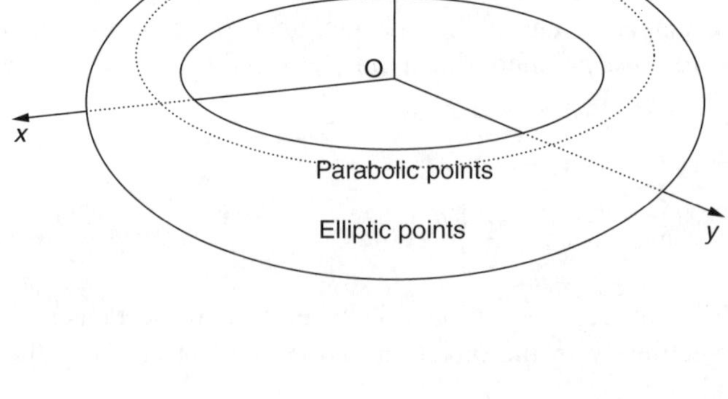

Fig. 3.4 Torus.

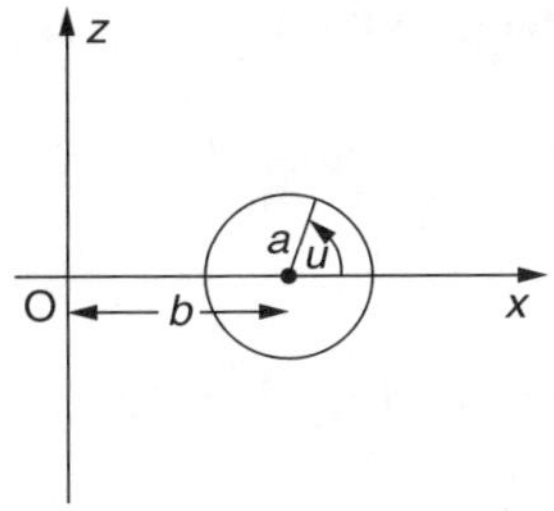

Fig. 3.5 Circle of radius a to be rotated around Oz to form torus.

elliptic points where $-\pi/2 < u < \pi/2$, hyperbolic points where $\pi/2 < u < 3\pi/2$, and a curve of parabolic points where $u = \pm\pi/2$.

If a surface curve C is a *normal section*, obtained from the intersection of S with a plane at P containing **N**, then $\boldsymbol{\kappa}_g = \mathbf{0}$ and $\kappa = \kappa_N$, given by eqns (3.91) and (3.16) as the ratio of quadratic forms

$$\kappa_N = \frac{b_{\alpha\beta}\,\mathrm{d}u^\alpha\,\mathrm{d}u^\beta}{a_{\gamma\delta}\,\mathrm{d}u^\gamma\,\mathrm{d}u^\delta} = \frac{L(\mathrm{d}u)^2 + 2M\,\mathrm{d}u\,\mathrm{d}v + N(\mathrm{d}v)^2}{E(\mathrm{d}u)^2 + 2F\,\mathrm{d}u\,\mathrm{d}v + G(\mathrm{d}v)^2}. \tag{3.101}$$

If C is not a normal section, with principal normal **n** making an angle ϕ with the surface normal **N**, then taking the scalar product of eqn (3.90) with **N** gives *Meusnier's Theorem*

$$\kappa\cos\phi = \kappa_N, \tag{3.102}$$

which, expressed in terms of the radius of curvature ρ of C and the radius of curvature ρ_N of the normal section with the same surface tangent at P, is equivalent to

$$\rho = \rho_N\cos\phi. \tag{3.103}$$

If C is a geodesic, we know from eqn (3.78) that $\boldsymbol{\kappa}$ has the direction of **N**. Hence for a geodesic, as for a normal section, $\boldsymbol{\kappa}_g = \mathbf{0}$; a geodesic through a point in a certain direction has the same curvature as a normal section through that point in the same direction. In the case of a spherical surface, normal sections at a point on the surface are the same as geodesics (great circles), but this is not generally the case.

In general, it can be shown that the magnitude κ_g of $\boldsymbol{\kappa}_g$ is the geodesic curvature defined in the last section. It gives a measure of how much the curvature of a surface curve differs at a point from that of a geodesic curve in the same direction passing through that point. By eqn (3.90) we have

$$\kappa^2 = \kappa_N^2 + \kappa_g^2. \tag{3.104}$$

3.7 Principal curvatures and lines of curvature

From eqn (3.101) the curvature κ_N of a normal section at a point P satisfies

$$(b_{\alpha\beta} - \kappa_N a_{\alpha\beta})\,\mathrm{d}u^\alpha\,\mathrm{d}u^\beta = 0,$$

or, in terms of the surface tangent vector $\lambda^\alpha = du^\alpha/ds$ where the normal section cuts the surface,

$$(b_{\alpha\beta} - \kappa_N a_{\alpha\beta})\lambda^\alpha \lambda^\beta = 0, \tag{3.105}$$

that is,

$$(b_{11} - \kappa_N a_{11}) + 2(b_{12} - \kappa_N a_{12})\left(\frac{dv}{du}\right) + (b_{22} - \kappa_N a_{22})\left(\frac{dv}{du}\right)^2 = 0, \tag{3.106}$$

writing (u, v) instead of (u^1, u^2). We may look for stationary values of κ_N as the directional parameter dv/du varies (or as the plane of the normal section at P is rotated about the normal **N**). Differentiation of eqn (3.106) with respect to this parameter, assuming that κ_N is stationary, gives

$$(b_{12} - \kappa_N a_{12}) + (b_{22} - \kappa_N a_{22})\left(\frac{dv}{du}\right) = 0,$$

and it follows, substituting this result in eqn (3.106), that we also have

$$(b_{11} - \kappa_N a_{11}) + (b_{12} - \kappa_N a_{12})\left(\frac{dv}{du}\right) = 0.$$

The last two equations may be summarized in the form

$$(b_{\alpha\beta} - \kappa_N a_{\alpha\beta})\lambda^\beta = 0, \quad \alpha = 1, 2. \tag{3.107}$$

This is a set of two homogeneous linear equations in the unknown quantities λ^α, with non-trivial solutions if κ_N satisfies the quadratic equation

$$\det(b_{\alpha\beta} - \kappa_N a_{\alpha\beta}) = 0, \tag{3.108}$$

which may be expressed as

$$\begin{aligned} 0 &= (b_{11} - \kappa_N a_{11})(b_{22} - \kappa_N a_{22}) - (b_{12} - \kappa_N a_{12})^2 \\ &= (b_{11}b_{22} - b_{12}^2) - \kappa_N(a_{11}b_{22} + a_{22}b_{11} - 2a_{12}b_{12}) + \kappa_N^2(a_{11}a_{22} - a_{12}^2). \end{aligned}$$

Using eqns (3.24) and (3.30), we obtain

$$0 = \det(b_{\alpha\beta}) - a\kappa_N(a^{22}b_{22} + a^{11}b_{11} + 2a^{12}b_{12}) + a\kappa_N^2.$$

In other words, the stationary curvatures are the roots of the quadratic equation

$$\kappa_N^2 - a^{\alpha\beta}b_{\alpha\beta}\kappa_N + \frac{1}{a}\det(b_{\alpha\beta}) = 0. \tag{3.109}$$

The roots must give the maximum and minimum values of κ_N:

$$\kappa_{\max,\min} = \frac{1}{2}a^{\alpha\beta}b_{\alpha\beta} \pm \sqrt{\left(\frac{1}{2}a^{\alpha\beta}b_{\alpha\beta}\right)^2 - \frac{1}{a}(b_{11}b_{22} - b_{12}^2)}. \tag{3.110}$$

These are called the *principal curvatures* of the surface at P. We can also define the *mean curvature* of the surface

$$\kappa_m = \tfrac{1}{2}(\kappa_{\max} + \kappa_{\min}) = \tfrac{1}{2}a^{\alpha\beta}b_{\alpha\beta}, \tag{3.111}$$

as well as the *Gaussian curvature*

$$\kappa_G = \kappa_{\max}\kappa_{\min} = \frac{1}{a}(b_{11}b_{22} - b_{12}^2) = \frac{\det(b_{\alpha\beta})}{\det(a_{\alpha\beta})} = \frac{LN - M^2}{EG - F^2}. \tag{3.112}$$

Exercise 10. Show that

$$\kappa_m = \frac{EN - 2FM + GL}{2(EG - F^2)}. \tag{3.113}$$

Exercise 11. For the surface $z = f(x, y)$, with covariant metric tensor components given by eqn (3.23), show that the unit surface normal, the coefficients of the second fundamental form, and the Gaussian curvature, are given by

$$\mathbf{N} = \left(-\frac{\partial f}{\partial x}\mathbf{i} - \frac{\partial f}{\partial y}\mathbf{j} + \mathbf{k}\right)\left[1 + \left(\frac{\partial f}{\partial x}\right)^2 + \left(\frac{\partial f}{\partial y}\right)^2\right]^{-\frac{1}{2}}, \tag{3.114}$$

$$L = \frac{1}{\sqrt{a}}\frac{\partial^2 f}{\partial x^2}, \quad M = \frac{1}{\sqrt{a}}\frac{\partial^2 f}{\partial x \partial y}, \quad N = \frac{1}{\sqrt{a}}\frac{\partial^2 f}{\partial y^2}, \tag{3.115}$$

where $a = \left(1 + \left[\frac{\partial f}{\partial x}\right]^2 + \left[\frac{\partial f}{\partial y}\right]^2\right)$, and

$$\kappa_G = \left[\left(\frac{\partial^2 f}{\partial x^2}\right)\left(\frac{\partial^2 f}{\partial y^2}\right) - \left(\frac{\partial^2 f}{\partial x \partial y}\right)^2\right]\left[1 + \left(\frac{\partial f}{\partial x}\right)^2 + \left(\frac{\partial f}{\partial y}\right)^2\right]^{-2}. \tag{3.116}$$

For elliptic points, as defined in the last section, $\kappa_{\max}$ and $\kappa_{\min}$ have the same sign, and κ_G is positive, whereas for hyberbolic points $\kappa_{\max}$ and $\kappa_{\min}$ have opposite signs and κ_G is negative. At parabolic points either $\kappa_{\max}$ or $\kappa_{\min}$ is zero, and so is κ_G.

If the surface tangents corresponding to these curvatures are $\lambda^\alpha_{\max}$ and $\lambda^\alpha_{\min}$, eqn (3.107) gives

$$b_{\alpha\beta}\lambda^\beta_{\max} = \kappa_{\max} a_{\alpha\beta}\lambda^\beta_{\max} \quad \text{and}$$
$$b_{\alpha\beta}\lambda^\beta_{\min} = \kappa_{\min} a_{\alpha\beta}\lambda^\beta_{\min}.$$

Hence

$$\begin{aligned}(\kappa_{\max} - \kappa_{\min})a_{\alpha\beta}\lambda^\alpha_{\max}\lambda^\beta_{\min} &= \kappa_{\max} a_{\alpha\beta}\lambda^\beta_{\max}\lambda^\alpha_{\min} - \kappa_{\min} a_{\alpha\beta}\lambda^\alpha_{\max}\lambda^\beta_{\min} \\ &= b_{\alpha\beta}\lambda^\beta_{\max}\lambda^\alpha_{\min} - b_{\alpha\beta}\lambda^\alpha_{\max}\lambda^\beta_{\min} = 0,\end{aligned}$$

where the symmetry of both $a_{\alpha\beta}$ and $b_{\alpha\beta}$ has been used. It follows that, provided $\kappa_{\max} \neq \kappa_{\min}$,

$$a_{\alpha\beta}\lambda^\alpha_{\max}\lambda^\beta_{\min} = 0. \tag{3.117}$$

We see from eqn (3.47) that the corresponding surface vectors $\boldsymbol{\lambda}_{\max}, \boldsymbol{\lambda}_{\min}$ must be orthogonal to each other, except when $\kappa_{\max} = \kappa_{\min}$, in which case all normal sections through P have the same curvature. The two orthogonal directions are called the *principal directions* at P. Surface curves to which the tangents are everywhere in the principal directions are called *lines of curvature*.

The case $\lambda_{max} = \lambda_{min}$ arises at points where the components $b_{\alpha\beta}$ are proportional to $a_{\alpha\beta}$, in other words there is a constant k such that $b_{\alpha\beta} = ka_{\alpha\beta}$ for all α, β ranging over the values 1, 2. In this case the normal curvature at the point must be equal to k for *all* normal sections, by eqn (3.101). Such points (for example, all points on the surface of a sphere) are called *umbilics*.

Note that from $b_{\alpha\beta}$ we can derive the associated mixed second-order surface tensor $b^{\alpha}_{\gamma} = a^{\alpha\beta}b_{\gamma\beta}$, raising one index by the application of the surface metric tensor $a^{\alpha\beta}$. Equations (3.111) and (3.112) can now be written as

$$\kappa_m = \tfrac{1}{2}b^{\alpha}_{\alpha} = \tfrac{1}{2}(b^1_1 + b^2_2) \quad \text{and } \kappa_G = \det(b^{\alpha}_{\beta}) = b^1_1 b^2_2 - b^1_2 b^2_1. \tag{3.118}$$

If we assume that we can choose our surface co-ordinates u^{α} such that the surface co-ordinate curves coincide with the lines of curvature, certain simplifications occur. Firstly, since the lines of curvature are orthogonal to each other, we have $a_{12} = F = 0$, by eqn (3.41). Secondly, writing the principal curvatures as κ_a and κ_b (without specifying which is the maximum or minimum), eqns (3.107) become, taking the unit vector λ^{α} in the u^1-direction, where the principal curvature is κ_a and $\lambda^1 = 1/\sqrt{a_{11}}$, $\lambda^2 = 0$ by eqn (3.40),

$$(b_{11} - \kappa_a a_{11})\frac{1}{\sqrt{a_{11}}} = 0 \quad \text{and } (b_{21} - \kappa_a a_{21})\frac{1}{\sqrt{a_{11}}} = 0.$$

Hence we also have $b_{12} = b_{21} = M = 0$, and the principal curvatures are

$$\kappa_a = \frac{b_{11}}{a_{11}} = \frac{L}{E} \quad \text{and } \kappa_b = \frac{b_{22}}{a_{22}} = \frac{N}{G} \tag{3.119}$$

similarly.

Example: For the surface of revolution as given by eqn (3.32), we have already seen that $F = M = 0$. Hence the co-ordinate curves are also lines of curvature in that case. From eqns (3.33) and (3.99) we immediately obtain the principal curvatures

$$\kappa_a = (f'^2 + g'^2)^{-\frac{3}{2}}(f'g'' - f''g') \quad \text{and } \kappa_b = \frac{(f'^2 + g'^2)^{-\frac{1}{2}}g'}{f}.$$

More generally, consider a unit directional vector μ^{α} making an angle ψ with the κ_a line of curvature. Then

$$\mu^1 = \frac{1}{\sqrt{a_{11}}}\cos\psi \quad \text{and } \mu^2 = \frac{1}{\sqrt{a_{22}}}\sin\psi$$

making use of eqns (3.39) and (3.40). The curvature of a normal section in this direction is, by eqn (3.93), given by

$$\kappa = b_{\alpha\beta}\mu^{\alpha}\mu^{\beta} = \frac{b_{11}}{a_{11}}\cos^2\psi + \frac{b_{22}}{a_{22}}\sin^2\psi = \kappa_a\cos^2\psi + \kappa_b\sin^2\psi, \tag{3.120}$$

and this is sometimes referred to as *Euler's Theorem*.

3.8 Weingarten, Gauss, and Gauss-Codazzi equations

Here we investigate the spatial derivatives of the surface covariant base vectors $\mathbf{a}_\alpha$ and the surface normal $\mathbf{N}$. Since $\mathbf{N}$ is a unit vector, we have $\mathbf{N}\cdot\mathbf{N} = \mathbf{1}$, and partial differentiation with respect to u^α gives immediately

$$\mathbf{N}\cdot\frac{\partial \mathbf{N}}{\partial u^\alpha} = 0,$$

which implies that the vectors $\partial\mathbf{N}/\partial u^\alpha$ lie in the tangent plane and must be linear combinations of $\mathbf{a}_1, \mathbf{a}_2$. We have already seen in eqn (3.91) that a definition of $b_{\alpha\beta}$ was

$$b_{\alpha\beta} = -\frac{\partial \mathbf{N}}{\partial u^\alpha}\cdot\mathbf{a}_\beta.$$

Consequently, if we write

$$\frac{\partial \mathbf{N}}{\partial u^\alpha} = c_\alpha^\gamma \mathbf{a}_\gamma$$

for some set of coefficients c_α^γ, taking scalar products of both sides with $\mathbf{a}_\beta$ gives

$$-b_{\alpha\beta} = c_\alpha^\gamma a_{\beta\gamma}.$$

It follows, using eqn (3.29), that

$$c_\alpha^\gamma = -b_{\alpha\beta}a^{\beta\gamma},$$

and hence

$$\frac{\partial \mathbf{N}}{\partial u^\alpha} = -b_{\alpha\beta}a^{\beta\gamma}\mathbf{a}_\gamma = -b_\alpha^\gamma \mathbf{a}_\gamma, \quad \alpha = 1, 2, \tag{3.121}$$

which are *Weingarten's equations.*

Exercise 12. Show that in an alternative notation Weingarten's equations can be written:

$$\frac{\partial \mathbf{N}}{\partial u} = \frac{MF - LG}{EG - F^2}\mathbf{a}_1 + \frac{LF - ME}{EG - F^2}\mathbf{a}_2,$$
$$\frac{\partial \mathbf{N}}{\partial v} = \frac{NF - MG}{EG - F^2}\mathbf{a}_1 + \frac{MF - NE}{EG - F^2}\mathbf{a}_2.$$

We have previously noted that the derivatives of the surface covariant base vectors with respect to the surface co-ordinates are not themselves surface vectors. However, recalling that $\partial\mathbf{a}_\alpha/\partial u^\beta = \partial^2\mathbf{r}/\partial u^\alpha\partial u^\beta$, it may be seen that eqns (3.50) and (3.94) immediately provide the required projections of $\partial\mathbf{a}_\alpha/\partial u^\beta$ in the directions $\mathbf{a}_1, \mathbf{a}_2$, and $\mathbf{N}$. Thus, as may be verified by taking scalar products of both sides of the following with $\mathbf{a}^\mu$ and $\mathbf{N}$,

$$\frac{\partial \mathbf{a}_\alpha}{\partial u^\beta} = \Gamma_{\alpha\beta}^\gamma \mathbf{a}_\gamma + b_{\alpha\beta}\mathbf{N}, \tag{3.122}$$

which is *Gauss's formula.*

Further important equations may be obtained by considering the consistency of different representations of Gauss's formula through the equations

$$\frac{\partial}{\partial u^\gamma}\left(\frac{\partial^2 \mathbf{r}}{\partial u^\alpha \partial u^\beta}\right) = \frac{\partial}{\partial u^\beta}\left(\frac{\partial^2 \mathbf{r}}{\partial u^\alpha \partial u^\gamma}\right),$$

or

$$\frac{\partial}{\partial u^\gamma}\left(\frac{\partial \mathbf{a}_\alpha}{\partial u^\beta}\right) = \frac{\partial}{\partial u^\beta}\left(\frac{\partial \mathbf{a}_\alpha}{\partial u^\gamma}\right),$$

which gives

$$\begin{aligned}
\mathbf{0} &= \frac{\partial}{\partial u^\gamma}(\Gamma^\mu_{\alpha\beta}\mathbf{a}_\mu + b_{\alpha\beta}\mathbf{N}) - \frac{\partial}{\partial u^\beta}(\Gamma^\mu_{\alpha\gamma}\mathbf{a}_\mu + b_{\alpha\gamma}\mathbf{N}) \\
&= \left(\frac{\partial \Gamma^\mu_{\alpha\beta}}{\partial u^\gamma} - \frac{\partial \Gamma^\mu_{\alpha\gamma}}{\partial u^\beta}\right)\mathbf{a}_\mu + \Gamma^\mu_{\alpha\beta}\frac{\partial \mathbf{a}_\mu}{\partial u^\gamma} - \Gamma^\mu_{\alpha\gamma}\frac{\partial \mathbf{a}_\mu}{\partial u^\beta} \\
&\quad + b_{\alpha\beta}\frac{\partial \mathbf{N}}{\partial u^\gamma} - b_{\alpha\gamma}\frac{\partial \mathbf{N}}{\partial u^\beta} + \left(\frac{\partial b_{\alpha\beta}}{\partial u^\gamma} - \frac{\partial b_{\alpha\gamma}}{\partial u^\beta}\right)\mathbf{N} \\
&= \left(\frac{\partial \Gamma^\mu_{\alpha\beta}}{\partial u^\gamma} - \frac{\partial \Gamma^\mu_{\alpha\gamma}}{\partial u^\beta} + \Gamma^\delta_{\alpha\beta}\Gamma^\mu_{\delta\gamma} - \Gamma^\delta_{\alpha\gamma}\Gamma^\mu_{\delta\beta} - b_{\alpha\beta}b_{\gamma\delta}a^{\delta\mu} + b_{\alpha\gamma}b_{\beta\delta}a^{\delta\mu}\right)\mathbf{a}_\mu \\
&\quad + \left(\Gamma^\mu_{\alpha\beta}b_{\mu\gamma} - \Gamma^\mu_{\alpha\gamma}b_{\mu\beta} + \frac{\partial b_{\alpha\beta}}{\partial u^\gamma} - \frac{\partial b_{\alpha\gamma}}{\partial u^\beta}\right)\mathbf{N},
\end{aligned}$$

using both eqns (3.121) and (3.122). Since $\mathbf{a}_1, \mathbf{a}_2$, and $\mathbf{N}$ are an independent set of vectors, it follows that we must have both

$$\frac{\partial \Gamma^\mu_{\alpha\beta}}{\partial u^\gamma} - \frac{\partial \Gamma^\mu_{\alpha\gamma}}{\partial u^\beta} + \Gamma^\delta_{\alpha\beta}\Gamma^\mu_{\delta\gamma} - \Gamma^\delta_{\alpha\gamma}\Gamma^\mu_{\delta\beta} - b_{\alpha\beta}b_{\gamma\delta}a^{\delta\mu} + b_{\alpha\gamma}b_{\beta\delta}a^{\delta\mu} = 0 \tag{3.123}$$

and

$$\Gamma^\mu_{\alpha\beta}b_{\mu\gamma} - \Gamma^\mu_{\alpha\gamma}b_{\mu\beta} + \frac{\partial b_{\alpha\beta}}{\partial u^\gamma} - \frac{\partial b_{\alpha\gamma}}{\partial u^\beta} = 0 \tag{3.124}$$

for all possible values of the free (unrepeated) indices.

We can appeal to eqn (3.64) to re-write eqn (3.123) as

$$R^\mu_{.\alpha\beta\gamma} = b_{\alpha\gamma}b_{\beta\delta}a^{\delta\mu} - b_{\alpha\beta}b_{\gamma\delta}a^{\delta\mu} \tag{3.125}$$

and multiplication through by $a_{\nu\mu}$ (with summation over μ), using eqn (3.65), gives

$$R_{\nu\alpha\beta\gamma} = b_{\alpha\gamma}b_{\beta\nu} - b_{\alpha\beta}b_{\gamma\nu}. \tag{3.126}$$

Since we know that the curvature tensor $R_{\nu\alpha\beta\gamma}$ has only one independent component R_{1212}, eqns (3.126) reduce to a single equation

$$R_{1212} = b_{11}b_{22} - (b_{12})^2 = \det(b_{\alpha\beta}), \tag{3.127}$$

and this is *Gauss's equation*. It has the consequence that since R_{1212} is an *intrinsic* quantity, derivable entirely in terms of the covariant metric tensor $a_{\alpha\beta}$, then so also must $\det(b_{\alpha\beta})$ be intrinsic. Moreover, by eqn (3.112) we obtain Gauss's result

$$\kappa_G = \frac{R_{1212}}{a}, \tag{3.128}$$

which shows that the Gaussian curvature is also an intrinsic quantity, that is, it depends only on the tensor components $a_{\alpha\beta}$ and their derivatives. Such a quantity is sometimes referred to as a *bending invariant*, as it remains unchanged by any deformation of the surface which involves pure bending without stretching, shrinking, or tearing, thus preserving distances between points, angles between directions at a point, and the coefficients of the first fundamental form and their derivatives.

Equation (3.124) may be written in the form

$$\frac{\partial b_{\alpha\beta}}{\partial u^\gamma} - \Gamma^\mu_{\alpha\gamma} b_{\beta\mu} - \Gamma^\mu_{\beta\gamma} b_{\alpha\mu} = \frac{\partial b_{\alpha\gamma}}{\partial u^\beta} - \Gamma^\mu_{\alpha\beta} b_{\gamma\mu} - \Gamma^\mu_{\beta\gamma} b_{\alpha\mu}$$

by subtracting the same term from each side, which gives covariant derivative identities

$$b_{\alpha\beta,\gamma} = b_{\alpha\gamma,\beta}. \tag{3.129}$$

This contains only two non-trivial results:

$$b_{11,2} = b_{12,1} \quad \text{and } b_{21,2} = b_{22,1}. \tag{3.130}$$

These are the *Codazzi equations*(or the *Mainardi-Codazzi equations*).

Exercise 13. Express the Codazzi equations in the form

$$\begin{aligned} \frac{\partial L}{\partial v} - \frac{\partial M}{\partial u} &= L\Gamma^1_{12} + M(\Gamma^2_{12} - \Gamma^1_{11}) - N\Gamma^2_{11} \\ \frac{\partial M}{\partial v} - \frac{\partial N}{\partial u} &= L\Gamma^1_{22} + M(\Gamma^2_{22} - \Gamma^1_{12}) - N\Gamma^2_{12}. \end{aligned} \tag{3.131}$$

Exercise 14. Deduce from eqns (3.131) and (3.52) that when the co-ordinate curves coincide with lines of curvature (so that $F = M = 0$), the Codazzi equations may be expressed as

$$\begin{aligned} \frac{\partial L}{\partial v} &= \frac{1}{2}\frac{\partial E}{\partial v}\left(\frac{L}{E} + \frac{N}{G}\right) = \frac{1}{2}\frac{\partial E}{\partial v}(\kappa_a + \kappa_b) \\ \frac{\partial N}{\partial u} &= \frac{1}{2}\frac{\partial G}{\partial u}\left(\frac{L}{E} + \frac{N}{G}\right) = \frac{1}{2}\frac{\partial G}{\partial u}(\kappa_a + \kappa_b). \end{aligned} \tag{3.132}$$

To derive the Gauss equation in the form due to Brioschi, we start with eqns (3.97) and

$$\begin{aligned} LN - M^2 &= \frac{1}{a}\left\{\left[\frac{\partial^2 \mathbf{r}}{\partial u^2}, \frac{\partial \mathbf{r}}{\partial u}, \frac{\partial \mathbf{r}}{\partial v}\right]\left[\frac{\partial^2 \mathbf{r}}{\partial v^2}, \frac{\partial \mathbf{r}}{\partial u}, \frac{\partial \mathbf{r}}{\partial v}\right] - \left[\frac{\partial^2 \mathbf{r}}{\partial u \partial v}, \frac{\partial \mathbf{r}}{\partial u}, \frac{\partial \mathbf{r}}{\partial v}\right]^2\right\} \\ &= \frac{1}{a}\left\{\begin{vmatrix} \mathbf{r}_{uu}\cdot\mathbf{r}_{vv} & \mathbf{r}_{uu}\cdot\mathbf{r}_u & \mathbf{r}_{uu}\cdot\mathbf{r}_v \\ \mathbf{r}_u\cdot\mathbf{r}_{vv} & \mathbf{r}_u\cdot\mathbf{r}_u & \mathbf{r}_u\cdot\mathbf{r}_v \\ \mathbf{r}_v\cdot\mathbf{r}_{vv} & \mathbf{r}_v\cdot\mathbf{r}_u & \mathbf{r}_v\cdot\mathbf{r}_v \end{vmatrix} - \begin{vmatrix} \mathbf{r}_{uv}\cdot\mathbf{r}_{uv} & \mathbf{r}_{uv}\cdot\mathbf{r}_u & \mathbf{r}_{uv}\cdot\mathbf{r}_v \\ \mathbf{r}_u\cdot\mathbf{r}_{uv} & \mathbf{r}_u\cdot\mathbf{r}_u & \mathbf{r}_u\cdot\mathbf{r}_v \\ \mathbf{r}_v\cdot\mathbf{r}_{uv} & \mathbf{r}_v\cdot\mathbf{r}_u & \mathbf{r}_v\cdot\mathbf{r}_v \end{vmatrix}\right\}, \end{aligned}$$

using properties of determinants and suffixes to denote partial differentiation. We deduce that

$$\kappa_G = \frac{LN - M^2}{EG - F^2}$$

$$= \frac{1}{a^2}\left\{\begin{vmatrix} \mathbf{r}_{uu}\cdot\mathbf{r}_{vv} & [11,1] & [11,2] \\ [22,1] & E & F \\ [22,2] & F & G \end{vmatrix} - \begin{vmatrix} \mathbf{r}_{uv}\cdot\mathbf{r}_{uv} & [12,1] & [12,2] \\ [12,1] & E & F \\ [12,2] & F & G \end{vmatrix}\right\}$$

$$= \frac{1}{a^2}\left\{\begin{vmatrix} \mathbf{r}_{uu}\cdot\mathbf{r}_{vv} - \mathbf{r}_{uv}\cdot\mathbf{r}_{uv} & \frac{1}{2}E_u & F_u - \frac{1}{2}E_v \\ F_v - \frac{1}{2}G_u & E & F \\ \frac{1}{2}G_v & F & G \end{vmatrix} - \begin{vmatrix} 0 & \frac{1}{2}E_v & \frac{1}{2}G_u \\ \frac{1}{2}E_v & E & F \\ \frac{1}{2}G_u & F & G \end{vmatrix}\right\},$$

again using simple properties of determinants and eqns (3.51). Moreover, it is straightforward to verify that

$$\mathbf{r}_{uu}\cdot\mathbf{r}_{vv} - \mathbf{r}_{uv}\cdot\mathbf{r}_{uv} = -\tfrac{1}{2}E_{vv} + F_{uv} - \tfrac{1}{2}G_{uu}. \tag{3.133}$$

Hence we obtain the equation

$$\kappa_G = \frac{1}{(EG - F^2)^2}\left\{\begin{vmatrix} F_{uv} - \frac{1}{2}E_{vv} - \frac{1}{2}G_{uu} & \frac{1}{2}E_u & F_u - \frac{1}{2}E_v \\ F_v - \frac{1}{2}G_u & E & F \\ \frac{1}{2}G_v & F & G \end{vmatrix} - \begin{vmatrix} 0 & \frac{1}{2}E_v & \frac{1}{2}G_u \\ \frac{1}{2}E_v & E & F \\ \frac{1}{2}G_u & F & G \end{vmatrix}\right\}. \tag{3.134}$$

Exercise 15. Verify that when the co-ordinate curves are orthogonal (so that $F = 0$), this formula may be expressed in the simpler form

$$\kappa_G = -\frac{1}{\sqrt{EG}}\left\{\frac{\partial}{\partial u}\left(\frac{1}{\sqrt{E}}\frac{\partial\sqrt{G}}{\partial u}\right) + \frac{\partial}{\partial v}\left(\frac{1}{\sqrt{G}}\frac{\partial\sqrt{E}}{\partial v}\right)\right\}. \tag{3.135}$$

The significance of the *Gauss-Codazzi* eqns (3.127) and (3.130) may be appreciated by referring to the *fundamental existence theorem for surfaces* (not proved here), which states that given two quadratic differential forms $a_{\alpha\beta}\,du^\alpha\,du^\beta$ and $b_{\alpha\beta}\,du^\alpha\,du^\beta$, such that $a_{\alpha\beta}\,du^\alpha\,du^\beta$ is positive definite and that the six coefficients $a_{11}, a_{12}, a_{22}, b_{11}, b_{12}, b_{22}$ satisfy the Gauss-Codazzi equations, then there exists a surface, uniquely determined apart from its precise position in space, whose first and second fundamental forms are given respectively by these forms.

3.9 Div, grad, and the Beltrami operator on surfaces

The purpose of generating grids is normally to solve a physical problem represented by a partial differential equation, the so-called *hosted* equation. In this section we present some formulas which are often useful in the transformation from cartesian to general surface co-ordinates of some of the standard vector operators which appear in hosted equations.

Here we begin by deriving the surface metric identity

$$\frac{\partial}{\partial u^1}\left(\frac{a_{22}\mathbf{a}_1 - a_{12}\mathbf{a}_2}{\sqrt{a}}\right) + \frac{\partial}{\partial u^2}\left(\frac{-a_{12}\mathbf{a}_1 + a_{11}\mathbf{a}_{11}}{\sqrt{a}}\right) = 2\kappa_m\sqrt{a}\mathbf{N}. \tag{3.136}$$

The left-hand side, by eqn (3.31), is equal to

$$\frac{\partial}{\partial u^1}(\sqrt{a}\mathbf{a}^1) + \frac{\partial}{\partial u^2}(\sqrt{a}\mathbf{a}^2). \tag{3.137}$$

Exercise 16. Show that

$$\mathbf{a}^1 = \frac{1}{\sqrt{a}}(\mathbf{a}_2 \times \mathbf{N}) \quad \text{and } \mathbf{a}^2 = \frac{1}{\sqrt{a}}(\mathbf{N} \times \mathbf{a}_1). \tag{3.138}$$

It follows that eqn (3.137) becomes

$$\begin{aligned}\frac{\partial}{\partial u^1}(\mathbf{a}_2 \times \mathbf{N}) - \frac{\partial}{\partial u^2}(\mathbf{a}_1 \times \mathbf{N}) &= \left[\frac{\partial}{\partial u^1}\left(\frac{\partial \mathbf{r}}{\partial u^2}\right) - \frac{\partial}{\partial u^2}\left(\frac{\partial \mathbf{r}}{\partial u^1}\right)\right] \\ &\quad \times \mathbf{N} + \mathbf{a}_2 \times \frac{\partial \mathbf{N}}{\partial u^1} - \mathbf{a}_1 \times \frac{\partial \mathbf{N}}{\partial u^2} \\ &= \mathbf{0} \times \mathbf{N} + \mathbf{a}_2 \times (-b_{1\beta}a^{\beta\gamma}\mathbf{a}_\gamma) - \mathbf{a}_1 \times (-b_{2\beta}a^{\beta\gamma}\mathbf{a}_\gamma),\end{aligned}$$

using the Weingarten eqns (3.121). The resulting expression is

$$-b_{1\beta}a^{\beta 1}\mathbf{a}_2 \times \mathbf{a}_1 + b_{2\beta}a^{\beta 2}\mathbf{a}_1 \times \mathbf{a}_2 = b_{\alpha\beta}a^{\beta\alpha}\mathbf{a}_1 \times \mathbf{a}_2 = 2\kappa_m\sqrt{a}\mathbf{N}$$

with the aid of eqns (3.96) and (3.111), thus proving the identity.

Now suppose there exists a surface scalar function $\phi(u^1, u^2)$. The gradient of this function is the surface vector

$$\nabla\varphi = \mathbf{a}^\alpha \frac{\partial\varphi}{\partial u^\alpha}, \tag{3.139}$$

where $\partial\varphi/\partial u^\alpha$ are the surface covariant components. Here we shall also be concerned with the background cartesian components of $\nabla\varphi$.

By eqn (3.138) we can write

$$\nabla\varphi = \frac{1}{\sqrt{a}}\left[(\mathbf{a}_2 \times \mathbf{N})\frac{\partial\varphi}{\partial u^1} + (\mathbf{N} \times \mathbf{a}_1)\frac{\partial\varphi}{\partial u^2}\right]. \tag{3.140}$$

If we now assemble the three background cartesian components of $(\mathbf{a}_2 \times \mathbf{N})$ and $(\mathbf{N} \times \mathbf{a}_1)$ into the two column vectors of a 3×2 matrix C, we can write the cartesian components of $\nabla\varphi$ as

$$(\nabla\varphi)_i = \frac{1}{\sqrt{a}}C_{i\alpha}\frac{\partial\varphi}{\partial u^\alpha}, \tag{3.141}$$

with C represented as

$$C = \left((\mathbf{a}_2 \times \mathbf{N}) \quad (\mathbf{N} \times \mathbf{a}_1) \right) = \sqrt{a}(\mathbf{a}^1, \mathbf{a}^2), \tag{3.142}$$

or, more explicitly, making use of eqn (3.31), and writing (x, y, z) for cartesians, (u, v) for surface co-ordinates, and partial derivatives x_u, etc.

$$C = \frac{1}{\sqrt{a}} \begin{pmatrix} a_{22}x_u - a_{12}x_v & -a_{12}x_u + a_{11}x_v \\ a_{22}y_u - a_{12}y_v & -a_{12}y_v + a_{11}y_v \\ a_{22}z_u - a_{12}z_v & -a_{12}z_u + a_{11}z_v \end{pmatrix}. \tag{3.143}$$

Equation (3.141) represents a non-conservative expression for $\nabla\varphi$. Summation is implied over α from 1 to 2, although we have been a little less than rigorous here in not writing the first α as a superscript.

Re-writing eqn (3.136) as

$$\frac{\partial}{\partial u^1}(\mathbf{a}_2 \times \mathbf{N}) + \frac{\partial}{\partial u^2}(\mathbf{N} \times \mathbf{a}_1) = 2\kappa_m\sqrt{a}\mathbf{N}, \tag{3.144}$$

it follows that we can express eqn (3.140) in the conservative form

$$\nabla\varphi = \frac{1}{\sqrt{a}}\left[\frac{\partial}{\partial u^1}((\mathbf{a}_2 \times \mathbf{N})\varphi) + \frac{\partial}{\partial u^2}((\mathbf{N} \times \mathbf{a}_1)\varphi)\right] - 2\kappa_m\varphi\mathbf{N}, \tag{3.145}$$

giving cartesian components

$$(\nabla\varphi)_i = \frac{1}{\sqrt{a}}\frac{\partial}{\partial u^\alpha}(C_{i\alpha}\varphi) - 2\kappa_m\varphi N_i. \tag{3.146}$$

Although C is not a square matrix, we can effectively define an inverse, at least on the left. To see this, consider the 3×2 (Jacobian) matrix $\mathcal{J}$ of the transformation $(u, v) \to (x, y, z)$, given by

$$\mathcal{J} = \begin{pmatrix} x_u & x_v \\ y_u & y_v \\ z_u & z_v \end{pmatrix} = \begin{pmatrix} \mathbf{a}_1 & \mathbf{a}_2 \end{pmatrix}. \tag{3.147}$$

Since eqn (3.27) implies the matrix identity

$$\begin{pmatrix} \mathbf{a}_1 \\ \mathbf{a}_2 \end{pmatrix}\begin{pmatrix} \mathbf{a}^1 & \mathbf{a}^2 \end{pmatrix} = \begin{pmatrix} 1 & 0 \\ 0 & 1 \end{pmatrix},$$

or, otherwise expressed, $\mathcal{J}^T\left(\frac{1}{\sqrt{a}}C\right) = I_2$, the 2×2 identity matrix, it follows that C has a *left* inverse

$$C^{-1} = \frac{1}{\sqrt{a}}\mathcal{J}^T = \frac{1}{\sqrt{a}}\begin{pmatrix} x_u & y_u & z_u \\ x_v & y_v & z_v \end{pmatrix}, \tag{3.148}$$

or $(C^{-1})_{\alpha i} = (a)^{-\frac{1}{2}}\partial y_i/\partial u^\alpha$.

We can now, by multiplication on the left, invert eqn (3.141) to obtain

$$\frac{\partial\phi}{\partial u^\alpha} = \sqrt{a}(C^{-1})_{\alpha i}(\nabla\varphi)_i \tag{3.149}$$

with summation over i from 1 to 3. This equation in fact gives the general relationship between the covariant components and the background cartesian components of a surface vector.

Exercise 17. Making use of eqn (3.26), show that if **A** is a surface vector with covariant components given by the 2×1 column vector $\tilde{\mathbf{A}}$ (with elements A_α) and background cartesian components given by the 3×1 column vector $\overline{\mathbf{A}}$ (with elements A_i), then

$$\tilde{\mathbf{A}} = \sqrt{a}C^{-1}\overline{\mathbf{A}} \tag{3.150}$$

or, equivalently,

$$A_\alpha = A_i \partial y_i/\partial u^\alpha. \tag{3.151}$$

Exercise 18. Show by direct matrix multiplication using eqns (3.143) and (3.147) that

$$C\left(\frac{1}{\sqrt{a}}\mathcal{J}^T\right) = I_3 - \mathbf{N}\mathbf{N}^T, \tag{3.152}$$

where here **N** stands for the column vector of cartesian components of the surface normal vector and I_3 is the unit 3×3 matrix, and hence that $\frac{1}{\sqrt{a}}\mathcal{J}^T$ is not a right inverse for C.

Now consider a surface vector field $\mathbf{V}(u^1, u^2)$, defined at all points of the surface and having the property that it is everywhere tangential to the surface. Then the divergence of **V** is given by an expression analogous to eqn (1.134):

$$\begin{aligned}\nabla\cdot\mathbf{V} &= \mathbf{a}^\alpha\cdot\frac{\partial\mathbf{V}}{\partial u^\alpha} = \mathbf{a}^1\cdot\frac{\partial\mathbf{V}}{\partial u^1} + \mathbf{a}^2\cdot\frac{\partial\mathbf{V}}{\partial u^2}\\ &= \frac{1}{\sqrt{a}}\left[(\mathbf{a}_2\times\mathbf{N})\cdot\frac{\partial\mathbf{V}}{\partial u^1} + (\mathbf{N}\times\mathbf{a}_1)\cdot\frac{\partial\mathbf{V}}{\partial u^2}\right] &(3.153)\\ &= \frac{1}{\sqrt{a}}C_{i\alpha}\frac{\partial V_i}{\partial u^\alpha} &(3.154)\end{aligned}$$

in terms of the background cartesian components V_i of **V**. (Here the index i is summed from 1 to 3, while α is summed from 1 to 2.) These expressions are non-conservative. A conservative form follows by using eqn (3.144):

$$\nabla\cdot\mathbf{V} = \frac{1}{\sqrt{a}}\left[\frac{\partial}{\partial u^1}((\mathbf{a}_2\times\mathbf{N})\cdot\mathbf{V}) + \frac{\partial}{\partial u^2}((\mathbf{N}\times\mathbf{a}_1)\cdot\mathbf{V})\right] - 2\kappa_m\mathbf{N}\cdot\mathbf{V},$$

but now the last term vanishes because **V** is a tangential vector. Hence we have

$$\begin{aligned}\nabla\cdot\mathbf{V} &= \frac{1}{\sqrt{a}}\left[\frac{\partial}{\partial u^1}((\mathbf{a}_2\times\mathbf{N})\cdot\mathbf{V}) + \frac{\partial}{\partial u^2}((\mathbf{N}\times\mathbf{a}_1)\cdot\mathbf{V})\right] &(3.155)\\ &= \frac{1}{\sqrt{a}}\frac{\partial}{\partial u^\alpha}(C_{i\alpha}V_i). &(3.156)\end{aligned}$$

For any surface scalar field $\varphi(u^1, u^2)$, the gradient $\nabla\varphi$ must be a surface vector field, according to eqn (3.139). Thus we can combine the results of eqns (3.141) and (3.156) to give a formula for the Laplacian $\nabla^2\varphi = \nabla\cdot\nabla\varphi$:

$$\nabla^2\varphi = \frac{1}{\sqrt{a}}\frac{\partial}{\partial u^\alpha}\left(\frac{1}{\sqrt{a}}C_{i\alpha}C_{i\beta}\frac{\partial\varphi}{\partial u^\beta}\right). \tag{3.157}$$

In fact, as the Laplacian operator associated with a particular surface, this second-order differential operator is known as the *Beltrami* operator of the surface, and given the special notation Δ_B. Since the 2×2 matrix represented by $C_{i\alpha}C_{i\beta}$, using eqn (3.142), is

$$C^T C = a \begin{pmatrix} (\mathbf{a}^1)^T \\ (\mathbf{a}^2)^T \end{pmatrix} \begin{pmatrix} \mathbf{a}^1 & \mathbf{a}^2 \end{pmatrix} = a \begin{pmatrix} a^{11} & a^{12} \\ a^{12} & a^{22} \end{pmatrix}, \tag{3.158}$$

we have

$$C_{i\alpha}C_{i\beta} = aa^{\alpha\beta}. \tag{3.159}$$

To make this identity appear consistent, we should have written the index α as a superscript in the definition of C in eqn (3.141). However, a consistent form for the Beltrami operator is now given, from eqn (3.157), by

$$\Delta_B\varphi = \frac{1}{\sqrt{a}}\frac{\partial}{\partial u^\alpha}\left(\sqrt{a}a^{\alpha\beta}\frac{\partial\varphi}{\partial u^\beta}\right), \tag{3.160}$$

which has precisely the same form as eqn (1.147). This can be written in terms of the covariant metric tensor as

$$\Delta_B\varphi = \frac{1}{\sqrt{a}}\left[\frac{\partial}{\partial u^1}\left(\frac{a_{22}\frac{\partial\varphi}{\partial u^1} - a_{12}\frac{\partial\varphi}{\partial u^2}}{\sqrt{a}}\right) + \frac{\partial}{\partial u^2}\left(\frac{a_{11}\frac{\partial\varphi}{\partial u^2} - a_{12}\frac{\partial\varphi}{\partial u^1}}{\sqrt{a}}\right)\right], \tag{3.161}$$

which leads to the identities

$$\Delta_B u = \frac{1}{\sqrt{a}}\left\{\frac{\partial}{\partial u}\left(\frac{a_{22}}{\sqrt{a}}\right) - \frac{\partial}{\partial v}\left(\frac{a_{12}}{\sqrt{a}}\right)\right\}, \quad \Delta_B v = \frac{1}{\sqrt{a}}\left\{\frac{\partial}{\partial v}\left(\frac{a_{11}}{\sqrt{a}}\right) - \frac{\partial}{\partial u}\left(\frac{a_{12}}{\sqrt{a}}\right)\right\},$$

writing u^α as (u, v).

Using eqns (3.56) and (3.61), we also have

$$\begin{aligned}
\Delta_B\varphi &= \frac{\partial}{\partial u^\alpha}\left(a^{\alpha\beta}\frac{\partial\varphi}{\partial u^\beta}\right) + \Gamma^\gamma_{\gamma\alpha}a^{\alpha\beta}\frac{\partial\varphi}{\partial u^\beta} \\
&= a^{\alpha\beta}\frac{\partial^2\varphi}{\partial u^\alpha\partial u^\beta} - (a^{\delta\alpha}\Gamma^\beta_{\delta\alpha} + a^{\delta\beta}\Gamma^\alpha_{\delta\alpha})\frac{\partial\varphi}{\partial u^\beta} + \Gamma^\gamma_{\gamma\alpha}a^{\alpha\beta}\frac{\partial\varphi}{\partial u^\beta} \\
&= a^{\alpha\beta}\frac{\partial^2\varphi}{\partial u^\alpha\partial u^\beta} - a^{\delta\alpha}\Gamma^\beta_{\delta\alpha}\frac{\partial\varphi}{\partial u^\beta} = a^{\alpha\beta}\left(\frac{\partial^2\varphi}{\partial u^\alpha\partial u^\beta} - \Gamma^\delta_{\alpha\beta}\frac{\partial\varphi}{\partial u^\delta}\right)
\end{aligned} \tag{3.162}$$

after some manipulation of indices.

Exercise 19. Show, using eqn (3.57), that $\Delta_B\varphi = a^{\alpha\beta}\varphi_{,\alpha\beta}$ in terms of covariant derivatives.

If we consider the special case $\varphi = u^\gamma$, we obtain

$$\Delta_B u^\gamma = -a^{\delta\alpha}\Gamma^\beta_{\delta\alpha}\delta^\gamma_\beta = -a^{\delta\alpha}\Gamma^\gamma_{\delta\alpha}, \tag{3.163}$$

which may be compared directly with eqn (1.111). Hence we can write

$$\Delta_B\varphi = a^{\alpha\beta}\frac{\partial^2\varphi}{\partial u^\alpha\partial u^\beta} + (\Delta_B u^\beta)\frac{\partial\varphi}{\partial u^\beta}. \tag{3.164}$$

We can establish a connection with the Gauss Formula, eqn (3.122), if we multiply that equation through by $a^{\alpha\beta}$ to get

$$a^{\alpha\beta}\frac{\partial^2 \mathbf{r}}{\partial u^\alpha \partial u^\beta} = a^{\alpha\beta}\Gamma^\gamma_{\alpha\beta}\frac{\partial \mathbf{r}}{\partial u^\gamma} + a^{\alpha\beta}b_{\alpha\beta}\mathbf{N}.$$

Hence, by eqns (3.111) and (3.163),

$$a^{\alpha\beta}\frac{\partial^2 \mathbf{r}}{\partial u^\alpha \partial u^\beta} + (\Delta_B u^\gamma)\frac{\partial \mathbf{r}}{\partial u^\gamma} = 2\kappa_m \mathbf{N}, \tag{3.165}$$

which may be written, in view of (3.164), as the identity

$$\Delta_B \mathbf{r} = 2\kappa_m \mathbf{N}. \tag{3.166}$$

However, if we use the expression eqn (3.161) for Δ_B, we see that we have again derived the surface metric identity (3.136).

4

Structured grid generation – algebraic methods

4.1 Co-ordinate transformations

Boundary-value problems of physics and engineering may usually be expressed in terms of partial differential equations (the *hosted* equations) to be solved for certain field quantities as functions of space and possibly time over some region of space, subject to certain specified boundary (and possibly initial) conditions. When the equations are expressed in terms of cartesian co-ordinates, a standard numerical method of solution is through *finite differences*, where a uniform rectangular grid of regularly-arranged points is constructed to cover the physical region of space (more precisely, its mathematical representation) and the partial derivatives in the equations are approximated in terms of the differences between values of the field quantities at adjacent points of the grid. The resulting equations may then in principle be solved by algebraic methods.

The essential ideas of this section may be illustrated in two dimensions, with the Oxy plane divided into equal rectangles with sides of length h, k parallel to the axes (Fig. 4.1), and grid points $x = ih$, $y = jk$, with i,j taking integer values. Denoting values of a field variable $f(x, y)$ at the i, j grid-point by $f_{i,j}$, we typically approximate first derivatives at this point by either forward differences

$$\frac{\partial f}{\partial x} \simeq \frac{1}{h}(f_{i+1,j} - f_{i,j}), \tag{4.1}$$

or backward differences

$$\frac{\partial f}{\partial x} \simeq \frac{1}{h}(f_{i,j} - f_{i-1,j}), \tag{4.2}$$

each with 'first-order' accuracy, or, with greater ('second-order') accuracy, by central differences

$$\frac{\partial f}{\partial x} \simeq \frac{1}{2h}(f_{i+1,j} - f_{i-1,j}), \tag{4.3}$$

and similarly for $\partial f/\partial y$. For second derivatives we have

$$\frac{\partial^2 f}{\partial x^2} \simeq \frac{1}{h^2}(f_{i+1,j} - 2f_{i,j} + f_{i-1,j}), \tag{4.4}$$

$$\frac{\partial^2 f}{\partial y^2} \simeq \frac{1}{k^2}(f_{i,j+1} - 2f_{i,j} + f_{i,j-1}), \tag{4.5}$$

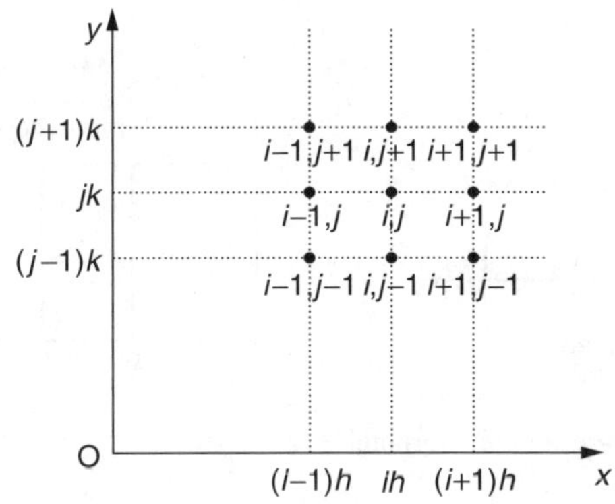

Fig. 4.1 Rectangular array of points for finite differences.

and

$$\frac{\partial^2 f}{\partial x \partial y} \simeq \frac{1}{4hk}(f_{i+1,j+1} + f_{i-1,j-1} - f_{i+1,j-1} - f_{i-1,j+1}), \tag{4.6}$$

all with second-order accuracy, as may be easily verified using Taylor Series expansions. A variety of methods will generally be available to solve the resulting algebraic equations for the grid-point values of the field quantities, provided that the boundary or initial conditions can be incorporated in some way. However, this may not be an easy task, particularly if the boundaries are not rectangular.

For boundary-value problems with relatively simple, non-rectangular boundaries, other co-ordinate systems than cartesian may suggest themselves. To take a basic example in two dimensions, consider the curved area bounded by the inequalities $r_1 \leqslant r \leqslant r_2$, $0 \leqslant \theta \leqslant \alpha$ in terms of polar co-ordinates r, θ (Fig. 4.2). The mapping

$$x = r\cos\theta, \quad y = r\sin\theta, \tag{4.7}$$

gives a 1-1 correspondence between points (r, θ) in a rectangular region in which r and θ are treated like cartesian co-ordinates and points (x, y) in 'physical space' (Fig. 4.3). One could imagine a sort of elastic sheet occupying the physical region

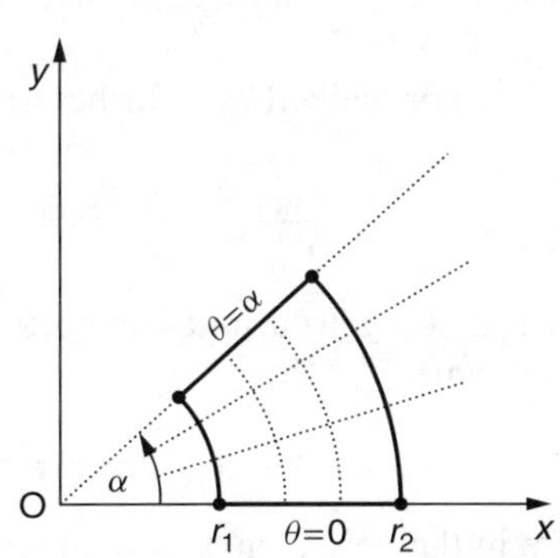

Fig. 4.2 Generating grids using a polar co-ordinate system.

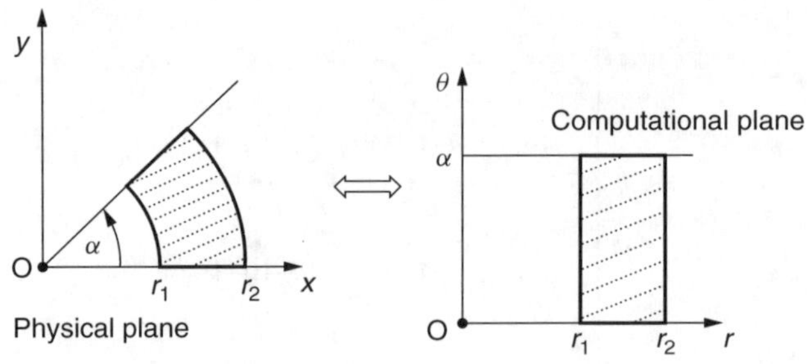

Fig. 4.3 Mapping a curved region onto a rectangle.

which is deformed by stretching, compression, and shearing into a rectangular shape in 'computational space' (the r, θ plane) under the (inverse) mapping

$$r = \sqrt{x^2 + y^2}, \quad \theta = \tan^{-1}(y/x). \tag{4.8}$$

Boundary conditions specified on the curved boundaries $r = r_1$, $r = r_2$ of physical space are then mapped to boundary conditions on straight boundaries in computational space, and a partial differential equation may be conveniently solved using a uniform rectangular grid in computational space. Solution values at a grid-point in computational space may then be associated with the corresponding grid-point in physical space.

The price to be paid for the geometric simplification is that the hosted equation, initially expressed in terms of cartesian co-ordinates, has itself to be transformed to the new co-ordinates, and this may result in a more complicated equation to solve. For example, as is well-known, Laplace's equation in two dimensions,

$$\frac{\partial^2 \varphi}{\partial x^2} + \frac{\partial^2 \varphi}{\partial y^2} = 0$$

becomes

$$\frac{\partial^2 \varphi}{\partial r^2} + \frac{1}{r}\frac{\partial \varphi}{\partial r} + \frac{\partial^2 \varphi}{\partial \theta^2} = 0$$

in polar co-ordinates, which has an extra term, although it is clearly not much more complicated in this case.

The mapping eqn (4.7) may be normalized by a further transformation of co-ordinates from r, θ to ξ, η, where

$$\xi = \frac{r - r_1}{r_2 - r_1}, \quad \eta = \frac{\theta}{\alpha}, \tag{4.9}$$

which maps the rectangle in Fig. 4.3 into a unit square $0 \leqslant \xi \leqslant 1, 0 \leqslant \eta \leqslant 1$ in the ξ, η plane. Then eqns (4.7) become

$$x = [(r_2 - r_1)\xi + r_1]\cos(\alpha\eta), \quad y = [(r_2 - r_1)\xi + r_1]\sin(\alpha\eta), \tag{4.10}$$

which now maps a unit square in the new computational space onto the original curved physical region. Moreover, a uniform rectangular grid in the unit square, where increments in ξ and η along grid lines between adjacent grid-points are constant, maps to a grid in the physical region where increments in r and θ (or ξ and η) along grid lines $\theta = const.$ and grid curves $r = const.$, respectively, are constant. This method of generating a grid in the physical region, using a single analytical transformation (4.10) or a combination of transformations (4.9) and (4.7), may be regarded as one form of an *algebraic* method of grid generation.

Note, however, that eqns (4.9) are not the only way to transform the rectangle in Fig. 4.3 into a unit square. Another one is given by the equations

$$\xi = \frac{\ln(r/r_1)}{\ln(r_2/r_1)}, \qquad \eta = \frac{\theta}{\alpha}. \tag{4.11}$$

It turns out that this transformation satisfies the requirements of one of the fundamental elliptic grid generation methods, as discussed in the first section of the next chapter, namely that each of ξ, η satisfies Laplace's equation (in two dimensions here):

$$\nabla^2\xi = 0, \quad \nabla^2\eta = 0.$$

A uniform grid in computational ξ, η space still maps to a grid in physical space, but note now that equal increments in ξ do not correspond to equal increments in r. The grid in physical space still consists of radial lines and concentric circles, but the distance between the concentric circles diminishes as the inner boundary $r = r_1$ is approached. If this is not regarded as a desirable feature of the grid, the spacing of grid lines can be adjusted using an additional transformation with stretching functions as shown in Section 4.4 below or through the use of control functions as described in Chapter 5.

When $\alpha = 2\pi$ the physical region becomes a complete annulus between the circles $r = r_1$ and $r = r_2$. The radial lines $\theta = 0$ and 2π (imagined slightly separated) may be regarded as forming a *branch cut* (Fig. 4.4), and the annulus can still be mapped into a unit square using eqns (4.10) with $\alpha = 2\pi$.

Of course there are many classical curvilinear co-ordinate systems which may be used to represent physical regions analytically. Even for essentially two dimensional problems we have, for example, elliptic cylindrical co-ordinates, parabolic cylindrical co-ordinates, and bipolar co-ordinates at our disposal. If the physical domain has a configuration which admits representation by a boundary-conforming system of this type, then grids can be easily generated. We refer to this here as *analytic grid generation*, and a number of examples are provided on the disk with this book (see Section 4.6.5). However, if the geometry differs in any significant way from such an ideal configuration, then analytic co-ordinate transformation becomes useless. A primary objective of structured grid generation is to obtain transformations between physical and computational domains which are not subject to this limitation.

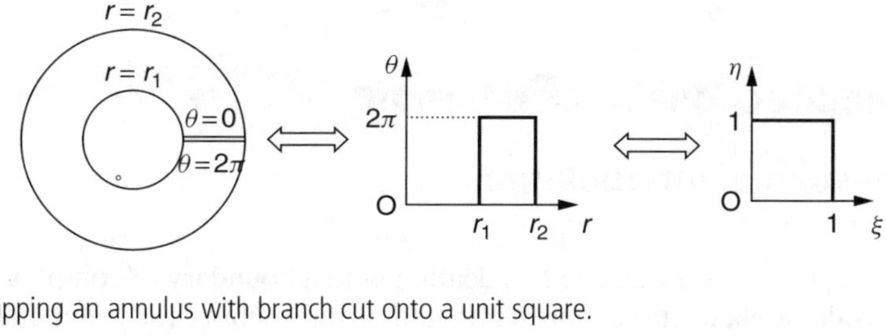

Fig. 4.4 Mapping an annulus with branch cut onto a unit square.

In general we seek to avoid mappings of a sheet in the physical region into a square in the computational region which would involve folding the sheet, resulting in a map which is not one–one. A simple example of such a transformation is the mapping of the rectangle $0 \leqslant x \leqslant 1$, $-1 \leqslant y \leqslant 1$ into the square $0 \leqslant \xi \leqslant 1$, $0 \leqslant \eta \leqslant 1$, with

$$\xi = x, \quad \eta = y^2, \tag{4.12}$$

which folds the sheet in the xy-plane along the x-axis. The Jacobian

$$\begin{vmatrix} \dfrac{\partial \xi}{\partial x} & \dfrac{\partial \xi}{\partial y} \\ \dfrac{\partial \eta}{\partial x} & \dfrac{\partial \eta}{\partial y} \end{vmatrix}$$

of the transformation vanishes when $y = 0$ (along the x-axis, where the fold is located), and this is the mathematical cause of the difficulty. Hence, in order to generate good grids, we seek to avoid mappings where the Jacobian is zero (or infinite, in which case the Jacobian of the inverse mapping would be zero) at points inside the physical region.

Similar considerations apply in three dimensions, where we seek transformations between a physical domain in which there is a cartesian co-ordinate system $Oxyz$ and a unit cube $0 \leqslant \xi \leqslant 1$, $0 \leqslant \eta \leqslant 1$, $0 \leqslant \varsigma \leqslant 1$ in computational space with ξ, η, ς as co-ordinates. The transformations can be imagined to deform a sponge-like object occupying the physical domain into the unit cube, or vice versa, with corresponding transformations of co-ordinate curves and co-ordinate surfaces. Again, the Jacobians of the transformations are required to be non-zero and finite.

Once we have recognized that the boundary curves (or boundary surfaces in three dimensions) in the physical space can be regarded as co-ordinate curves which map onto the sides of a square in computational space (or co-ordinate surfaces which map onto the faces of a cube in three dimensions), certain simple methods of interpolating a set of grid points may suggest themselves. These algebraic methods based on interpolation are extensively used in computational fluid dynamics, exploiting their advantages of ease of computation (compared with differential models involving the solution of partial differential equations) and the capability of direct control over grid node location. On the other hand, algebraic methods may not generate smooth grids; in particular, they tend to preserve the features of boundaries, and any discontinuities in the slope of boundary curves will generally propagate into the interior region. A common use of *algebraic* methods is to generate a first attempt at a grid, which may then be used as a starting point in the iterative implementation of differential models of grid generation, to be considered in Chapter 5.

In the following section we review some basic methods of interpolation.

4.2 Unidirectional interpolation

4.2.1 Polynomial interpolation

A need for interpolation may arise when dealing with a boundary of complex shape. For example, consider a plane curve representing an airfoil section, for which measurement gives a finite number of points with cartesian co-ordinates (x_0, y_0), (x_1, y_1), (x_2, y_2), $\dots$, (x_n, y_n). It may be necessary to be able to represent the set of data mathematically in terms of a functional relationship $y = f(x)$, in order to enable us to perform mathematical operations such as differentiation, integration, and also interpolation to specify other boundary points. Curve-fitting from a discrete set of points may be carried out by standard interpolation methods, in which a curve is found which passes through all data points. Also well-developed are *approximation* methods, in which a curve of a given type, say a polynomial of a certain degree, is obtained which passes as close as possible, in some sense, to all points. Here we concentrate on interpolation methods.

It is easy to show that there is a unique polynomial of degree n which passes through the above points, and it is convenient for some purposes to represent this polynomial in terms of the well-known *Lagrange basis polynomials*

$$L_i(x) = \frac{(x - x_0)(x - x_1)\dots(x - x_{i-1})(x - x_{i+1})\dots(x - x_n)}{(x_i - x_0)(x_i - x_1)\dots(x_i - x_{i-1})(x_i - x_{i+1})\dots(x_i - x_n)}, \quad i = 0, 1, \dots n, \tag{4.13}$$

of degree n, where the numerators omit the linear factors $(x - x_i)$. These functions may be written as

$$L_i(x) = \prod_{\substack{j=0 \\ j \neq i}}^{n} \frac{(x - x_j)}{(x_i - x_j)}.$$

By inspection the Lagrange basis polynomials have the property that

$$L_i(x_j) = 0 \quad \text{for } j \neq i \quad \text{and } L_i(x_i) = 1,$$

or, using the Kronecker delta with both indices written as suffixes (and with 0 included as a possible value of the indices),

$$L_i(x_j) = \delta_{ij}. \tag{4.14}$$

The polynomial of degree n which passes through the given points is then clearly

$$p(x) = \sum_{i=0}^{n} y_i L_i(x). \tag{4.15}$$

Here we shall be mainly concerned with the simplest case in which a polynomial of degree 1 (a straight line) may be constructed to pass through two points (x_0, y_0), (x_1, y_1). In this case we have linear Lagrange basis polynomials

$$L_0(x) = \frac{(x - x_1)}{(x_0 - x_1)}, \quad L_1(x) = \frac{(x - x_0)}{(x_1 - x_0)}, \tag{4.16}$$

(Fig. 4.5) and the resulting straight line

$$y = y_0\frac{(x - x_1)}{(x_0 - x_1)} + y_1\frac{(x - x_0)}{(x_1 - x_0)} = y_0(1 - \xi) + y_1\xi \tag{4.17}$$

with a change of variable on the x-axis, where

$$x = x_0 + \xi(x_1 - x_0),$$

so that $\xi = 0, 1$ when $x = x_0, x_1$, respectively.

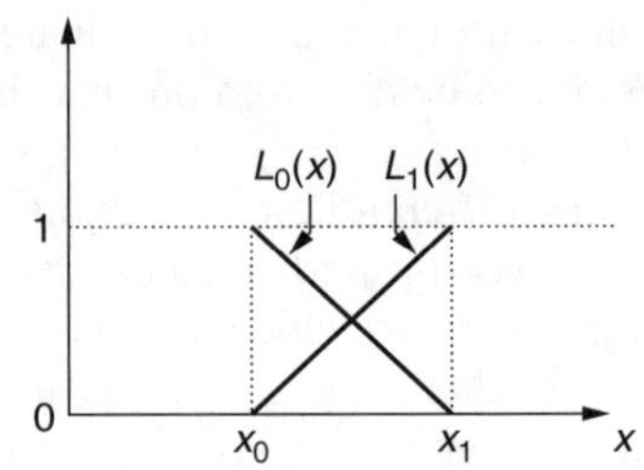

Fig. 4.5 Linear Lagrange basis polynomials.

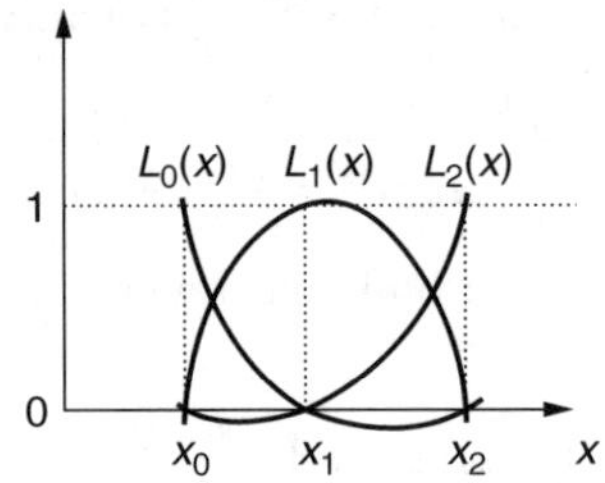

Fig. 4.6 Quadratic Lagrange basis polynomials.

In the case where three points (x_0, y_0), (x_1, y_1), (x_2, y_2) are given, the three quadratic Lagrange basis polynomials (Fig. 4.6) are

$$L_0(x) = \frac{(x - x_1)(x - x_2)}{(x_0 - x_1)(x_0 - x_2)}, \qquad L_1(x) = \frac{(x - x_0)(x - x_2)}{(x_1 - x_0)(x_1 - x_2)},$$

$$L_2(x) = \frac{(x - x_0)(x - x_1)}{(x_2 - x_0)(x_2 - x_1)}, \tag{4.18}$$

and the quadratic function passing through the three points is

$$p(x) = y_0 \frac{(x - x_1)(x - x_2)}{(x_0 - x_1)(x_0 - x_2)} + y_1 \frac{(x - x_0)(x - x_2)}{(x_1 - x_0)(x_1 - x_2)} + y_2 \frac{(x - x_0)(x - x_1)}{(x_2 - x_0)(x_2 - x_1)}.$$

The situation could arise in principle, of course, that the three points lie on a straight line, in which case the coefficient of x^2 in this expression vanishes, and the quadratic reduces to a linear function.

Unidirectional interpolation for algebraic grid generation may be carried out between selected grid-points on opposite boundary curves (or surfaces) of a physical region. With $\mathbf{r}_0$ the position vector of a chosen point on one boundary and $\mathbf{r}_1$ the position vector of another on the opposite boundary, the simplest approach, taking our cue from eqn (4.17), is to construct a straight line between the points, on which the parameter ξ varies, with parametric representation

$$\mathbf{r} = (1 - \xi)\mathbf{r}_0 + \xi \mathbf{r}_1, \tag{4.19}$$

with $0 \leqslant \xi \leqslant 1$. Grid-points may then be selected on this line at uniformly-spaced ξ values (Fig. 4.7), or in some other way, if uniform spacing is not desirable.

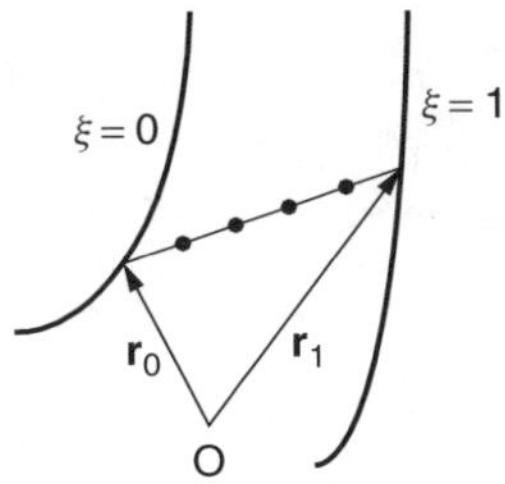

Fig. 4.7 Linear interpolation between curves.

We may wish to interpolate using selected boundary points $\mathbf{r}_0$, $\mathbf{r}_2$, and an intermediate point $\mathbf{r}_1$, in which case we can make use of quadratic Lagrange polynomials. Taking $x_0 = 0$, $x_1 = \frac{1}{2}$, $x_2 = 1$ in eqn (4.18) gives

$$L_0(x) = 2\left(x - \tfrac{1}{2}\right)(x-1), \quad L_1(x) = 4x(1-x), \quad L_2(x) = 2x\left(x - \tfrac{1}{2}\right) \tag{4.20}$$

and it follows that a parametric representation of a curve (not now a straight line in general) passing through the three points is

$$\mathbf{r} = 2\left(\xi - \tfrac{1}{2}\right)(\xi - 1)\mathbf{r}_0 + 4\xi(1-\xi)\mathbf{r}_1 + 2\xi\left(\xi - \tfrac{1}{2}\right)\mathbf{r}_2, \tag{4.21}$$

on which ξ may be regarded as a curvilinear co-ordinate, taking the values 0 and 1 at the end-points $\mathbf{r}_0$, $\mathbf{r}_2$, and $\frac{1}{2}$ at the intermediate point $\mathbf{r}_1$.

Given a set of $n+1$ points with position vectors $\mathbf{r}_0, \mathbf{r}_1, \ldots, \mathbf{r}_n$, the general form of the interpolating curve is given by

$$\mathbf{r}(\xi) = \sum_{i=0}^{n} L_i(\xi)\mathbf{r}_i, \tag{4.22}$$

where, just as in eqn (4.13),

$$L_i(\xi) = \frac{(\xi - \xi_0)(\xi - \xi_1)\ldots(\xi - \xi_{i-1})(\xi - \xi_{i+1})\ldots(\xi - \xi_n)}{(\xi_i - \xi_0)(\xi_i - \xi_1)\ldots(\xi_i - \xi_{i-1})(\xi_i - \xi_{i+1})\ldots(\xi_i - \xi_n)}$$
$$= \prod_{\substack{j=0 \\ j \neq i}}^{n} \frac{(\xi - \xi_j)}{(\xi_i - \xi_j)},$$

so that ξ takes the values ξ_i at the points $\mathbf{r}_i$, $i = 0, 1, \ldots, n$. Functions of a single variable ξ appearing in interpolation expressions such as eqn (4.22) are often called *blending functions*. Here we use blending functions to make the grid distribution match the distribution of end-points $\mathbf{r}_0$, $\mathbf{r}_n$, and interior points $\mathbf{r}_1, \ldots, \mathbf{r}_{n-1}$. In the next section we see that blending functions can also be chosen to provide matching for grid directions at given points.

To show the generation of a two-dimensional plane grid using unidirectional interpolation, we consider a physical domain ABCD (Fig. 4.8) in which only the boundaries AB and CD are specified at the outset. We shall take the curves AB and CD to be

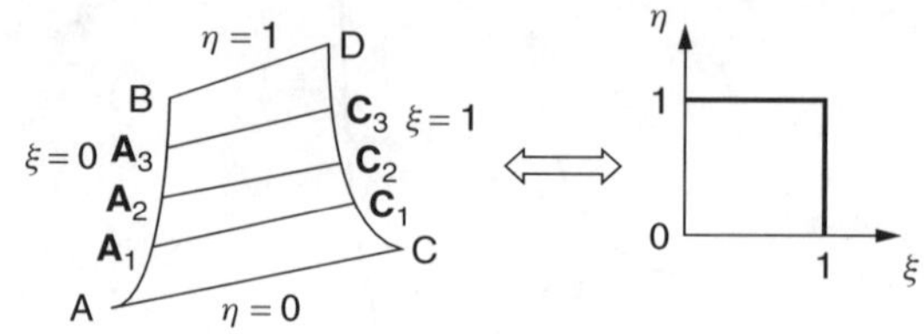

Fig. 4.8 Linear interpolation between two curves.

co-ordinate lines for the η co-ordinate, and assume that we have already selected sets of points on these curves in some way. As shown in Fig. 4.8, we have five points $AA_1A_2A_3B$, $CC_1C_2C_3D$ on each curve corresponding to certain values of η, varying between $\eta = 0$ at A and C and $\eta = 1$ at B and D. Suppose we take $\eta = 0.25$ at A_1, C_1, $\eta = 0.5$ at A_2, C_2, and $\eta = 0.75$ at A_3, C_3. Moreover we can take the co-ordinate ξ to be constant on AB and CD; to be specific, we take $\xi = 0$ on AB and $\xi = 1$ on CD. Points in the physical domain will have cartesian co-ordinates (x, y) which are functions of ξ and η. Unidirectional interpolation between the points on AC and corresponding points on BD will map the unit square in the $\xi\eta$ plane onto the domain shown in Fig. 4.8.

Employing linear interpolation between A and C according to eqn (4.19) then gives

$$\mathbf{r}(\xi, 0) = (1 - \xi)\mathbf{r}(0, 0) + \xi\mathbf{r}(1, 0),$$

while between A_1 and C_1 we obtain

$$\mathbf{r}(\xi, 0.25) = (1 - \xi)\mathbf{r}(0, 0.25) + \xi\mathbf{r}(1, 0.25).$$

Thus the parametric equation of the interpolating line is

$$\mathbf{r}(\xi, \eta_j) = (1 - \xi)\mathbf{r}(0, \eta_j) + \xi\mathbf{r}(1, \eta_j), \tag{4.23}$$

where $0 \leqslant \eta_j = \frac{j-1}{\tilde{j}-1} \leqslant 1$, $j = 1, 2, \ldots, \tilde{j}$, and we have taken $\tilde{j} = 5$ here.

Marking off equal divisions along these straight lines then produces a grid. The general grid-point corresponds to $\xi = \xi_i$, $\eta = \eta_j$, $i = 1, 2, \ldots, \tilde{\imath}$, $j = 1, 2, \ldots, \tilde{j}$, where $\tilde{\imath} = 5$ also and

$$0 \leqslant \xi_i = \frac{i-1}{\tilde{\imath}-1} \leqslant 1.$$

Thus

$$\mathbf{r}(\xi_i, \eta_j) = (1 - \xi_i)\mathbf{r}(0, \eta_j) + \xi_i\mathbf{r}(1, \eta_j). \tag{4.24}$$

Of course, the physical domain mapped out by this process will not coincide with the actual physical domain, unless the physical boundaries AC and BD are straight.

The same process can be used to carry out a unidirectional linear interpolation in the η-direction, starting with given curved boundaries AC and BD on which we take $\eta = 0$ and 1, respectively. This will give a set of grid-points with

$$\mathbf{r}(\xi_i, \eta_j) = (1 - \eta_j)\mathbf{r}(\xi_i, 0) + \eta_j\mathbf{r}(\xi_i, 1). \tag{4.25}$$

4.2.2 Hermite interpolation polynomials

While the Lagrange interpolation polynomials match function values provided by data-points, it is possible to generate a polynomial which matches first derivative values as well as function values at a given set of points. Suppose we have $n + 1$ data points $(x_0, y_0), (x_1, y_1), \ldots, (x_n, y_n)$ as at the beginning of the previous section, together with $n + 1$ corresponding values $y'_0, y'_1, \ldots, y'_n$ of the derivatives of y with respect to x at these points. It is clear that a polynomial of degree $2n + 1$ in general will be required. We would like to be able to write the matching polynomial, in comparison with eqns (4.15) and (4.14), as

$$p(x) = \sum_{i=0}^{n} y_i H_i(x) + \sum_{i=0}^{n} y'_i \tilde{H}_i(x), \tag{4.26}$$

where $H_i(x)$ and $\tilde{H}_i(x)$ are polynomials of degree $2n + 1$ satisfying

$$H_i(x_j) = \delta_{ij}, \quad H'_i(x_j) = 0, \quad \tilde{H}_i(x_j) = 0, \quad \tilde{H}_i{}'(x_j) = \delta_{ij}. \tag{4.27}$$

A convenient set of formulas defines the *Hermite interpolating polynomials* $H_i(x)$, $\tilde{H}_i(x)$ in terms of the Lagrange polynomials as follows:

$$H_i(x) = \{1 - 2L'_i(x_i)(x - x_i)\}[L_i(x)]^2, \tag{4.28}$$

$$\tilde{H}_i(x) = (x - x_i)[L_i(x)]^2. \tag{4.29}$$

It is straightforward to verify, using eqn (4.14), that these definitions satisfy eqn (4.27).

The most commonly used form of Hermite interpolation makes use of the *cubic Hermite polynomial*, for which $n = 1$; the corresponding Lagrange polynomials are linear and given by eqn (4.16). Taking $x_0 = 0$ and $x_1 = 1$ for clarity, we have $L_0(x) = 1 - x$, $L_1(x) = x$, and hence

$$H_0(x) = \{1 - 2L'_0(0)x\}[L_0(x)]^2 = (1 + 2x)(1 - x)^2 = 2x^3 - 3x^2 + 1 \tag{4.30}$$

$$H_1(x) = \{1 - 2L'_1(1)(x - 1)\}[L_1(x)]^2 = (3 - 2x)x^2 = 3x^2 - 2x^3 \tag{4.31}$$

$$\tilde{H}_0(x) = x[L_0(x)]^2 = x(1 - x)^2 = x^3 - 2x^2 + x \tag{4.32}$$

$$\tilde{H}_1(x) = (x - 1)[L_1(x)]^2 = (x - 1)x^2 = x^3 - x^2. \tag{4.33}$$

The graph of these polynomials is shown in Fig. 4.9

A unidirectional interpolation, now with some control over the gradient of the interpolating curve, can be carried out between points $\mathbf{r}_0$ and $\mathbf{r}_n$, with intermediate points $\mathbf{r}_1, \ldots, \mathbf{r}_{n-1}$, according to the formula

$$\mathbf{r}(\xi) = \sum_{i=0}^{n} \mathbf{r}_i H_i(\xi) + \sum_{i=0}^{n} \mathbf{r}'_i \tilde{H}_i(\xi), \tag{4.34}$$

rather than eqn (4.22), where $\mathbf{r}'_i$ is the value of the derivative $\mathrm{d}\mathbf{r}/\mathrm{d}\xi$ at the point $\mathbf{r}_i$.

Once again we can perform unidirectional interpolation starting with boundaries AB and CD which we take as co-ordinate curves $\xi = 0, 1$ as in the previous section

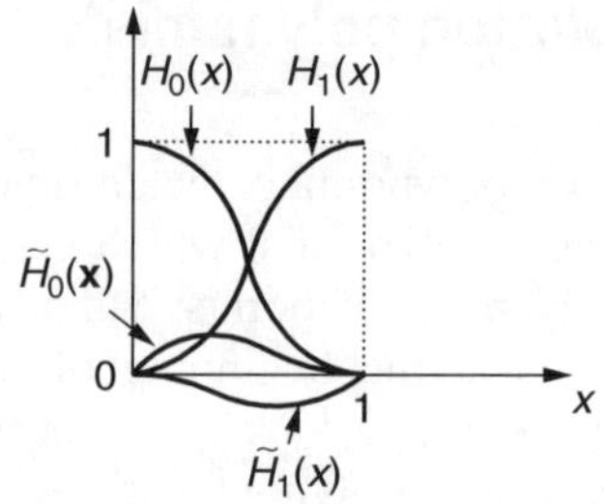

Fig. 4.9 Hermite cubic polynomials.

with sets of corresponding points on each where $\eta = \eta_1, \eta_2, \ldots, \eta_{\bar{J}}$. The parametric equation of the interpolating curve between corresponding points with $\eta = \eta_j$ is now

$$\begin{aligned}\mathbf{r}(\xi, \eta_j) &= \mathbf{r}(0, \eta_j)(2\xi^3 - 3\xi^2 + 1) + \mathbf{r}(1, \eta_j)(3\xi^2 - 2\xi^3) \\ &\quad + \mathbf{r}'(0, \eta_j)(\xi^3 - 2\xi^2 + \xi) + \mathbf{r}'(1, \eta_j)(\xi^3 - \xi^2),\end{aligned} \tag{4.35}$$

where the dash now denotes *partial* differentiation with respect to ξ. This equation may be compared with eqn (4.23). By appropriate choice of $\mathbf{r}'$, whose direction is tangential to the interpolating curve, at the end-points, we are able to force the curve to cut the boundary curves orthogonally.

Equation (4.35) may be written as

$$\mathbf{r} = \Psi_1(\xi)\mathbf{r}_{AB} + \Psi_2(\xi)\mathbf{r}_{CD} + \Psi_3(\xi)\mathbf{r}'_{AB} + \Psi_4(\xi)\mathbf{r}'_{CD}, \tag{4.36}$$

where the Hermite cubic polynomials, or blending functions, have been written as $\Psi_i(\xi)$, and are given by

$$\begin{cases} \Psi_1(\xi) = (\xi^3, \xi^2, \xi, 1)(2, -3, 0, 1)^T \\ \Psi_2(\xi) = (\xi^3, \xi^2, \xi, 1)(-2, 3, 0, 0)^T \\ \Psi_3(\xi) = (\xi^3, \xi^2, \xi, 1)(1, -2, 1, 0)^T \\ \Psi_4(\xi) = (\xi^3, \xi^2, \xi, 1)(1, -1, 0, 0)^T, \end{cases} \tag{4.37}$$

so that we have, using matrices,

$$\mathbf{r} = \Psi(\xi)\begin{pmatrix} \mathbf{r}_{AB} \\ \mathbf{r}_{CD} \\ \mathbf{r}'_{AB} \\ \mathbf{r}'_{CD} \end{pmatrix}, \tag{4.38}$$

where

$$\begin{aligned}\Psi(\xi) &= (\Psi_1(\xi) \quad \Psi_2(\xi) \quad \Psi_3(\xi) \quad \Psi_4(\xi)) \\ &= (\xi^3, \xi^2, \xi, 1)\begin{pmatrix} 2 & -2 & 1 & 1 \\ -3 & 3 & -2 & -1 \\ 0 & 0 & 1 & 0 \\ 1 & 0 & 0 & 0 \end{pmatrix}.\end{aligned}$$

We also have, in the obvious notation, $\mathbf{r}_{AB} = \mathbf{r}(0, \eta_j)$, $\mathbf{r}_{CD} = \mathbf{r}(1, \eta_j)$, $\mathbf{r}'_{AB} = \mathbf{r}'(0, \eta_j)$, $\mathbf{r}'_{CD} = \mathbf{r}'(1, \eta_j)$.

4.2.3 Cubic splines

Fitting a single polynomial to a set of data points $(x_0, y_0), (x_1, y_1), \ldots, (x_n, y_n)$ is often unsatisfactory, even for relatively low values of n, due to the fact that a polynomial of degree N can have $(N-1)$ relative maxima and minima, so that the interpolating curve may oscillate, or wiggle, excessively between data points and hence may not appear to be a good fit. This difficulty may be overcome by generating a 'composite' interpolation curve, constructed out of low-degree polynomials fitted together in a piecewise manner. Such curves are called *splines*. So when in a process of algebraic grid generation we want to control grid distribution by prescribing a large number of interior points, splines may be used as blending functions.

There are many ways in which piecewise interpolation may be carried out. Here we concentrate on one of the commonest methods, that of *cubic splines*. In a cubic spline fit the interpolating function between any two adjacent points is a third-degree polynomial. For the $(n+1)$ data points above there are n intervals between the points, in each of which a cubic polynomial is required. We can write these as

$$\phi_i(x) = a_i + b_i x + c_i x^2 + d_i x^3 \quad \text{for } x_{i-1} \leqslant x \leqslant x_i, \quad i = 1, 2, \ldots, n, \tag{4.39}$$

for some constants a_i, b_i, c_i, d_i, to be found. Differentiation gives

$$\begin{aligned} \phi_i'(x) &= b_i + 2c_i x + 3d_i x^2, \\ \phi_i''(x) &= 2c_i + 6d_i x. \end{aligned} \tag{4.40}$$

We denote the overall piecewise-cubic interpolating function by $y(x)$; the smoothness of this function is made possible by arranging that its first and second derivatives are continuous at the interior points $x_1, x_2, \ldots, x_{n-1}$. So, in addition to the basic continuity requirements

$$\begin{aligned} y(x_i) &= \phi_{i+1}(x_i) = y_i, \quad i = 0, 1, \ldots, (n-1), \\ y(x_i) &= \phi_i(x_i) = y_i, \quad i = 1, 2, \ldots, n, \end{aligned} \tag{4.41}$$

the cubic spline must also satisfy (Fig. 4.10)

$$\phi_i'(x_i) = \phi_{i+1}'(x_i) = y_i', \quad i = 1, 2, \ldots, (n-1) \tag{4.42}$$

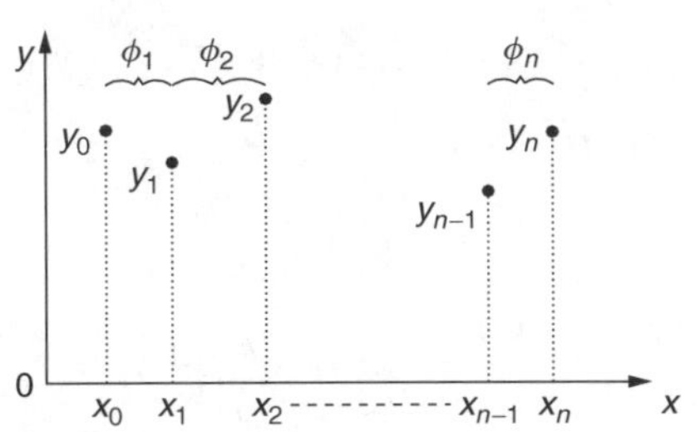

Fig. 4.10 Cubic splines.

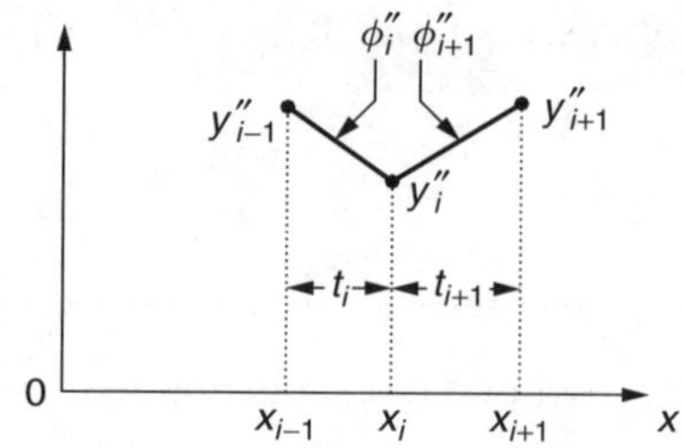

Fig. 4.11 Second derivatives of cubic splines.

and

$$\phi_i''(x_i) = \phi_{i+1}''(x_i) = y_i'', \quad i = 1, 2, \ldots, (n-1), \tag{4.43}$$

where the values of y_i' and y_i'', $i = 1, 2, \ldots, (n-1)$, are not prescribed.

The basic equations for the cubic spline may be derived starting with the observation that from eqns (4.40) and (4.43) y'' must be a continuous piecewise-linear function (Fig. 4.11). Thus we can immediately write

$$\phi_{i+1}''(x) = y_i'' + \frac{(x - x_i)}{(x_{i+1} - x_i)}(y_{i+1}'' - y_i''), \quad x_i \leqslant x \leqslant x_{i+1} \tag{4.44}$$

$$= y_{i+1}'' \frac{(x - x_i)}{(x_{i+1} - x_i)} + y_i'' \frac{(x_{i+1} - x)}{(x_{i+1} - x_i)}, \quad i = 0, 1, \ldots, (n-1), \tag{4.45}$$

where two more unprescribed quantities y_0'' and y_n'' are needed.

Successive direct integrations now give

$$\phi_{i+1}'(x) = \frac{1}{2} y_{i+1}'' \frac{(x - x_i)^2}{t_{i+1}} - \frac{1}{2} y_i'' \frac{(x_{i+1} - x)^2}{t_{i+1}} + C_{i+1}, \tag{4.46}$$

where $t_{i+1} = (x_{i+1} - x_i)$ is an interval width, and

$$\phi_{i+1}(x) = \frac{1}{6} y_{i+1}'' \frac{(x - x_i)^3}{t_{i+1}} + \frac{1}{6} y_i'' \frac{(x_{i+1} - x)^3}{t_{i+1}} + C_{i+1}x + D_{i+1}, \tag{4.47}$$

where C_{i+1} and D_{i+1} are constants of integration.

Substituting $x = x_i$ and $x = x_{i+1}$ into eqn (4.47) gives the simultaneous equations

$$\tfrac{1}{6} y_i'' t_{i+1}^2 + C_{i+1} x_i + D_{i+1} = y_i,$$
$$\tfrac{1}{6} y_{i+1}'' t_{i+1}^2 + C_{i+1} x_{i+1} + D_{i+1} = y_{i+1},$$

which may be solved for C_{i+1} and D_{i+1} to give

$$C_{i+1} = \frac{(y_{i+1} - y_i)}{t_{i+1}} - \frac{1}{6} t_{i+1}(y_{i+1}'' - y_i''), \tag{4.48}$$

$$D_{i+1} = \frac{(x_{i+1} y_i - x_i y_{i+1})}{t_{i+1}} + \frac{1}{6} t_{i+1}(x_i y_{i+1}'' - x_{i+1} y_i''). \tag{4.49}$$

Substituting into eqn (4.47), we obtain the basic equations of the cubic spline:

$$\phi_{i+1}(x) = \frac{1}{6}y_i''\left[\frac{(x_{i+1}-x)^3}{t_{i+1}} - t_{i+1}(x_{i+1}-x)\right] + \frac{1}{6}y_{i+1}''\left[\frac{(x-x_i)^3}{t_{i+1}} - t_{i+1}(x-x_i)\right]$$
$$+y_i\frac{(x_{i+1}-x)}{t_{i+1}} + y_{i+1}\frac{(x-x_i)}{t_{i+1}}, \quad i=0,1,\ldots,(n-1). \qquad (4.50)$$

The second derivatives y_i'', $i = 0, 1, \ldots, n$, however, appear as undetermined quantities in these equations. To proceed further, we still have the continuity condition (4.42) to apply. Equations (4.46) and (4.48) give

$$\phi_{i+1}'(x) = \frac{1}{2}y_{i+1}''\frac{(x-x_i)^2}{t_{i+1}} - \frac{1}{2}y_i''\frac{(x_{i+1}-x)^2}{t_{i+1}} + \frac{(y_{i+1}-y_i)}{t_{i+1}} - \frac{1}{6}t_{i+1}(y_{i+1}''-y_i''). \qquad (4.51)$$

Changing i to $i-1$ gives

$$\phi_i'(x) = \frac{1}{2}y_i''\frac{(x-x_{i-1})^2}{t_i} - \frac{1}{2}y_{i-1}''\frac{(x_i-x)^2}{t_i} + \frac{(y_i-y_{i-1})}{t_i} - \frac{1}{6}t_i(y_i''-y_{i-1}'').$$

Consequently, eqn (4.42) gives

$$-\frac{1}{2}y_i''t_{i+1} + \frac{(y_{i+1}-y_i)}{t_{i+1}} - \frac{1}{6}t_{i+1}(y_{i+1}''-y_i'') = \frac{1}{2}y_i''t_i + \frac{(y_i-y_{i-1})}{t_i} - \frac{1}{6}t_i(y_i''-y_{i-1}''),$$

which may be written as

$$y_{i-1}''t_i + 2y_i''(t_i+t_{i+1}) + y_{i+1}''t_{i+1} = 6\left[\frac{(y_{i+1}-y_i)}{t_{i+1}} - \frac{(y_i-y_{i-1})}{t_i}\right],$$
$$i = 1, 2, \ldots, (n-1). \qquad (4.52)$$

Here we have a set of $(n-1)$ linear equations for the $(n+1)$ quantities y_i'', so clearly there is still some indeterminacy in the system. To resolve the problem we need to specify two more conditions. There are a number of standard ways of doing this.

Method 1 (Natural spline fit) Here we put $y_0'' = y_n'' = 0$. This means that the curvature of the spline is zero at the end-points. Equations (4.52) can be expressed in matrix form as

$$A\begin{pmatrix} y_1'' \\ y_2'' \\ y_3'' \\ - \\ - \\ - \\ y_{n-1}'' \end{pmatrix} = \begin{pmatrix} 6\left(\frac{y_2-y_1}{t_2} - \frac{y_1-y_0}{t_1}\right) \\ 6\left(\frac{y_3-y_2}{t_3} - \frac{y_2-y_1}{t_2}\right) \\ - \\ - \\ - \\ - \\ 6\left(\frac{y_n-y_{n-1}}{t_n} - \frac{y_{n-1}-y_{n-2}}{t_{n-1}}\right) \end{pmatrix}, \qquad (4.53)$$

where A is the $(n-1) \times (n-1)$ symmetric tridiagonal matrix

$$A = \begin{pmatrix} 2\,(t_1+t_2) & t_2 & 0 & 0 & - & - & 0 \\ t_2 & 2\,(t_2+t_3) & t_3 & 0 & - & - & - \\ 0 & t_3 & 2\,(t_3+t_4) & t_4 & - & - & - \\ 0 & 0 & t_4 & - & - & - & - \\ - & - & - & - & - & - & - \\ - & - & - & - & - & - & t_{n-1} \\ 0 & - & - & - & - & t_{n-1} & 2(t_{n-1}+t_n) \end{pmatrix} \tag{4.54}$$

Solution of the system (4.53) may result in a cubic spline that is flatter near the end-points than we might want, given that y_0'' and y_n'' are both zero.

Method 2 Here we make the assumption that the second derivative at each end of the set of data points is the same at the two points adjacent to the ends. That is, we take $y_0'' = y_1''$ and $y_n'' = y_{n-1}''$. This hypothesis generally results in greater curvature in the interpolating curve near the end-points than in the previous method.

We again have a matrix equation of the form (4.53), and the matrix A, still symmetric and tridiagonal, is given by

$$A = \begin{pmatrix} (3t_1+2t_2) & t_2 & 0 & 0 & - & - & 0 \\ t_2 & 2(t_2+t_3) & t_3 & 0 & - & - & - \\ 0 & t_2 & 2(t_3+t_4) & t_4 & - & - & - \\ 0 & 0 & t_4 & - & - & - & - \\ - & - & - & - & - & - & - \\ - & - & - & - & - & 2(t_{n-2}+t_{n-1}) & t_{n-1} \\ 0 & - & - & - & - & t_{n-1} & 2t_{n-1}+3t_n \end{pmatrix}. \tag{4.55}$$

Method 3 Here it is assumed that y_0'' and y_n'' are linear extrapolations of the values of y'' at the two nearest data points at each end. Thus at the left-hand end we put

$$\frac{y_1''-y_0''}{t_1} = \frac{y_2''-y_1''}{t_2}, \tag{4.56}$$

or, rearranging,

$$y_0'' = y_1''\frac{t_1+t_2}{t_2} - y_2''\frac{t_1}{t_2}. \tag{4.57}$$

Similarly at the right-hand end we obtain the equation

$$y_n'' = y_{n-1}''\frac{t_{n-1}+t_n}{t_{n-1}} - y_{n-2}''\frac{t_n}{t_{n-1}}. \tag{4.58}$$

It is easy to show that the matrix equation again has the form (4.53) with the same matrix A as in (4.55), except that the first row is changed to

$$\begin{pmatrix} \dfrac{(t_1+t_2)(t_1+2t_2)}{t_2} & \dfrac{(t_s^2-t_1^2)}{t_2} & 0 & 0 & - & - & 0 \end{pmatrix} \tag{4.59}$$

and the last row becomes

$$\left(\begin{array}{cccccc} 0 & - & - & - & 0 & \dfrac{(t_{n-1}^2 - t_n^2)}{t_{n-1}} \quad \dfrac{(t_{n-1}+t_n)(2t_{n-1}+t_n)}{t_{n-1}} \end{array}\right). \tag{4.60}$$

Method 4 Here we prescribe the gradients y_0', y_n' of the interpolating curve at the end-points. If these gradients are known, we obtain the best cubic spline fit of all these methods. However, we commonly have only estimates of their values at our disposal.

Using eqn (4.51), we obtain

$$y_0' = \phi_1'(x_0) = -\frac{1}{2}y_0''t_1 + \frac{y_1 - y_0}{t_1} - \frac{1}{6}t_1(y_1'' - y_0'') = -\frac{1}{3}y_0''t_1 - \frac{1}{6}y_1''t_1 + \frac{y_1 - y_0}{t_1},$$

and hence

$$y_0''t_1 = -\frac{1}{2}y_1''t_1 - 3y_0' + 3\frac{y_1 - y_0}{t_1}. \tag{4.61}$$

Substituting from (4.61) into the first equation ($i = 1$) of eqn (4.52) gives

$$y_1''\left(\frac{3}{2}t_1 + 2t_2\right) + y_2''t_2 = 6\frac{y_2 - y_1}{t_2} - 9\frac{y_1 - y_0}{t_1} + 3y_0' \tag{4.62}$$

Similarly we can show that the last equation ($i = n - 1$) of eqn (4.52) becomes

$$y_{n-2}''t_{n-2} + y_{n-1}''\left(2t_{n-1} + \frac{3}{2}t_n\right) = 9\frac{y_n - y_{n-1}}{t_n} - 6\frac{y_{n-1} - y_{n-2}}{t_{n-1}} - 3y_n'. \tag{4.63}$$

The other equations of (4.52) are unchanged, and hence we obtain the matrix equation

$$A\begin{pmatrix} y_1'' \\ y_2'' \\ y_3'' \\ - \\ - \\ - \\ y_{n-1}'' \end{pmatrix} = \begin{pmatrix} 6\dfrac{y_2 - y_1}{t_2} - 9\dfrac{y_1 - y_0}{t_1} + 3y_0' \\ 6\dfrac{y_3 - y_2}{t_3} - 6\dfrac{y_2 - y_1}{t_2} \\ 6\dfrac{y_4 - y_3}{t_4} - 6\dfrac{y_3 - y_2}{t_3} \\ - \\ - \\ 6\dfrac{y_{n-1} - y_{n-2}}{t_{n-1}} - 6\dfrac{y_{n-2} - y_{n-3}}{t_{n-2}} \\ 9\dfrac{y_n - y_{n-1}}{t_n} - 6\dfrac{y_{n-1} - y_{n-2}}{t_{n-1}} - 3y_n' \end{pmatrix}, \tag{4.64}$$

where again the matrix A is the same as in the above sections, except for the first row, which becomes

$$\left(\tfrac{3}{2}t_1 + 2t_2 \quad t_2 \quad 0 \quad 0 \quad - \quad - \quad 0\right), \tag{4.65}$$

and the last row, which is

$$\left(0 \quad - \quad - \quad - \quad 0 \quad t_{n-1} \quad 2t_{n-1} + \tfrac{3}{2}t_n\right). \tag{4.66}$$

In all the above approaches a matrix equation has to be solved for the values of $y_1'', y_2'', \ldots, y_{n-1}''$. The matrix A is tridiagonal and diagonally dominant in each case, and possesses a large degree of sparseness. There are standard numerical methods of solving such a system, such as the Thomas Algorithm (see Section 5.2). It is then straightforward to calculate y_0'' and y_n'', and we are then in a position to calculate the interpolating cubic spline itself from eqn (4.50).

The same remarks hold for the following.

Method 5 Here we may consider a mixture of end-conditions of the type considered in the previous four approaches. For example, one might have a natural spline at the left-hand end (Method 1) with a specified slope at the right-hand end (Method 3). In this case we simply have to take the matrix A as in eqn (4.54) but change the last row in accordance with eqn (4.66). Moreover, the column vector on the right-hand side of the matrix equation must agree with eqn (4.53), except that the last entry will have to agree with the last entry in eqn (4.64). Other combinations of end-conditions can be handled in a similar way.

4.3 Multidirectional interpolation and TFI

4.3.1 Projectors and bilinear mapping in two dimensions

Suppose there exists a transformation $\mathbf{r} = \mathbf{r}(\xi, \eta)$ (or $x = x(\xi, \eta)$, $y = y(\xi, \eta)$) which maps the unit square $0 < \xi < 1$, $0 < \eta < 1$ onto the interior of the region $ABDC$ in the xy (physical) plane (Fig. 4.12), such that the edges $\xi = 0, 1$ map to the boundaries AB, CD, respectively, which we can formulate as $\mathbf{r}(0, \eta)$ and $\mathbf{r}(1, \eta)$, the boundaries AC, BD being similarly given by $\mathbf{r}(\xi, 0)$, $\mathbf{r}(\xi, 1)$. We can write down another transformation $\mathbf{P}_\xi$, called a *projector*, which maps points in computational space to points (or position vectors) in physical space, defined by

$$\mathbf{P}_\xi(\xi, \eta) = (1 - \xi)\mathbf{r}(0, \eta) + \xi\mathbf{r}(1, \eta). \tag{4.67}$$

As we have seen in Section 4.2.1 this maps the unit square in the $\xi\eta$ plane onto the region shown in Fig. 4.8, in which the boundaries AC, BD are replaced by straight lines. The sides $\xi = 0, 1$ are mapped onto AB, CD respectively, and the sides $\eta = 0, 1$ are mapped onto the straight lines AC, BD. Furthermore, co-ordinate lines of constant η are mapped into straight lines rather than co-ordinate curves in the physical plane.

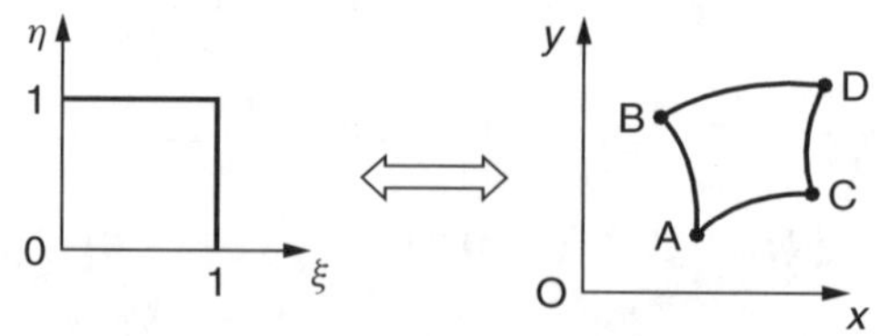

Fig. 4.12 Mapping unit square onto curved four-sided figure.

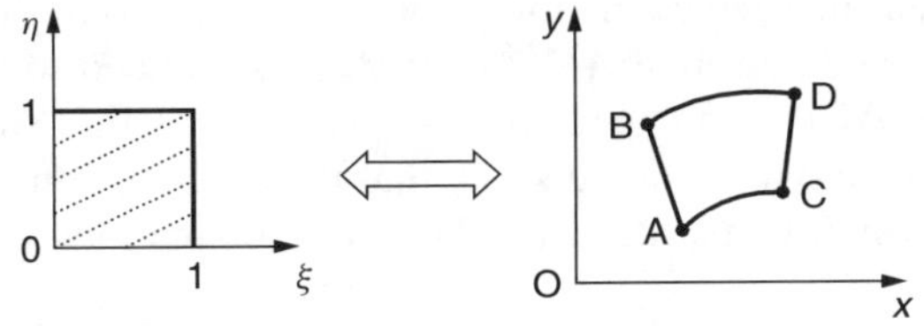

Fig. 4.13 Projector $\mathbf{P}_\eta$.

Similarly we can define the projector

$$\mathbf{P}_\eta(\xi, \eta) = (1 - \eta)\mathbf{r}(\xi, 0) + \eta\mathbf{r}(\xi, 1) \tag{4.68}$$

which maps the unit square onto a region which preserves the boundaries AC, BD, but replaces the boundaries AB, CD with straight lines (Fig. 4.13).

We can form the composite mapping $\mathbf{P}_\xi\mathbf{P}_\eta$, such that

$$\begin{aligned}\mathbf{P}_\xi(\mathbf{P}_\eta(\xi, \eta)) &= \mathbf{P}_\xi((1 - \eta)\mathbf{r}(\xi, 0) + \eta\mathbf{r}(\xi, 1)) \\ &= (1 - \xi)[(1 - \eta)\mathbf{r}(0, 0) + \eta\mathbf{r}(0, 1)] + \xi[(1 - \eta)\mathbf{r}(1, 0) + \eta\mathbf{r}(1, 1)] \\ &= (1 - \xi)(1 - \eta)\mathbf{r}(0, 0) \\ &\quad +(1 - \xi)\eta\mathbf{r}(0, 1) + \xi(1 - \eta)\mathbf{r}(1, 0) + \xi\eta\mathbf{r}(1, 1).\end{aligned} \tag{4.69}$$

This *bilinear* transformation has the property that the four vertices A, B, C, D are preserved, but the boundaries are all replaced by straight lines; that is, the unit square is mapped onto a quadrilateral $ABDC$ (Fig. 4.14). Moreover, straight lines $\xi = const.$ and $\eta = const.$ in computational space are mapped onto straight lines in physical space.

It is easy to show that this composition of projectors, often referred to as the *tensor product*of $\mathbf{P}_\xi$ and $\mathbf{P}_\eta$, is commutative; that is,

$$\mathbf{P}_\xi\mathbf{P}_\eta = \mathbf{P}_\eta\mathbf{P}_\xi. \tag{4.70}$$

The accompanying disk contains a program, listed in Section 4.6.3, to generate a grid in a straight-sided quadrilateral using bilinear transformation.

Note also that we can form the composite map $\mathbf{P}_\xi\mathbf{P}_\xi$; we obtain

$$\mathbf{P}_\xi(\mathbf{P}_\xi(\xi, \eta)) = \mathbf{P}_\xi[(1 - \xi)\mathbf{r}(0, \eta) + \xi\mathbf{r}(1, \eta)] = (1 - \xi)\mathbf{r}(0, \eta) + \xi\mathbf{r}(1, \eta) = \mathbf{P}_\xi(\xi, \eta).$$

Hence we can write

$$\mathbf{P}_\xi\mathbf{P}_\xi = \mathbf{P}_\xi, \tag{4.71}$$

which is the usual defining property of *projection operators*.

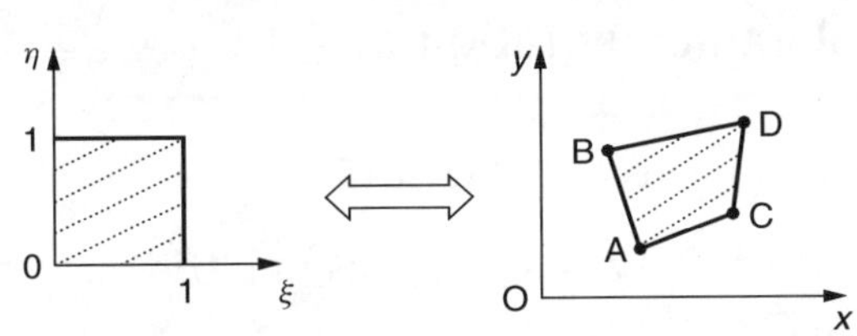

Fig. 4.14 Bilinear transformation $\mathbf{P}_\xi\mathbf{P}_\eta$.

Let us now consider the various mappings of the side $\eta = 0$ of the unit square. Under $\mathbf{P}_\xi$ it is mapped to the straight line AC; under $\mathbf{P}_\eta$ it is mapped to the curved boundary AC; finally under $\mathbf{P}_\xi \mathbf{P}_\eta$ it is mapped to the straight line AC. Similar considerations applied to each side of the unit square show that the composite map $(\mathbf{P}_\xi + \mathbf{P}_\eta - \mathbf{P}_\xi \mathbf{P}_\eta)$ is a transformation which maps the entire boundary of the unit square onto the entire curved boundary $ABDC$.

This map is called the *Boolean sum* of the transformations $\mathbf{P}_\xi$ and $\mathbf{P}_\eta$, and denoted by $\mathbf{P}_\xi \oplus \mathbf{P}_\eta$. Thus

$$\mathbf{P}_\xi \oplus \mathbf{P}_\eta = \mathbf{P}_\xi + \mathbf{P}_\eta - \mathbf{P}_\xi \mathbf{P}_\eta. \tag{4.72}$$

It is clear that $\mathbf{P}_\xi \oplus \mathbf{P}_\eta = \mathbf{P}_\eta \oplus \mathbf{P}_\xi$. The complete formulation is

$$\begin{aligned}(\mathbf{P}_\xi \oplus \mathbf{P}_\eta)(\xi, \eta) &= \mathbf{P}_\xi(\xi, \eta) + \mathbf{P}_\eta(\xi, \eta) - \mathbf{P}_\xi \mathbf{P}_\eta(\xi, \eta) \\ &= (1-\xi)\mathbf{r}(0, \eta) + \xi \mathbf{r}(1, \eta) + (1-\eta)\mathbf{r}(\xi, 0) + \eta \mathbf{r}(\xi, 1) \\ &\quad -(1-\xi)(1-\eta)\mathbf{r}(0,0) - (1-\xi)\eta \mathbf{r}(0,1) \\ &\quad -(1-\eta)\xi \mathbf{r}(1,0) - \xi\eta \mathbf{r}(1,1). \end{aligned} \tag{4.73}$$

This transformation is the basis of *transfinite interpolation (TFI)* in two dimensions. A grid will be generated by eqn (4.73) by taking discrete values ξ_i, η_j of ξ and η with

$$0 \leqslant \xi_i = \frac{i-1}{\tilde{\imath}-1} \leqslant 1 \text{ and } 0 \leqslant \eta_j = \frac{j-1}{\tilde{\jmath}-1} \leqslant 1, \quad i = 1, 2, \ldots, \tilde{\imath}, \quad j = 1, 2, \ldots, \tilde{\jmath},$$

for some choice of $\tilde{\imath}$ and $\tilde{\jmath}$.

Transfinite interpolation is the most common approach to algebraic grid generation. It can produce excellent grids quickly in situations where other methods would be difficult to apply, and it also allows for direct control of the location of grid nodes. Many two-dimensional regions are easy to grid accurately using TFI. However, there are some geometries, such as the airfoil, 'backstep', and C-grids, where TFI proves to be unsatisfactory. The main disadvantages are (1) a lack of smoothness in the generated grids, with any discontinuities in gradient in the boundary curves tending to propagate into the interior, and (2) a tendency to fold when the geometries are complex.

The method can be extended in many ways. For example, the physical region can be divided into several parts, with grids being generated in each separate part and then matched together at the interfaces. This results in discontinuities of slope at the interfaces, and Hermite polynomial interpolation may be exploited to match slopes and thus remove the discontinuities. It is also possible to use TFI with higher-order polynomials as blending functions.

4.3.2 Numerical implementation of TFI

We write eqn (4.73), with reference to Fig. 4.15, as

$$\begin{aligned}\mathbf{r}(\xi.\eta) &= (1-\xi)\mathbf{r}_l(\eta) + \xi \mathbf{r}_r(\eta) + (1-\eta)\mathbf{r}_b(\xi) + \eta \mathbf{r}_t(\xi) - (1-\xi)(1-\eta)\mathbf{r}_b(0) \\ &\quad -(1-\xi)\eta \mathbf{r}_t(0) - (1-\eta)\xi \mathbf{r}_b(1) - \xi\eta \mathbf{r}_t(1), \end{aligned} \tag{4.74}$$

where the abbreviations l, r, b, t stand for 'left', 'right', 'bottom', 'top'.

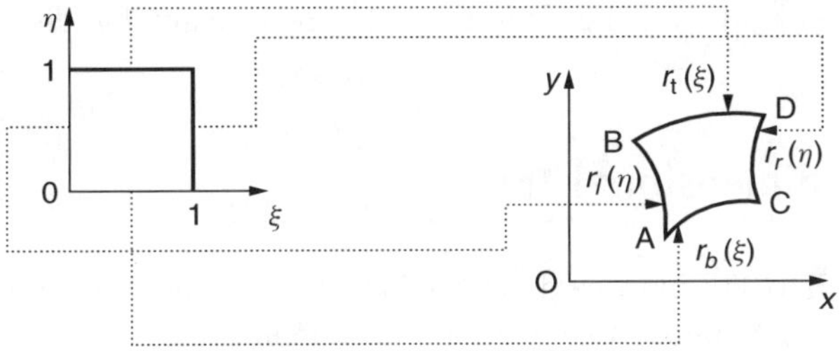

Fig. 4.15 Mapping of boundary curves.

At the four vertices of the physical domain we need the consistency conditions

$$\mathbf{r}_b(0) = \mathbf{r}_l(0), \quad \mathbf{r}_b(1) = \mathbf{r}_r(0), \quad \mathbf{r}_r(1) = \mathbf{r}_t(1), \quad \mathbf{r}_l(1) = \mathbf{r}_t(0). \tag{4.75}$$

Equation (4.74) is equivalent to the two component equations

$$x(\xi.\eta) = (1-\xi)x_l(\eta) + \xi x_r(\eta) + (1-\eta)x_b(\xi) + \eta x_t(\xi) - (1-\xi)(1-\eta)x_b(0)$$
$$-(1-\xi)\eta x_t(0) - (1-\eta)\xi x_b(1) - \xi\eta x_t(1) \tag{4.76}$$

and

$$y(\xi.\eta) = (1-\xi)y_l(\eta) + \xi y_r(\eta) + (1-\eta)y_b(\xi) + \eta y_t(\xi) - (1-\xi)(1-\eta)y_b(0)$$
$$-(1-\xi)\eta y_t(0) - (1-\eta)\xi y_b(1) - \xi\eta y_t(1). \tag{4.77}$$

These equations can be discretized and evaluated through a 'nested DO loop'. Suppose we choose $(m+1)$ grid nodes on the bottom and top boundaries in the computational plane, with equal increments $\Delta\xi = 1/m$ in ξ between nodes; similarly, $(n+1)$ nodes on left and right, with equal increments $\Delta\eta = 1/n$ in η. We need the boundary data for the functions $\mathbf{r}_b$, $\mathbf{r}_t$, $\mathbf{r}_l$, $\mathbf{r}_r$, i.e. the values of the (x, y) co-ordinates at the selected points corresponding to the chosen values of ξ and η on each part of the boundary. This data can be made available to the main routine through a data-file. Or, if the boundaries can be calculated according to some analytical expression, then this can be done in a subroutine.

A basic program with a 'double loop' to compute eqns (4.76) and (4.77), setting $\xi = s$, $\eta = t$, $\Delta\xi = \mathrm{d}X = 1/m$, $\Delta\eta = \mathrm{d}Y = 1/n$, would then take the form:

```
DO J=2,n
    t=(J-1)*dY
DO 2 I=2,m
    s=(I-1)*dX
    X(I,J)=(1.0-s)*Xl(J)+s*Xr(J)+(1.0-t)*Xb(I)+t*Xt(I)
           -(1.0-s)*(1.0-t)*Xb(1)-(1.0-s)*t*Xt(1)
           -s*(1.0-t)*Xb(m+1)-s*t*Xt(m+1)
    Y(I,J)=(1.0-s)*Yl(J)+s*Yr(J)+(1.0-t)*Yb(I)+t*Yt(I)
           -(1.0-s)*(1.0-t)*Yb(1)-(1.0-s)*t*Yt(1)
           -s*(1.0-t)*Yb(m+1)-s*t*Yt(m+1)
2 Continue
1 Continue
```

A complete program may be found on the accompanying disk as described in Section 4.6.4.

4.3.3 Three-dimensional TFI

A simple approach to TFI in three dimensions is through the extension of the definition of projectors to 3D. Suppose that we have a mapping $\mathbf{r}(\xi, \eta, \varsigma)$ from the unit cube $0 \leqslant \xi \leqslant 1, 0 \leqslant \eta \leqslant 1, 0 \leqslant \varsigma \leqslant 1$ to a six-sided volume R of physical space. The opposite planar faces of the cube given by $\xi = 0, 1$ map onto the (in general, curved) opposite faces $\mathbf{r}(0, \eta, \varsigma), \mathbf{r}(1, \eta, \varsigma)$ of R. On these faces there are curvilinear co-ordinate systems with η and ς as co-ordinates. Edges of the cube such as that given by $\eta = \varsigma = 0$ (with $0 \leqslant \xi \leqslant 1$) map into edges of R such as $\mathbf{r}(\xi, 0, 0)$, which is a ξ-co-ordinate curve.

Using linear Lagrange polynomials as blending functions, the following projectors may be defined:

$$\mathbf{P}_\xi(\xi, \eta, \varsigma) = (1 - \xi)\mathbf{r}(0, \eta, \varsigma) + \xi\mathbf{r}(1, \eta, \varsigma) \tag{4.78}$$

$$\mathbf{P}_\eta(\xi, \eta, \varsigma) = (1 - \eta)\mathbf{r}(\xi, 0, \varsigma) + \eta\mathbf{r}(\xi, 1, \varsigma) \tag{4.79}$$

$$\mathbf{P}_\varsigma(\xi, \eta, \varsigma) = (1 - \varsigma)\mathbf{r}(\xi, \eta, 0) + \varsigma\mathbf{r}(\xi, \eta, 1). \tag{4.80}$$

Now the projector $\mathbf{P}_\xi$ still maps the opposite faces $\xi = 0, 1$ of the cube onto the opposite faces $\mathbf{r}(0, \eta, \varsigma), \mathbf{r}(1, \eta, \varsigma)$ of R. It also maps all the vertices of the cube, $(0, 0, 0)$, $(1, 0, 0)$, etc., onto the vertices $\mathbf{r}(0, 0, 0)$, $\mathbf{r}(1, 0, 0)$, etc., of R. However, the four edges of the cube which connect opposite vertices of the faces $\xi = 0$ and $\xi = 1$ are mapped onto straight lines connecting corresponding vertices of R.

For example, $(\xi, 0, 0) \to (1 - \xi)\mathbf{r}(0, 0, 0) + \xi\mathbf{r}(1, 0, 0)$, $0 \leqslant \xi \leqslant 1$.

Clearly the other projectors $\mathbf{P}_\eta, \mathbf{P}_\varsigma$ have similar properties. Moreover they all satisfy the basic projection property given by eqn (4.71).

If we started out with only two opposite faces of R specified and were able to construct curvilinear co-ordinate systems on these surfaces with η and ς as co-ordinates (Fig. 4.16), we could then have a grid on these faces corresponding to discrete

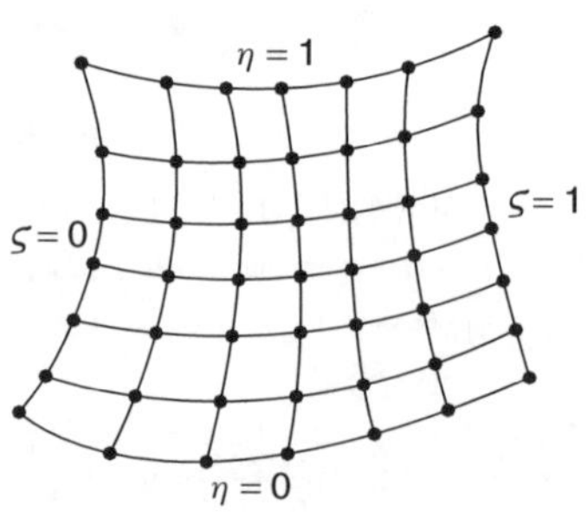

Fig. 4.16 Surface grid.

values η_j, ς_k with

$$0 \leqslant \eta_j = \frac{j-1}{\tilde{j}-1} \leqslant 1, \quad 0 \leqslant \varsigma_k = \frac{k-1}{\tilde{k}-1} \leqslant 1, \quad j = 1, 2, \ldots, \tilde{j}, \quad k = 1, 2, \ldots, \tilde{k}$$

for some $\tilde{j}, \tilde{k}$. We could then use $\mathbf{P}_\xi$, through eqn (4.78), to interpolate a grid between these faces, taking discrete values of ξ also, with $0 \leqslant \xi_i = \frac{i-1}{\tilde{i}-1} \leqslant 1$, $i = 1, 2, \ldots, \tilde{i}$.

The bilinear 'tensor product' $\mathbf{P}_\xi \mathbf{P}_\eta$ may be expressed in full as

$$\begin{aligned} \mathbf{P}_\xi \mathbf{P}_\eta(\xi, \eta, \varsigma) = {} & (1-\xi)(1-\eta)\mathbf{r}(0, 0, \varsigma) + (1-\xi)\eta\mathbf{r}(0, 1, \varsigma) \\ & + \xi(1-\eta)\mathbf{r}(1, 0, \varsigma) + \xi\eta\mathbf{r}(1, 1, \varsigma). \end{aligned} \tag{4.81}$$

The effect of this transformation on the unit cube is to map all four straight edges parallel to the ς direction onto the corresponding four curved edges $\mathbf{r}(0, 0, \varsigma)$, etc., of R. Between these curved edges we then have linear interpolation in both ξ and η directions. This map could be used for linear interpolation if we started with just those four edges of R. The other bilinear products have similar properties, and are given by

$$\begin{aligned} \mathbf{P}_\eta \mathbf{P}_\varsigma(\xi, \eta, \varsigma) = {} & (1-\eta)(1-\varsigma)\mathbf{r}(\xi, 0, 0) \\ & + (1-\eta)\varsigma\mathbf{r}(\xi, 0, 1) + \eta(1-\varsigma)\mathbf{r}(\xi, 1, 0) + \eta\varsigma\mathbf{r}(\xi, 1, 1), \end{aligned} \tag{4.82}$$

$$\begin{aligned} \mathbf{P}_\xi \mathbf{P}_\varsigma(\xi, \eta, \varsigma) = {} & (1-\xi)(1-\varsigma)\mathbf{r}(0, \eta, 0) \\ & + (1-\xi)\varsigma\mathbf{r}(0, \eta, 1) + \xi(1-\varsigma)\mathbf{r}(1, \eta, 0) + \xi\varsigma\mathbf{r}(1, \eta, 1). \end{aligned} \tag{4.83}$$

Clearly these products all have the property of commutativity.

We can also formulate the 'trilinear' transformation $\mathbf{P}_\xi \mathbf{P}_\eta \mathbf{P}_\varsigma$, which may be expressed in full as

$$\begin{aligned} \mathbf{P}_\xi \mathbf{P}_\eta \mathbf{P}_\varsigma(\xi, \eta, \varsigma) = {} & (1-\xi)(1-\eta)(1-\varsigma)\mathbf{r}(0, 0, 0) \\ & + \xi(1-\eta)(1-\varsigma)\mathbf{r}(1, 0, 0) + (1-\xi)\eta(1-\varsigma)\mathbf{r}(0, 1, 0) \\ & + (1-\xi)(1-\eta)\varsigma\mathbf{r}(0, 0, 1) + \xi\eta(1-\varsigma)\mathbf{r}(1, 1, 0) \\ & + \xi(1-\eta)\varsigma\mathbf{r}(1, 0, 1) + (1-\xi)\eta\varsigma\mathbf{r}(0, 1, 1) + \xi\eta\varsigma\mathbf{r}(1, 1, 1). \end{aligned} \tag{4.84}$$

This *trilinear interpolant* maps the unit cube onto a region of physical space with the same vertices as R but with straight lines connecting the vertices.

The Boolean sum $\mathbf{P}_\xi \oplus \mathbf{P}_\eta \oplus \mathbf{P}_\varsigma$ may be formulated in terms of the above mappings by successively applying the definition (4.72). We have

$$\mathbf{P}_\xi \oplus (\mathbf{P}_\eta \oplus \mathbf{P}_\varsigma) = \mathbf{P}_\xi \oplus (\mathbf{P}_\eta + \mathbf{P}_\varsigma - \mathbf{P}_\eta \mathbf{P}_\varsigma) = \mathbf{P}_\xi + \mathbf{P}_\eta + \mathbf{P}_\varsigma - \mathbf{P}_\eta \mathbf{P}_\varsigma - \mathbf{P}_\xi \mathbf{P}_\eta - \mathbf{P}_\xi \mathbf{P}_\varsigma + \mathbf{P}_\xi \mathbf{P}_\eta \mathbf{P}_\varsigma. \tag{4.85}$$

It is straightforward to show that the same result emerges from evaluating $(\mathbf{P}_\xi \oplus \mathbf{P}_\eta) \oplus \mathbf{P}_\varsigma$, which means that Boolean summation is *associative*. Putting $\xi = 0$ in the expressions (4.78), (4.79), (4.80), (4.81), (4.82), (4.83), and (4.84), and combining the results according to the vector sums in (4.85) shows that the face $\xi = 0$ of the unit cube maps onto the curved face $\mathbf{r}(0, \eta, \varsigma)$ of R under the Boolean sum (4.85). In fact each face of the cube maps onto a face of R.

From the above discussion it is clear that, in terms of projectors, the product $\mathbf{P}_\xi \mathbf{P}_\eta \mathbf{P}_\varsigma$ is 'algebraically minimal', in that it is the weakest member of the set of projectors to generate a grid (based on TFI) in R, given that it interpolates only from the eight vertices of R. The Boolean sum $\mathbf{P}_\xi \oplus \mathbf{P}_\eta \oplus \mathbf{P}_\varsigma$, on the other hand, is 'algebraically maximal' and the strongest member of the projector set. To use it we need boundary data on all six faces of R (including the twelve edges and eight vertices). Then eqn (4.85) will generate a grid within R by trilinear interpolation, taking discrete values of ξ, η, ς.

In practice, however, we may not have a complete set of boundary data. Suppose, for example, that we have only boundary data pertaining to the twelve edges of the physical region R. Since eqn (4.81) showed that the product $\mathbf{P}_\xi \mathbf{P}_\eta$ interpolates linearly from four edges of R, we might expect the appropriate grid generation formula to be given by the Boolean product $(\mathbf{P}_\xi \mathbf{P}_\eta \oplus \mathbf{P}_\eta \mathbf{P}_\varsigma \oplus \mathbf{P}_\varsigma \mathbf{P}_\xi)$. This can be easily evaluated in terms of the 'tensor products' above with use of commutativity and the basic projection property (4.71). We have

$$
\begin{aligned}
\mathbf{P}_\xi \mathbf{P}_\eta \oplus (\mathbf{P}_\eta \mathbf{P}_\varsigma \oplus \mathbf{P}_\varsigma \mathbf{P}_\xi) &= \mathbf{P}_\xi \mathbf{P}_\eta \oplus (\mathbf{P}_\eta \mathbf{P}_\varsigma + \mathbf{P}_\varsigma \mathbf{P}_\xi - \mathbf{P}_\eta \mathbf{P}_\varsigma \mathbf{P}_\varsigma \mathbf{P}_\xi) \\
&= \mathbf{P}_\xi \mathbf{P}_\eta \oplus (\mathbf{P}_\eta \mathbf{P}_\varsigma + \mathbf{P}_\varsigma \mathbf{P}_\xi - \mathbf{P}_\eta \mathbf{P}_\varsigma \mathbf{P}_\xi) \\
&= \mathbf{P}_\xi \mathbf{P}_\eta + (\mathbf{P}_\eta \mathbf{P}_\varsigma + \mathbf{P}_\varsigma \mathbf{P}_\xi - \mathbf{P}_\eta \mathbf{P}_\varsigma \mathbf{P}_\xi) \\
&\quad - \mathbf{P}_\xi \mathbf{P}_\eta (\mathbf{P}_\eta \mathbf{P}_\varsigma + \mathbf{P}_\varsigma \mathbf{P}_\xi - \mathbf{P}_\eta \mathbf{P}_\varsigma \mathbf{P}_\xi) \\
&= \mathbf{P}_\xi \mathbf{P}_\eta + \mathbf{P}_\eta \mathbf{P}_\varsigma + \mathbf{P}_\varsigma \mathbf{P}_\xi - \mathbf{P}_\eta \mathbf{P}_\varsigma \mathbf{P}_\xi \\
&\quad - \mathbf{P}_\xi \mathbf{P}_\eta \mathbf{P}_\varsigma - \mathbf{P}_\xi \mathbf{P}_\eta \mathbf{P}_\varsigma + \mathbf{P}_\xi \mathbf{P}_\eta \mathbf{P}_\varsigma \\
&= \mathbf{P}_\xi \mathbf{P}_\eta + \mathbf{P}_\eta \mathbf{P}_\varsigma + \mathbf{P}_\varsigma \mathbf{P}_\xi - 2\mathbf{P}_\xi \mathbf{P}_\eta \mathbf{P}_\varsigma. \qquad (4.86)
\end{aligned}
$$

An explicit expression for this transfinite interpolation (based on twelve edges of boundary data) may be written down by combining eqns (4.81), (4.82), (4.83), and (4.84) according to eqn (4.86).

4.4 Stretching transformations

Algebraic grid generation may be used in combination with univariate stretching transformations to control grid density. For example, in fluid dynamics it is essential to increase the density of grid points near solid boundaries so that boundary layer behaviour, involving sharp variations in flow properties, can be realistically simulated. The stretching transformations discussed below are just a few of a family of general stretching transformations proposed by Roberts (1971). We present the transformations initially in the simple context of a two-dimensional rectangular physical domain $0 \leqslant x \leqslant L$, $0 \leqslant y \leqslant h$ being mapped directly onto a rectangular computational domain, such that a non-uniform grid with the desired clustering of grid points transforms to a uniform grid in the computational plane (Fig. 4.17), on which the 'hosted' partial differential equations can be solved. Stretching functions can also be applied when the physical region is non-rectangular, as mappings between the square computational domain and an intermediate rectangular parametric domain, as in the example (4.11) above, where the (r, θ) rectangular domain serves as the intermediate parameter

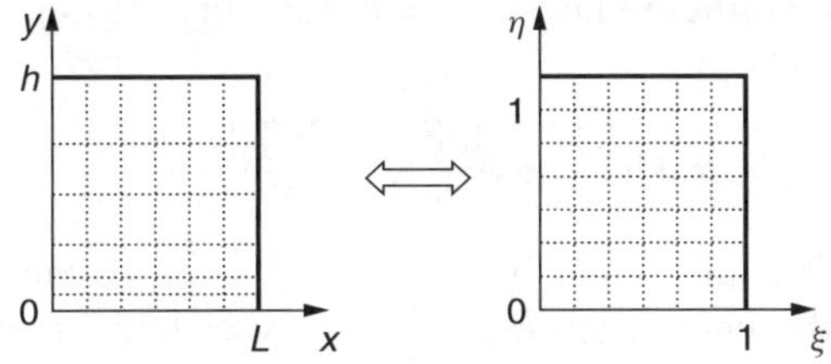

Fig. 4.17 Mapping non-uniform grid in physical plane onto uniform grid in computational plane.

space. A further mapping between the parameter space and the physical domain is then involved. We return to this subject in the next chapter.

Stretching transformations involve positive monotonic univariate functions, here given by $x = x(\xi)$ and $y = y(\eta)$, with inverses $\xi = \xi(x)$ and $\eta = \eta(y)$. A suitable transformation for dealing with two-dimensional boundary-layer flow along a wall at $y = 0$ will concentrate grid lines in the neighbourhood of $y = 0$. Thus we require the derivative $\mathrm{d}y/\mathrm{d}\eta$ to take smaller values (and the derivative $\mathrm{d}\eta/\mathrm{d}y$ larger values) near $y = 0$ than in the rest of the region, so that uniform increments $\delta\eta$ in the computational domain will correspond to smaller increments δy in the physical domain.

One possible transformation is given by

$$\begin{cases} \xi = x \\ \eta = 1 - \dfrac{\ln\{[\beta + 1 - y/h]/[\beta - 1 + y/h]\}}{\ln\{(\beta + 1)/(\beta - 1)\}} \end{cases} \tag{4.87}$$

for some chosen constant β with $\beta > 1$. This maps $y = 0$ directly to $\eta = 0$ and $y = h$ onto $\eta = 1$. While the grid spacing in the x-direction is unaffected, the variation in spacing of grid lines in the y-direction, given a uniform spacing in the η-direction, is governed by the derivative $\mathrm{d}\eta/\mathrm{d}y$, which increases, together with the grid-density near the wall $y = 0$, as the parameter β approaches the limiting value 1.

Any such transformation has implications for the hosted equations, which would need to be transformed in terms of ξ, η if they are to be solved on a uniform grid in the $\xi\eta$-plane. For example, the two-dimensional steady-state incompressible continuity equation in fluid dynamics would become

$$\frac{\partial u}{\partial x} + \frac{\partial v}{\partial y} = \left(\frac{\partial u}{\partial \xi}\frac{\partial \xi}{\partial x} + \frac{\partial u}{\partial \eta}\frac{\partial \eta}{\partial x}\right) + \left(\frac{\partial v}{\partial \xi}\frac{\partial \xi}{\partial y} + \frac{\partial v}{\partial \eta}\frac{\partial \eta}{\partial y}\right) = \frac{\partial u}{\partial \xi} + \frac{\partial v}{\partial \eta}\frac{\partial \eta}{\partial y} = 0,$$

where $\partial\eta/\partial y$ may be obtained in terms of y from (4.87), or in terms of η from the inverse relationship

$$\begin{cases} x = \xi \\ y = h\left\{\dfrac{(\beta + 1) - (\beta - 1)[(\beta + 1)/(\beta - 1)]^{1-\eta}}{[(\beta + 1)/(\beta - 1)]^{1-\eta} + 1}\right\}. \end{cases} \tag{4.88}$$

A similar stretching transformation with an additional parameter α is

$$\begin{cases} \xi = x \\ \eta = \alpha + (1 - \alpha)\dfrac{\ln\{[\beta + y(1 + 2\alpha)/h - 2\alpha]/[\beta - y(1 + 2\alpha)/h + 2\alpha]\}}{\ln[(\beta + 1)/(\beta - 1)]}. \end{cases} \tag{4.89}$$

Note that we still have the boundary $y = h$ mapping into the boundary $\eta = 1$. But the boundary $y = 0$ maps to

$$\eta = \alpha + (1-\alpha)\frac{\ln\{[\beta - 2\alpha]/[\beta + 2\alpha]\}}{\ln[(\beta+1)/(\beta-1)]},$$

so the computational domain is not in general the same rectangle as in the previous example, except for the case when $\alpha = \frac{1}{2}$. The variation of grid spacing in the y-direction is again governed by the derivative $\mathrm{d}\eta/\mathrm{d}y$, which with $\alpha = \frac{1}{2}$ is given by

$$\frac{\mathrm{d}\eta}{\mathrm{d}y} = \frac{2\beta}{h\left\{\beta^2 - \left(\frac{2y}{h} - 1\right)^2\right\}\ln[(\beta+1)/(\beta-1)]},$$

which takes its maximum values in the range $0 \leqslant y \leqslant h$ when $y = 0$ and $y = h$. Thus clustering of grid lines occurs *both* near $y = 0$ and near $y = h$. An example of such a grid for the case $\alpha = 0.5$, $\beta = 1.07$, is shown in Fig. 4.18.

A univariate stretching transformation which gives a clustering of grid lines around the line $y = y_0$ is given by

$$\begin{cases} \xi = x \\ \eta = B + \dfrac{1}{r}\sinh^{-1}\left\{\left(\dfrac{y}{y_0} - 1\right)\sinh(rB)\right\} \end{cases} \tag{4.90}$$

where

$$B = \frac{1}{2r}\ln\left\{\frac{1 + (e^r - 1)y_0/h}{1 - (1 - e^{-r})y_0/h}\right\} \tag{4.91}$$

and r is the 'stretching' parameter. As r approaches zero, eqns (4.90) approach the zero-stretching case $\eta = y/h$. Larger values of r are required to give clustering around $y = y_0$.

Exercise 1. Verify that $y = 0$ maps to $\eta = 0$ and $y = h$ to $\eta = 1$ under eqn (4.90).

The clustering around $y = y_0$ is evident from the derivative

$$\frac{\mathrm{d}\eta}{\mathrm{d}y} = \frac{\sinh(rB)}{ry_0\left\{1 + [(y/y_0) - 1]^2\sinh^2(rB)\right\}^{1/2}}.$$

which takes its maximum value at $y = y_0$.

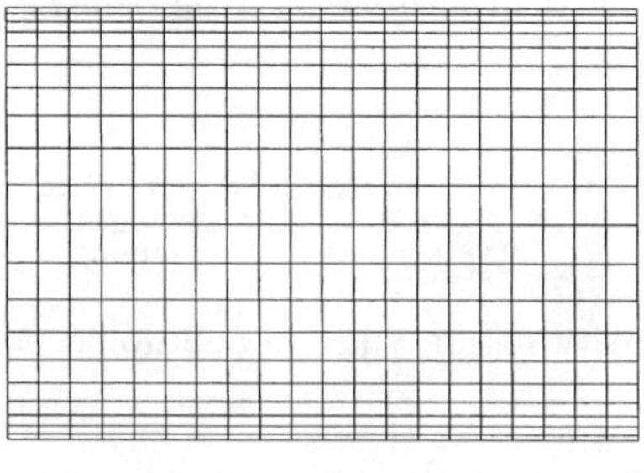

Fig. 4.18 Algebraic grid with grid clustering at both boundaries for beta = 1.07, alpha = 0.5.

The inverse transformation is given by

$$\begin{cases} x = \xi \\ y = y_0 \left\{ 1 + \dfrac{\sinh[r(\eta - B)]}{\sinh(rB)} \right\} \end{cases}. \tag{4.92}$$

Other possible stretching functions are given in Section 4.6.1.

The Eriksson function

The following stretching transformation, due to Eriksson (1982), also involves exponential functions, but has the simpler form

$$y = h \left[\frac{e^{\alpha\eta} - 1}{e^{\alpha} - 1} \right] \tag{4.93}$$

for some constant α, with inverse

$$\eta = \frac{1}{\alpha} \ln \left[1 + \frac{y}{h}(e^{\alpha} - 1) \right]. \tag{4.94}$$

Here $\mathrm{d}y/\mathrm{d}\eta$ takes its lowest values, thereby increasing grid density, as we approach $y = 0$. Denoting the function in (4.93) by $f(\eta)$, we can move the clustering to $y = h$ by forming the function $\{h - f(1 - \eta)\}$, which gives

$$y = h \left\{ \frac{e^{\alpha} - e^{\alpha(1-\eta)}}{e^{\alpha} - 1} \right\}. \tag{4.95}$$

It is straightforward to verify that, if we wish to create a clustering of grid lines near the *interior* line $y = y_0$ (corresponding to $\eta = \eta_0 = y_0/h$ in the computational plane), the above functions can be fitted together after appropriate scaling, with

$$y = \begin{cases} h\eta_0[(e^{\alpha} - e^{\alpha(1-\eta/\eta_0)})/(e^{\alpha} - 1)], & 0 \leqslant \eta \leqslant \eta_0 \\ h\eta_0 + h(1 - \eta_0)[(e^{\alpha(\eta-\eta_0)/(1-\eta_0)} - 1)/(e^{\alpha} - 1)], & \eta_0 \leqslant \eta \leqslant 1. \end{cases} \tag{4.96}$$

This function is monotonically increasing, and has a continuous derivative at $\eta = \eta_0$.

The functions (4.93) and (4.95), suitably scaled, can also be fitted together in the opposite order at an arbitrary interior value $y = y_1$ (with corresponding $\eta = \eta_1 = y_1/h$) to give a stretching function which gives grid clustering near *both* boundaries $y = 0$ and $y = h$. With $f(\eta)$ again as given by eqn (4.93), the two parts of the function are $y = y_1 f(\eta/\eta_1)$ for $0 \leqslant \eta \leqslant \eta_1$ and $y = h - (h - y_1) f((1 - \eta)/(1 - \eta_1))$ for $\eta_1 \leqslant \eta \leqslant 1$.

Exercise 2. Show that the required function is given by

$$y = \begin{cases} h\eta_1(e^{\alpha\eta/\eta_1} - 1)/(e^{\alpha} - 1), & 0 \leqslant \eta \leqslant \eta_1 \\ h - h(1 - \eta_1)(e^{\alpha(1-\eta)/(1-\eta_1)} - 1)/(e^{\alpha} - 1), & \eta_1 \leqslant \eta \leqslant 1, \end{cases} \tag{4.97}$$

which has a continuous first derivative at $y = y_1$.

Example – Flow in a Divergent Nozzle:

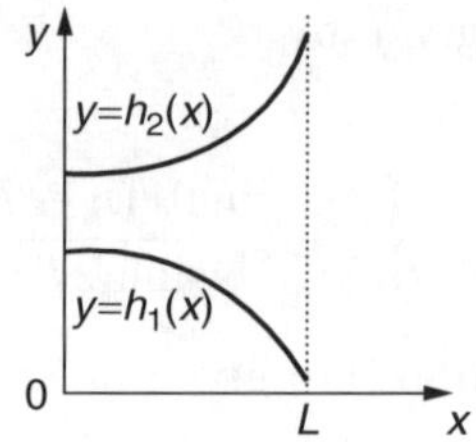

Fig. 4.19 Divergent nozzle.

Here we extend the use of stretching functions to a non-rectangular physical domain. Fig. 4.19 shows a two-dimensional divergent nozzle bounded by the curves $y = h_1(x)$ and $y = h_2(x)$. The physical domain defined by $0 \leqslant x \leqslant L$, $h_1(x) \leqslant y \leqslant h_2(x)$ may be mapped directly onto computational space $0 \leqslant \xi, \eta \leqslant 1$ through

$$\begin{cases} \xi = x \\ \eta = [y - h_1(x)]/[h_2(x) - h_1(x)] \end{cases} \tag{4.98}$$

with inverse

$$\begin{cases} x = \xi \\ y = h_1(\xi) + [h_2(\xi) - h_1(\xi)]\eta. \end{cases} \tag{4.99}$$

The accompanying disk contains a program for directly generating a grid using this transformation, and is listed at Section 4.6.2.

We can concentrate the grid-lines near the boundaries by adapting eqn (4.97) in an obvious way so that the mapping becomes (again with $x = \xi$)

$$y = \begin{cases} [h_2(\xi) - h_1(\xi)]\eta_1(e^{\alpha\eta/\eta_1} - 1)/(e^\alpha - 1) + h_1(\xi), & 0 \leqslant \eta \leqslant \eta_1 \\ [h_2(\xi) - h_1(\xi)][1 - (1 - \eta_1) & \\ \quad \times (e^{\alpha(1-\eta)/(1-\eta_1)} - 1)/(e^\alpha - 1)] + h_1(\xi), & \eta_1 \leqslant \eta \leqslant 1. \end{cases} \tag{4.100}$$

A grid generated by this transformation (using a uniform rectangular grid in the computational plane) is shown in Fig. 4.20.

Returning to our rectangular physical domain $0 \leqslant x \leqslant L$, $0 \leqslant y \leqslant h$, we note that similar piecewise fitting together of Eriksson functions can give stretching functions

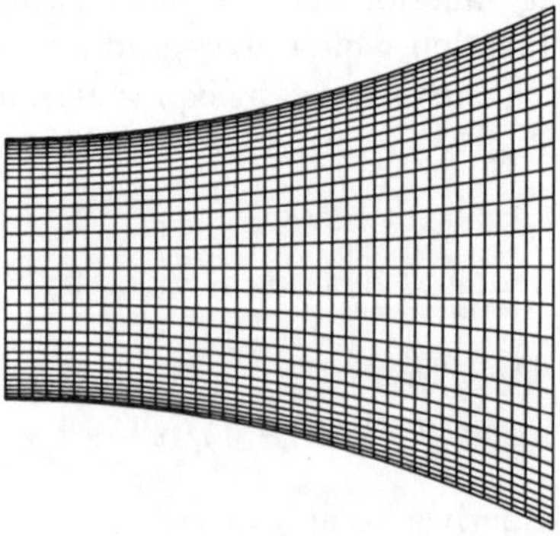

Fig. 4.20 Algebraic grid with grid-clustering near boundaries.

which cluster grids near any number of stipulated lines. For example, clustering near the two lines $y = y_1$ and $y = y_2$ is achieved by the combination

$$y = \begin{cases} \eta_1[h - f(1 - \eta/\eta_1)], & 0 \leqslant \eta \leqslant \eta_1 \\ h\eta_1 + (\eta_0 - \eta_1)f((\eta - \eta_1)/(\eta_0 - \eta_1)), & \eta_1 \leqslant \eta \leqslant \eta_0 \\ h\eta_0 + (\eta_2 - \eta_0)[h - f((\eta_2 - \eta)/(\eta_2 - \eta_0))], & \eta_0 \leqslant \eta \leqslant \eta_2 \\ h\eta_2 + (1 - \eta_2)f((\eta - \eta_2)/(1 - \eta_2)), & \eta_2 \leqslant \eta \leqslant 1. \end{cases} \quad (4.101)$$

where $f(\eta)$ is again given by eqn (4.93), $y = y_0$ is any value intermediate to y_1 and y_2, and $\eta_j = y_j/h$, $j = 0, 1, 2$. Thus we have, explicitly,

$$y = \begin{cases} h\eta_1[e^{\alpha} - e^{\alpha(1-\eta/\eta_1)}]/(e^{\alpha} - 1), & 0 \leqslant \eta \leqslant \eta_1 \\ h\eta_1 + (\eta_0 - \eta_1)h(e^{\alpha(\eta-\eta_1)/(\eta_0-\eta_1)} - 1)/(e^{\alpha} - 1), & \eta_1 \leqslant \eta \leqslant \eta_0 \\ h\eta_2 - (\eta_2 - \eta_0)h(e^{\alpha(\eta_2-\eta)/(\eta_2-\eta_0)} - 1)/(e^{\alpha} - 1), & \eta_0 \leqslant \eta \leqslant \eta_2 \\ h\eta_2 + (1 - \eta_2)h(e^{\alpha(\eta-\eta_2)/(1-\eta_2)} - 1)/(e^{\alpha} - 1), & \eta_2 \leqslant \eta \leqslant 1. \end{cases} \quad (4.102)$$

Similar expressions can be formulated on the basis of the stretching functions defined in eqns (4.88) and (4.92), and similar monotonically increasing functions can be defined to locate grid clustering near an arbitrary number of lines $y = const.$.

4.5 Two-boundary and multisurface methods

4.5.1 Two-boundary technique

The example of the divergent nozzle in the previous section shows how a two-dimensional grid can be generated in the physical domain between two boundaries, using stretching functions to control grid-density. The techniques described here start from these basic ideas, and incorporate Hermite interpolation to produce orthogonality at the boundaries. Moreover, stretching functions are used along the curved boundaries to produce the required position of grid-nodes.

Suppose we have to generate a grid between the two curves AB ($\eta = \eta_1$), CD ($\eta = \eta_2$), shown in Fig. 4.21, consisting of curves $\xi = const.$, $\eta = const.$ The parameters will be normalized so that $\eta_1 = 0$ and $\eta_2 = 1$; the curves connecting A to D and B to C will be $\xi = 0$ and $\xi = 1$, respectively. The parameter ξ could represent a normalized arc-length along the curves, and numerical integration will generally be required to

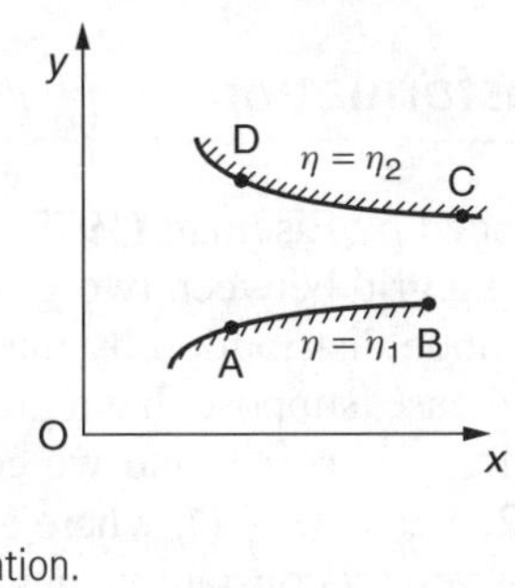

Fig. 4.21 Two-boundary grid generation.

calculate it. Stretching functions $h_{AB}(\xi)$ and $h_{DC}(\xi)$ can then be used to control the location of grid-nodes along AB and DC. Here we present the stretching function

$$h = P\xi + (1-P)\left\{1 - \frac{\tanh[Q(1-\xi)]}{\tanh Q}\right\}, \quad 0 \leqslant \xi \leqslant 1, \tag{4.103}$$

originally due to Roberts (1971) and later modified by Eiseman (1979), where P and Q are parameters to be chosen (for more details, see Section 4.6.1 below) on each boundary. In principle we can then calculate the cartesian co-ordinates of grid-nodes on the boundaries $(x_{AB}(h_{AB}), y_{AB}(h_{AB}))$ and $(x_{DC}(h_{DC}), y_{DC}(h_{DC}))$.

Stretching can also be used in the η-direction with a choice of similar stretching functions $h_{AD}(\eta)$ and $h_{BC}(\eta)$. Linear interpolation could then lead to the following expressions for cartesian co-ordinates:

$$\begin{cases} x(\xi,\eta) = \left\{1 - \tilde{h}(\xi,\eta)\right\} x_{AB}(h_{AB}(\xi)) + \tilde{h}(\xi,\eta) x_{DC}(h_{DC}(\xi)) \\ y(\xi,\eta) = \left\{1 - \tilde{h}(\xi,\eta)\right\} y_{AB}(h_{AB}(\xi)) + \tilde{h}(\xi,\eta) y_{DC}(h_{DC}(\xi)), \end{cases} \tag{4.104}$$

where $\tilde{h}(\xi,\eta) = h_{AD}(\eta) + \xi(h_{BC}(\eta) - h_{AD}(\eta))$. To achieve orthogonality at the two boundaries, however, it is possible to use the Hermite interpolation formula (4.36). Since tangent vectors at the boundaries have direction $\mathrm{d}\mathbf{r}/\mathrm{d}h$, with components $\left(\frac{\mathrm{d}x_{AB}}{\mathrm{d}h_{AB}}, \frac{\mathrm{d}y_{AB}}{\mathrm{d}h_{AB}}\right)$ and $\left(\frac{\mathrm{d}x_{DC}}{\mathrm{d}h_{DC}}, \frac{\mathrm{d}y_{DC}}{\mathrm{d}h_{DC}}\right)$, orthogonal directions will have components $\left(\frac{\mathrm{d}y_{AB}}{\mathrm{d}h_{AB}}, -\frac{\mathrm{d}x_{AB}}{\mathrm{d}h_{AB}}\right)$ and $\left(\frac{\mathrm{d}y_{DC}}{\mathrm{d}h_{DC}}, -\frac{\mathrm{d}x_{DC}}{\mathrm{d}h_{DC}}\right)$. Thus we can write the Hermite interpolation formula in component form as

$$\begin{cases} x(\xi,\eta) = \Psi_1(\eta)x_{AB}(h_{AB}) + \Psi_2(\eta)x_{DC}(h_{DC}) + T_1\Psi_3(\eta)\dfrac{\mathrm{d}y_{AB}}{\mathrm{d}h_{AB}} + T_2\Psi_4(\eta)\dfrac{\mathrm{d}y_{DC}}{\mathrm{d}h_{DC}} \\ y(\xi,\eta) = \Psi_1(\eta)y_{AB}(h_{AB}) + \Psi_2(\eta)y_{DC}(h_{DC}) - T_1\Psi_3(\eta)\dfrac{\mathrm{d}x_{AB}}{\mathrm{d}h_{AB}} - T_2\Psi_4(\eta)\dfrac{\mathrm{d}x_{DC}}{\mathrm{d}h_{DC}}, \end{cases} \tag{4.105}$$

where the two parameters T_1, T_2 that have been introduced, while not affecting orthogonality at the boundaries, can be used to control the extent to which the grid in the interior domain is orthogonal. In fact these parameters may have to be tuned to avoid folding of the grid in the interior.

The accompanying floppy disk contains a numerical code for generating a grid around an NACA-0012 airfoil using the two-boundary method. This may be found in the file *Two-boundary.f* listed in Section 4.6.4 below.

4.5.2 Multisurface transformation

The multisurface method introduced by Eiseman (1979) is another unidirectional interpolation technique for generating a grid between two given curves (or surfaces), allowing additional control over grid-node distribution by the use of intermediate curves (or surfaces). In the two-dimensional case, suppose that a given inner boundary is the curve $\mathbf{r} = \mathbf{r}_1(\xi)$, the given outer boundary is $\mathbf{r}_n(\xi)$, and we construct (n-2) non-intersecting intermediate curves $\mathbf{r}_i(\xi)$, $i = 2, 3, \ldots, (n-1)$, where ξ is a parameter for each curve. We assume that each surface is given by constant values of the independent co-ordinate

η which can be used to give an interpolation formula $\mathbf{r}(\xi, \eta)$ between inner and outer curves, with

$$\mathbf{r}(\xi, \eta_i) = \mathbf{r}_i(\xi), \quad i = 1, 2, 3, \ldots, n. \tag{4.106}$$

Piecewise-linear curves from inner curve $\mathbf{r}_1(\xi)$ to outer curve $\mathbf{r}_n(\xi)$ may be constructed by linking together points on all n curves corresponding to the same value of ξ. We assume that these piecewise-linear curves do not intersect each other. A segment of such a curve between the curves $\mathbf{r}_i(\xi)$ and $\mathbf{r}_{i+1}(\xi)$ must be the vector $\mathbf{r}_{i+1}(\xi) - \mathbf{r}_i(\xi)$. The multisurface method produces smooth co-ordinate curves $\mathbf{r}(\xi, \eta)$ with ξ fixed and η varying from η_1 to η_n by matching tangent vectors $\partial\mathbf{r}/\partial\eta$ at each surface $\mathbf{r}_i(\xi)$, $i = 1, 2, 3, \ldots, n-1$, to the directions of the line segment $\mathbf{r}_{i+1}(\xi) - \mathbf{r}_i(\xi)$. Thus we take

$$\frac{\partial \mathbf{r}}{\partial \eta} = \sum_{i=1}^{n-1} \psi_i(\eta) T_i \{\mathbf{r}_{i+1}(\xi) - \mathbf{r}_i(\xi)\}, \tag{4.107}$$

where the T_is are positive scalars associated with the surfaces $\mathbf{r}_1, \mathbf{r}_2, \ldots, \mathbf{r}_{n-1}$, and the ψ_is satisfy

$$\psi_i(\eta_k) = \delta_{ik}.$$

These could be Lagrange polynomials, given by eqn (4.13).

Integration then yields

$$\mathbf{r}(\xi, \eta) = \mathbf{r}_1(\xi) + \sum_{i=1}^{n-1} \left\{ \int_{\eta_1}^{\eta} \psi_i(\eta')\,\mathrm{d}\eta' \right\} T_i \{\mathbf{r}_{i+1}(\xi) - \mathbf{r}_i(\xi)\}. \tag{4.108}$$

Note that if we put

$$T_i = \left\{ \int_{\eta_1}^{\eta_n} \psi_i(\eta')\,\mathrm{d}\eta' \right\}^{-1}, \tag{4.109}$$

we obtain an equation which is consistent at $\eta = \eta_n$, since the right-hand side of eqn (4.108) reduces to

$$\mathbf{r}_1(\xi) + \sum_{i=1}^{n-1} \{\mathbf{r}_{i+1}(\xi) - \mathbf{r}_i(\xi)\},$$

which clearly telescopes down to the left-hand side $\mathbf{r}(\xi, \eta_n) = \mathbf{r}_n(\xi)$. Equation (4.108) is, of course, also consistent at $\eta = \eta_1$.

Thus the multisurface equation is taken to be

$$\mathbf{r}(\xi, \eta) = \mathbf{r}_1(\xi) + \sum_{i=1}^{n-1} \frac{\left\{ \int_{\eta_1}^{\eta} \psi_i(\eta')\,\mathrm{d}\eta' \right\}}{\left\{ \int_{\eta_1}^{\eta_n} \psi_i(\eta')\,\mathrm{d}\eta' \right\}} \{\mathbf{r}_{i+1}(\xi) - \mathbf{r}_i(\xi)\}. \tag{4.110}$$

In the case $n = 3$, where there is one intermediate surface $\mathbf{r}_2(\xi)$, the approach we take here is to choose parameter values $\eta_1 = 0$, $\eta_2 = 1$, and η_3 initially unspecified, with value greater than one. We can then use the linear Lagrange polynomials

$$\psi_1 = 1 - \eta, \quad \psi_2 = \eta.$$

The multisurface formula becomes

$$\mathbf{r}(\xi,\eta) = \mathbf{r}_1 + \frac{\eta - \frac{1}{2}\eta^2}{\eta_3 - \frac{1}{2}\eta_3^2}(\mathbf{r}_2 - \mathbf{r}_1) + \frac{\frac{1}{2}\eta^2}{\frac{1}{2}\eta_3^2}(\mathbf{r}_3 - \mathbf{r}_2)$$
$$= \frac{(\eta_3 - \eta)(2 - \eta_3 - \eta)}{\eta_3(2-\eta_3)}\mathbf{r}_1 + \frac{2\eta(\eta_3 - \eta)}{\eta_3^2(2-\eta_3)}\mathbf{r}_2 + \frac{\eta^2}{\eta_3^2}\mathbf{r}_3. \qquad (4.111)$$

Now if we take the value of η_3 to be very close to 1, this reduces (approximately) to the interpolation formula:

$$\mathbf{r}(\xi,\eta) = (1-\eta)^2\mathbf{r}_1(\xi) + 2\eta(1-\eta)\mathbf{r}_2(\xi) + \eta^2\mathbf{r}_3(\xi). \qquad (4.112)$$

When $n = 4$ and there are two intermediate surfaces, a similar method involves taking parameter values $\eta_1 = 0$, $\eta_2 = a$, $\eta_3 = 1$, and η_4 unspecified, with $0 < a < 1$ and $\eta_4 > 1$. Then we can take second degree Lagrange polynomials

$$\psi_1 = \frac{1}{a}(a-\eta)(1-\eta), \quad \psi_2 = \frac{\eta(1-\eta)}{a(1-a)}, \quad \psi_3 = \frac{\eta(\eta-a)}{(1-a)}.$$

Equation (4.110) now yields the formula

$$\mathbf{r} = \mathbf{r}_1 + \frac{\left\{\frac{1}{3}\eta^3 - \frac{1}{2}(1+a)\eta^2 + a\eta\right\}}{\left\{\frac{1}{3}\eta_4^3 - \frac{1}{2}(1+a)\eta_4^2 + a\eta_4\right\}}(\mathbf{r}_2 - \mathbf{r}_1)$$
$$+ \frac{\left(\frac{1}{2}\eta^2 - \frac{1}{3}\eta^3\right)}{\left(\frac{1}{2}\eta_4^2 - \frac{1}{3}\eta_4^3\right)}(\mathbf{r}_3 - \mathbf{r}_2) + \frac{\left(\frac{1}{3}\eta^3 - \frac{1}{2}a\eta^2\right)}{\left(\frac{1}{3}\eta_4^3 - \frac{1}{2}a\eta_4^2\right)}(\mathbf{r}_4 - \mathbf{r}_3). \qquad (4.113)$$

Again, if we take η_4 to be very close to 1, this expression reduces (approximately) to the formula:

$$\mathbf{r}(\xi,\eta) = (1-\eta)^2(1-a_1\eta)\mathbf{r}_1(\xi) + (2+a_1)\eta(1-\eta)^2\mathbf{r}_2(\xi)$$
$$+\eta^2(1-\eta)(2+a_2)\mathbf{r}_3(\xi) + \eta^2(1-a_2+a_2\eta)\mathbf{r}_4(\xi), \qquad (4.114)$$

where

$$a_1 = \frac{2}{3a-1} \quad \text{and} \quad a_2 = \frac{2}{2-3a}. \qquad (4.115)$$

Clearly we must avoid taking the values $\frac{1}{3}$ or $\frac{2}{3}$ for a.

Note that eqn (4.110) can also be applied in the trivial case where $n = 2$, when there are no intermediate curves. Taking ψ_1 to be a constant then leads to simple linear interpolation between inner and outer curves:

$$\mathbf{r}(\xi,\eta) = \mathbf{r}_1(\xi) + \eta\{\mathbf{r}_2(\xi) - \mathbf{r}_1(\xi)\}. \qquad (4.116)$$

4.5.3 Numerical implementation

Here we describe briefly the numerical implementation of the above methods for the case of an airfoil NACA-0012. The relevant programs, *Two-boundary.f* and *Multisurface.f*, are listed at Section 4.6.4 and may be found in the directory *Book/tfi.gds* on

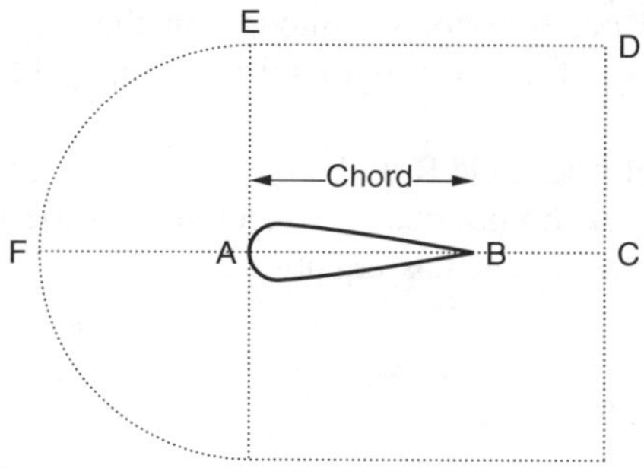

Fig. 4.22 Airfoil profile.

the companion website (www.bh.com/companions/0750650583). The geometry of the domain is shown in Fig. 4.22, with the curve AB representing the profile of an airfoil, one of a family of airfoils with the generic name NACA-00t, where t indicates the thickness as a percentage of the chord length as a two-digit number. Thus an NACA-0012 airfoil has 12% thickness.

The profiles of this family of airfoils are given (with origin at A and chord length unity) by the analytic expression

$$y = t\left(a_1 x^{\frac{1}{2}} + a_2 x + a_3 x^2 + a_4 x^3 + a_5 x^4\right), \tag{4.117}$$

where $a_1 = 1.4779155$, $a_2 = -0.624424$, $a_3 = 1.727016$, $a_4 = 1.384087$, $a_5 = -0.489769$.

The airfoil has an axis of symmetry, and it is sufficient to generate a grid for the upper half of the domain. We therefore consider the domain with inner boundary ABC, consisting of the profile AB and the straight line CD, and outer boundary FED, where FE is a quarter-circle of radius AE and ED is a straight line parallel to BC. The left-hand boundary is taken as AF and the right-hand boundary as CD. Normalized length parameters ξ can then be defined along the boundaries ABC and FED ($0 \leqslant \xi \leqslant 1$), and similarly η along AF and CD ($0 \leqslant \eta \leqslant 1$). For the profile AB this requires integration based on the length formula for plane curves and the analytical expression (4.117).

The first step of the program *Multisurface.f* is to use the univariate stretching functions given by eqn (4.103) with appropriate values of P and Q to generate grid-nodes with the required clustering along the two boundaries AF and CD. The stretching functions could be denoted by $h_{AF}(\eta)$ and $h_{CD}(\eta)$. The inner surface (curve) ABC can be denoted $\mathbf{r}_1$.

The program now has three options. The first is the 'no intermediate curve' option, in which the simple linear interpolation expression eqn (4.116) is used, subject to the stretching used in step 1. The second option is the 'one intermediate curve' option, in which eqn (4.112) is used. In this option, control of orthogonality of the grid is possible at either the inner boundary or the outer boundary.

The third option, that of 'two intermediate curves', is the one we pursue here. This allows control of orthogonality at both inner and outer boundaries. The first attempt at constructing the intermediate curves uses linear interpolation between the inner curve $\mathbf{r}_1$ and the outer one, denoted by $\mathbf{r}_4$. A subroutine 'orthogonality' is then called to adjust the positions of grid-nodes on the intermediate curves, denoted by $\mathbf{r}_2$ and $\mathbf{r}_3$, so as to enforce orthogonality of the grid at the inner and outer boundaries.

If we consider a typical grid node R with co-ordinates (x_1, y_1) on the inner boundary $\mathbf{r}_1$, at which the gradient of the boundary is calculated to be m_1, then the straight line normal to the boundary through R has equation

$$y - y_1 = -\frac{1}{m_1}(x - x_1). \tag{4.118}$$

Suppose that the corresponding node S on the linearly interpolated curve $\mathbf{r}_2^i$ has co-ordinates (x_2^i, y_2^i) and that we can estimate the gradient of the curve $\mathbf{r}_2^i$ at this point as m_2. The tangent to the curve $\mathbf{r}_2^i$ at this node has equation

$$y - y_2 = m_2(x - x_2). \tag{4.119}$$

A better position for S should be at the intersection of these two straight lines; the subroutine calculates the new co-ordinates. A similar procedure is applied to points on the outer boundary $\mathbf{r}_4$ and the linearly interpolated curve $\mathbf{r}_3$. Having now constructed new intermediate curves $\mathbf{r}_2$ and $\mathbf{r}_3$ such that grid-lines are orthogonal at inner and outer boundaries, we can employ the multisurface transformation formula (4.114), with a chosen value of a, in the form

$$\begin{aligned}\mathbf{r}(\xi, \eta) &= (1-h)^2(1-a_1 h)\mathbf{r}_1(\xi) + (2+a_1)h(1-h)^2\mathbf{r}_2(\xi)\\ &\quad + h^2(1-h)(2+a_2)\mathbf{r}_3(\xi) + h^2(1-a_2+a_2 h)\mathbf{r}_4(\xi),\end{aligned} \tag{4.120}$$

where we define h in terms of a linear interpolation between the two stretching functions,

$$h = h_{AF} + \xi(h_{CD} - h_{AF}). \tag{4.121}$$

The program *Two-boundary.f* is essentially a modification of *Multisurface.f*. The subroutine *orthogonality* is removed, and instead Hermite interpolation is used in a modified form of eqn (4.105),

$$\left\{\begin{aligned} x(\xi, \eta) &= \Psi_1(h)x_{AB}(\xi) + \Psi_2(h)x_{DC}(\xi) + T_1\Psi_3(h)\frac{\mathrm{d}y_{AB}}{\mathrm{d}\xi} + T_2\Psi_4(h)\frac{\mathrm{d}y_{DC}}{\mathrm{d}\xi}\\ y(\xi, \eta) &= \Psi_1(h)y_{AB}(\xi) + \Psi_2(h)y_{DC}(\xi) - T_1\Psi_3(h)\frac{\mathrm{d}x_{AB}}{\mathrm{d}\xi} - T_2\Psi_4(h)\frac{\mathrm{d}x_{DC}}{\mathrm{d}\xi}, \end{aligned}\right. \tag{4.122}$$

with h given by (4.121).

4.6 Website programs

Here we list the programs in the directory *Book* contained on the companion website (www.bh.com/companions/0750650583) that are relevant to the material in this chapter. First the subdirectories are specified and then the names of the files with a description are given.

4.6.1 Subdirectory: Book/univariate.gds

1. File: linear.f
   This program simply generates a one-dimensional grid on a straight line based on the linear interpolation

$$x = (1 - \xi)x_0 + \xi x_1, \quad 0 \leqslant \xi \leqslant 1.$$

2. File: Eriksson.f
   This generates a one-dimensional grid on a straight line using the Eriksson stretching function

$$x = (e^{\beta\xi} - 1)/(e^{\beta} - 1), \quad 0 \leqslant \xi \leqslant 1.$$

   Here the constant β controls the degree of clustering near $x = 0$.
3. File: Compress1.f
   This generates another grid on a straight line using the stretching function

$$x = (p^{kb} + c\xi)^{1/b} - p^{k}, \quad 0 \leqslant \xi \leqslant 1, \tag{4.123}$$

   where p, k, and b are constants, with $c = (p^k + 1)^b - p^{kb}$.
4. File: Compress2.f
   This generates a grid on a straight line with the stretching function

$$x = P\xi + (1 - P)\left\{1 - \frac{\tanh[Q(1 - \xi)]}{\tanh Q}\right\}, \quad 0 \leqslant \xi \leqslant 1, \tag{4.124}$$

   mentioned above. For example, when the parameters P and Q take the values $P = 1.8$, $Q = 2.0$, an equally spaced set of points in the ξ computational domain maps into a set of points in the physical x domain with clustering near $x = 1$. When $P = 0.9$, $Q = 2.0$, a fairly equally spaced set of points in physical space is obtained, while the choice $P = 0.1$, $Q = 2.0$ gives clustering near $x = 0$.

4.6.2 Subdirectory: Book/Algebra

1. File: Algebraic1.f
   This generates a two-dimensional planar boundary-conforming grid using the coordinate transformation given by eqn (4.98), where the functions $h_1(x)$ and $h_2(x)$ are specified by the user. The program calculates the so-called *metrics* of the transformation, given by the partial derivatives $\partial\xi/\partial x$, $\partial\xi/\partial y$, $\partial\eta/\partial x$, $\partial\eta/\partial y$, which are required in transforming the hosted partial differential equations to be solved.
   In this example we can show that

$$\frac{\partial\xi}{\partial x} = 1, \quad \frac{\partial\xi}{\partial y} = 0, \quad \frac{\partial\eta}{\partial x} = [h_2'(x)h_1(x) - h_1'(x)h_2(x) - yh_2'(x)] \times [h_2(x) - h_1(x)]^{-2},$$

$$\frac{\partial\eta}{\partial y} = [h_2(x) - h_1(x)]^{-1}.$$

We also need in numerical calculations the partial derivatives

$$\frac{\partial x}{\partial \xi} = J\frac{\partial \eta}{\partial y}, \; \frac{\partial x}{\partial \eta} = -J\frac{\partial \xi}{\partial y}, \; \frac{\partial y}{\partial \xi} = -J\frac{\partial \eta}{\partial x}, \; \frac{\partial y}{\partial \eta} = J\frac{\partial \xi}{\partial x},$$

where the Jacobian J is given by

$$J = \frac{\partial x}{\partial \xi}\frac{\partial y}{\partial \eta} - \frac{\partial x}{\partial \eta}\frac{\partial y}{\partial \xi} = h_2(\xi) - h_1(\xi).$$

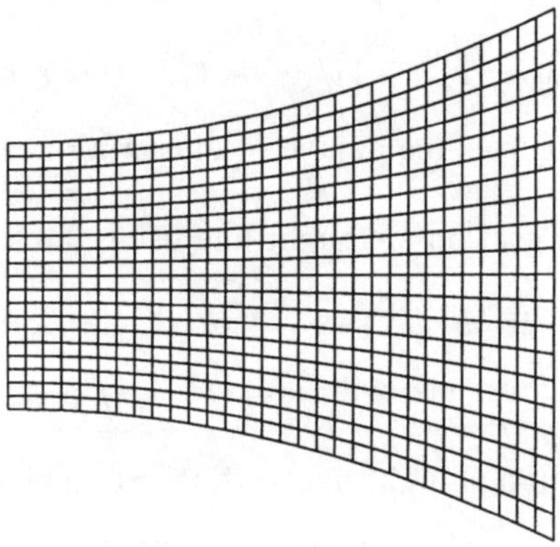

Fig. 4.23 Boundary-conforming algebraic grid.

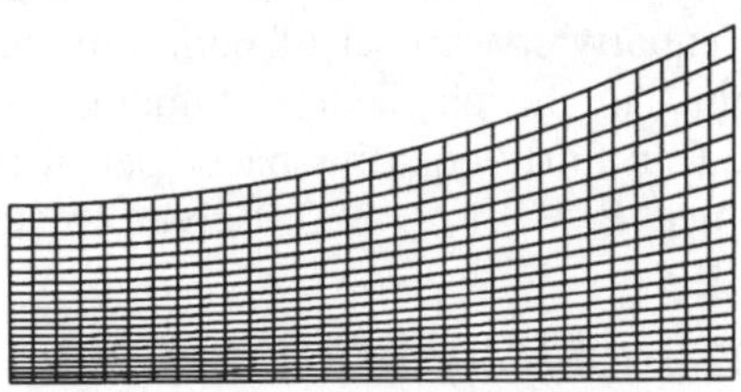

Fig. 4.24 Algebraic grid with grid-clustering near lower boundary.

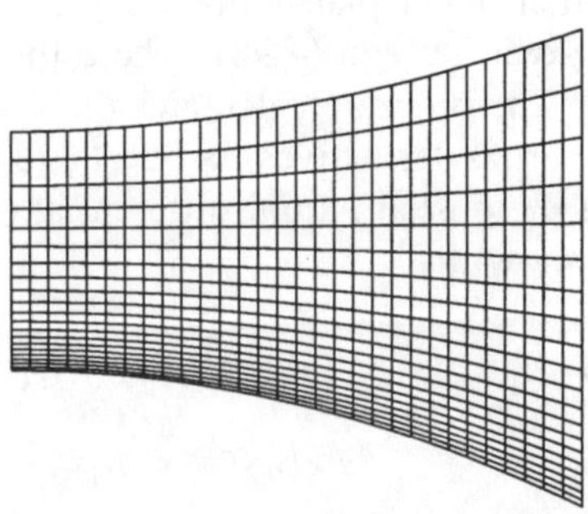

Fig. 4.25 Algebraic grid with grid-clustering near lower boundary.

2. File: Algebraic2.f
3. File: Algebraic3.f
4. File: Algebraic4.f
5. File: Algebraic5.f
   Some typical grids in divergent nozzles calculated using these programs, both without clustering and with clustering, are shown in figs 4.23–4.26.

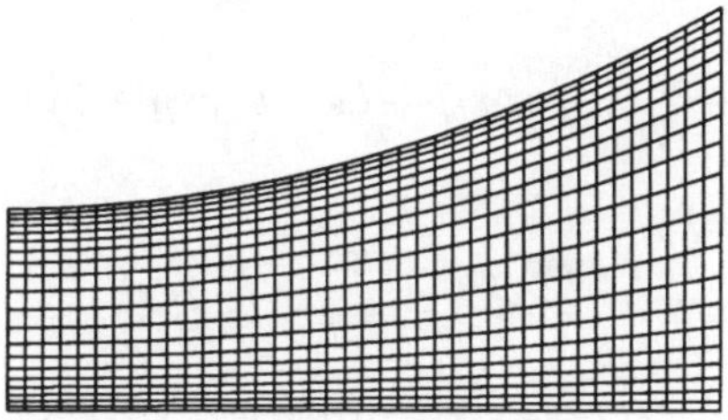

Fig. 4.26 Algebraic grid with grid-clustering near boundaries.

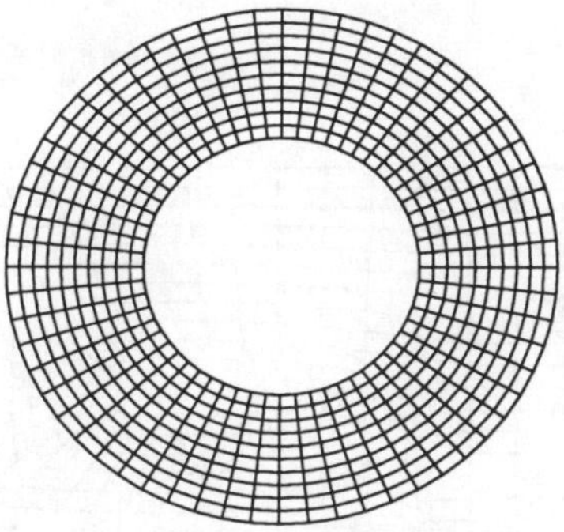

Fig. 4.27 Transfinite interpolation.

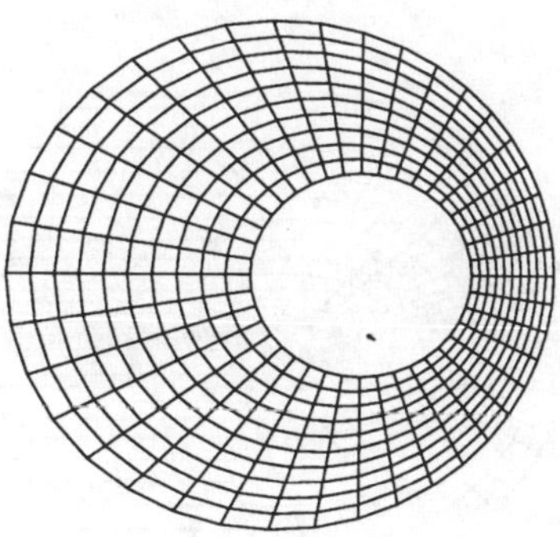

Fig. 4.28 Transfinite interpolation.

4.6.3 Subdirectory: Book/bilinear.gds

This subdirectory contains only one file.

1. File: bilinear.f
   This generates a two-dimensional planar grid in a straight-sided quadrilateral using bilinear interpolation given by

$$\mathbf{r}(\xi, \eta) = (1-\xi)(1-\eta)\mathbf{r}_{bl} + (1-\xi)\eta\mathbf{r}_{tl} + \xi(1-\eta)\mathbf{r}_{br} + \xi\eta\mathbf{r}_{tr},$$

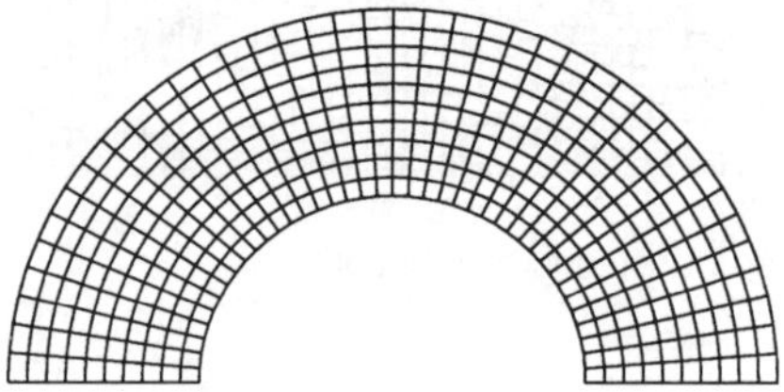

Fig. 4.29 Transfinite interpolation.

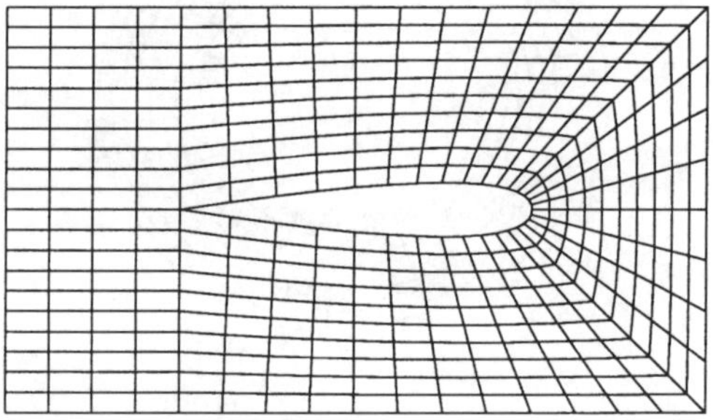

Fig. 4.30 Transfinite interpolation.

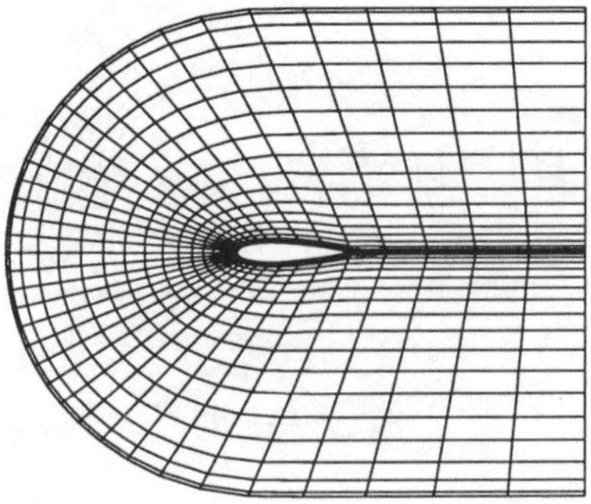

Fig. 4.31 Two-boundary technique.

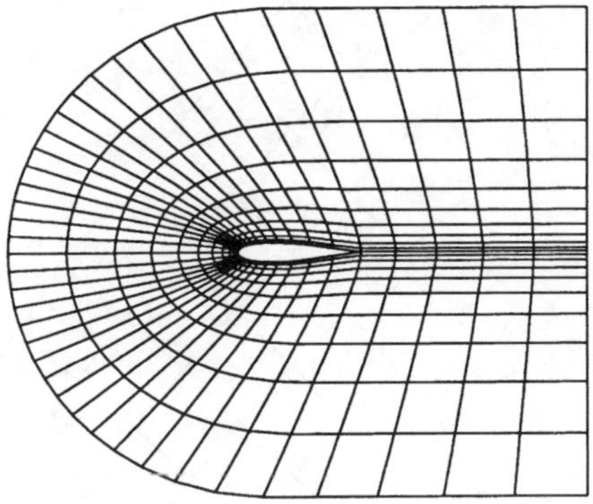

Fig. 4.32 Multisurface grid generation with one intermediate curve.

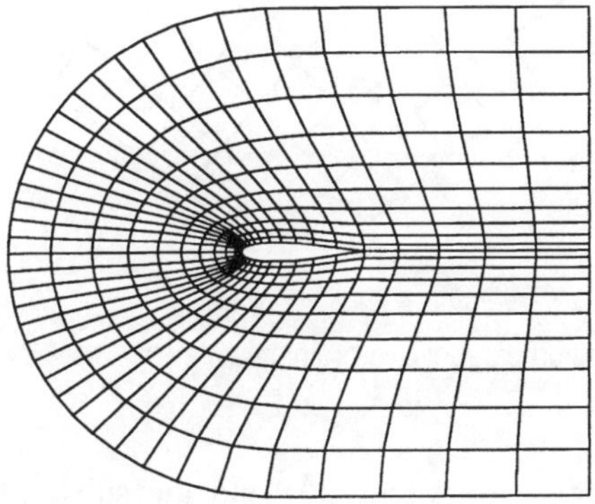

Fig. 4.33 Multisurface grid generation with two intermediate curves.

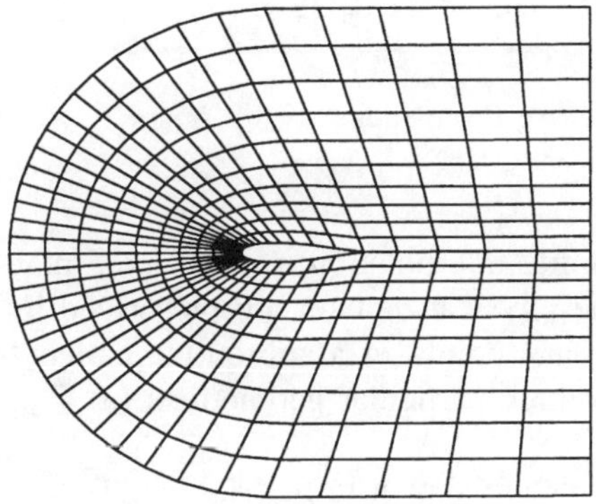

Fig. 4.34 Multisurface grid generation with no intermediate curve.

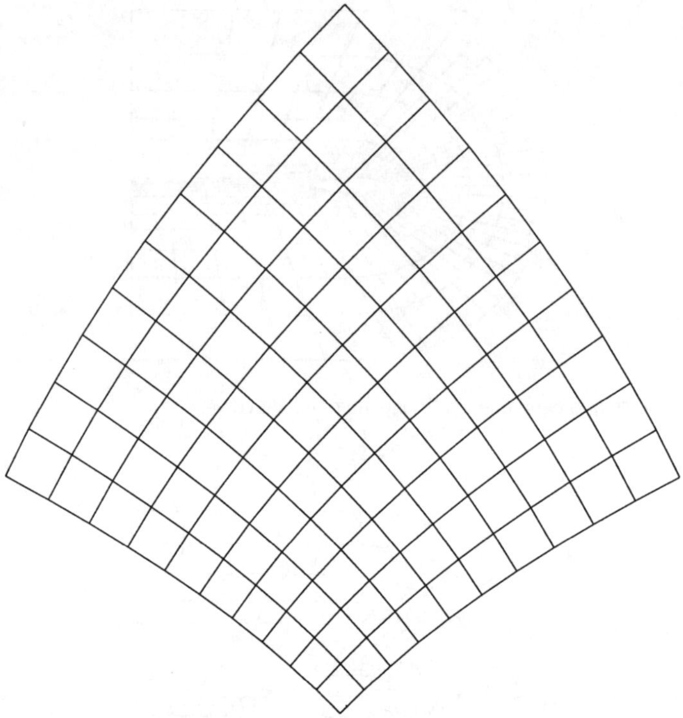

Fig. 4.35 Algebraic grid based on parabolic co-ordinates.

where **r** denotes the position vector $\binom{x}{y}$ and the subscripts bl, tl, br, tr stand for 'bottom left', 'top left', 'bottom right', and 'top right', respectively. The cartesian co-ordinates of the four corners of the quadrilateral must be specified by the user.

4.6.4 Subdirectory: Book/tfi.gds

In this subdirectory there are three files.

1. File: Transfinite.f
   This code uses transfinite interpolation to generate a planar two-dimensional grid. The boundaries of the physical domain are prescribed by specifying the cartesian co-ordinates of boundary points in a subroutine called 'boundary'. An option is included to deal with three particular geometries: (a) a square (b) an annulus (c) a trapezium.
   Some examples of grids are shown in figs 4.27–4.30.
2. File: Two-boundary.f
   A typical grid around an airfoil is shown in Fig. 4.31.
3. File: Multisurface.f
   Examples are shown in figs 4.32–4.34.

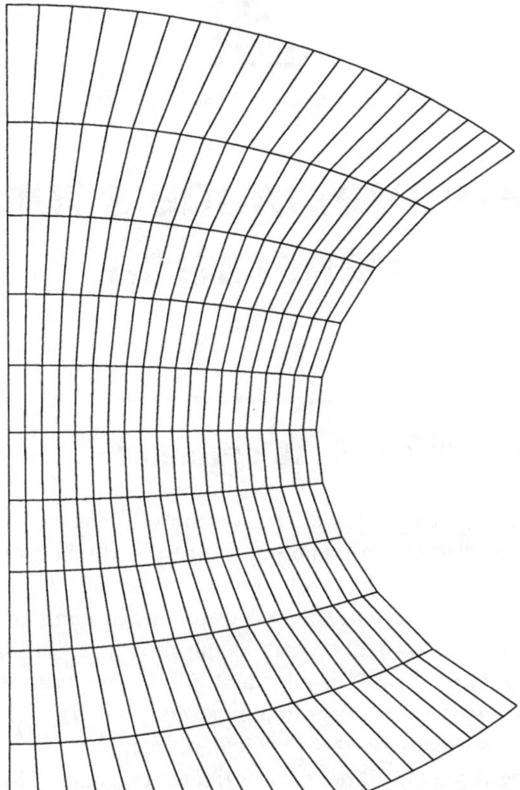

Fig. 4.36 Algebraic grid based on bipolar co-ordinates.

4.6.5 Subdirectory: Book/analytic.gds

This subdirectory contains three files. In the first two, planar two-dimensional grids are generated by discretizing classical co-ordinate transformations.

1. File: parabolic.f
   An example is shown in Fig. 4.35.
2. File: bipolar.f
   An example is shown in Fig. 4.36.

5

Differential models for grid generation

5.1 The direct and inverse problems

In Section 4.1 a number of examples were given in two dimensions of the way in which grids could be generated in a non-rectangular physical domain by utilizing a transformation between the domain and a rectangular (or square) domain in computational space, whereby the rectangular cartesian co-ordinates in computational space become curvilinear co-ordinates in physical space. As curvilinear co-ordinates, moreover, they are *boundary-conforming*, so that the physical boundaries become co-ordinate curves on which one of the curvilinear co-ordinates is constant. In three dimensions, in the same way, boundary surfaces of the physical domain become co-ordinate surfaces, again with one of the three curvilinear co-ordinates constant. The principal advantage of such transformations is naturally that physical boundary conditions expressed in terms of the new curvilinear co-ordinates are simplified and thus easier to incorporate in numerical work. This advantage comes at the cost of increasing the complexity of the partial differential equations to be solved.

Thus the 'direct problem' in two dimensions, given a physical domain R in the plane Oxy bounded by four segments of curves, is to determine two functions $\xi(x, y)$, $\eta(x, y)$ for x and y within the domain, such that ξ is constant (say $\xi = 0$ and 1, although other choices might be convenient) on two opposite boundaries and η is constant (0 and 1 again, say) on the other two boundaries. Furthermore, on each boundary where ξ is constant, η must increase monotonically (from 0 to 1) so as to make $\eta(x, y)$ continuous with a one–one relationship between points (x, y) and η on that boundary curve; similarly on the other two boundaries, where η is constant and ξ varies monotonically. The details of precisely how ξ and η vary monotonically on the boundaries remain at our disposal. A structured grid can then be generated in the physical domain by selecting a network of ξ and η co-ordinate curves. The same considerations apply to a doubly-connected physical domain (Fig. 5.1) once we have made a branch-cut, except that the cut introduces two artificial boundaries on which any boundary conditions must coincide, since they correspond to the same points in physical space.

Analysis of the accuracy of the numerical solution of the hosted partial differential equations based on a grid indicates as one would expect that grid spacing needs to be small to obtain accurate resolution where gradients in the solution are large. Moreover,

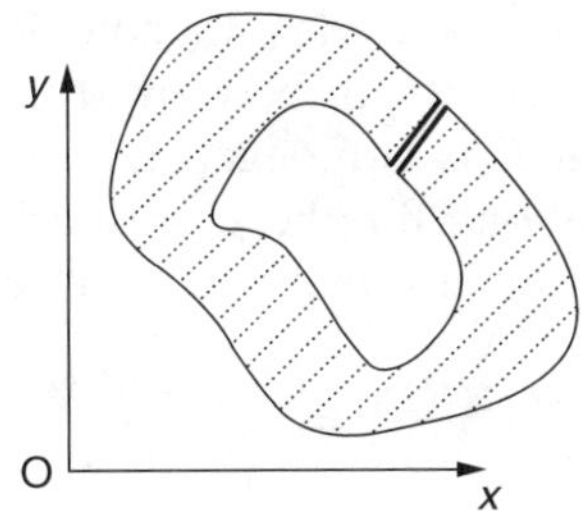

Fig. 5.1 Doubly-connected region with branch cut.

too high a rate of change of grid spacing and too large a departure from orthogonality result in errors. For a given grid spacing, smoothness and orthogonality usually result in smaller errors. However, it is not always possible to generate an orthogonal grid.

There will in principle be an infinite number of solutions to the direct problem, and an infinite number of possible grids. We can single out one particular solution by requiring the functions $\xi(x, y)$, $\eta(x, y)$ to satisfy certain partial differential equations in R. These equations could be *elliptic*, *parabolic*, or *hyperbolic*, but with boundary conditions specified over the entire boundary, as described in the previous paragraph, they must be chosen to be elliptic, the most common example of which is Laplace's equation. The unknown functions are then solutions to a well-posed boundary-value problem with Dirichlet-type boundary conditions (i.e. with function values specified on the entire boundary).

In practice, however, it is usually more convenient to solve the 'inverse problem' for the cartesian co-ordinates x, y as functions of ξ, η. That is, we set up a boundary-value problem in the transformed (computational) plane $O\xi\eta$ in which the domain is a rectangle (or square), on the sides of which the values of $x(\xi, \eta)$ and $y(\xi, \eta)$ are given by the distribution of cartesian co-ordinates on the corresponding boundaries of R. (These are again *Dirichlet* boundary conditions, in which values of the variables which we are attempting to solve for are prescribed over the whole boundary.) The selected partial differential equations are inverted so that x and y are expressed in terms of ξ and η, and can then be solved on a simple rectangular grid constructed in the $\xi\eta$ plane, using the finite difference approximations given in Section 4.1. Values of x and y given by the solution at the grid nodes in the computational plane now directly give the cartesian co-ordinates of the grid nodes in the physical plane.

The most widely used of such elliptic grid generators is the pair of Laplace's equations

$$\nabla^2\xi = 0 \quad \text{and } \nabla^2\eta = 0, \tag{5.1}$$

familiar in fluid dynamics as generating networks of mutually orthogonal streamlines and equipotential lines in two-dimensional ideal fluid flows. An example of a boundary-conforming co-ordinate system satisfying these equations has already been given by eqn (4.11).

The system (5.1) has certain desirable features, including an *extremum principle*, which guarantees that neither maxima nor minima in ξ, η can occur in the interior of R. Given the imposed smooth monotonic variation of these variables on the boundaries of R, this suggests that they will also vary monotonically along the grid-lines, and that folding will not occur. In fact it can be proved theoretically that the curvilinear co-ordinate system generated in R is non-degenerate.

The partial differential equations for the corresponding inverse problem may be obtained directly from the two-dimensional version of eqn (1.114):

$$(\nabla^2 x^i)\frac{\partial y_k}{\partial x^i} = -g^{ij}\frac{\partial^2 y_k}{\partial x^i \partial x^j}.$$

Using eqn (5.1), we obtain

$$g^{ij}\frac{\partial^2 y_k}{\partial x^i \partial x^j} = 0, \quad k = 1, 2. \tag{5.2}$$

Hence, substituting $x^1 = \xi$, $x^2 = \eta$, $y_1 = x$, $y_2 = y$, and using eqn (1.163), we have

$$g_{22}\frac{\partial^2 x}{\partial \xi^2} - 2g_{12}\frac{\partial^2 x}{\partial \xi \partial \eta} + g_{11}\frac{\partial^2 x}{\partial \eta^2} = 0,$$

$$g_{22}\frac{\partial^2 y}{\partial \xi^2} - 2g_{12}\frac{\partial^2 y}{\partial \xi \partial \eta} + g_{11}\frac{\partial^2 y}{\partial \eta^2} = 0, \tag{5.3}$$

for x and y as functions of ξ, η, with g_{11}, g_{12}, and g_{22} given by eqn (1.158). These equations, which here we call the *Winslow equations*, Winslow (1967), are also elliptic. However, although eqns (5.1) are linear and uncoupled in ξ and η, eqns (5.3) are in general non-linear and coupled in x and y through the coefficients g_{ij}.

An intrinsic property of elliptic grid generators is that the grids generated are always smooth. This means that discontinuities in slope at the boundaries are not propagated into the interior of the physical domain. In fact, as will be shown in Chapter 6, the grid generator based on Laplace's eqns (5.1) produces a grid which minimizes (for all admissible functions $\xi(x, y)$, $\eta(x, y)$) the integral

$$I = \iint_R (|\nabla\xi|^2 + |\nabla\eta|^2)\, dx\, dy \tag{5.4}$$

evaluated over the physical domain R. Given that the grid in the computational plane will be associated with equal increments in ξ and η, we might expect $|\nabla\xi|$ and $|\nabla\eta|$ to give measures of grid-point density in the physical plane, with large gradients in ξ and η where points are closely clustered together. Minimizing the integral I may then be regarded as equivalent to generating the (in some sense) smoothest possible grid.

Since we have a choice of models of grid generation for any given problem, the possibility of selecting the pair of Laplace equations

$$\frac{\partial^2 x}{\partial \xi^2} + \frac{\partial^2 x}{\partial \eta^2} = 0,$$

$$\frac{\partial^2 y}{\partial \xi^2} + \frac{\partial^2 y}{\partial \eta^2} = 0, \tag{5.5}$$

instead of the more complex system (5.3) suggests itself. However, although this model may generate satisfactory grids in some cases, in general the non-vanishing of the Jacobian of the transformation from the computational to the physical domain cannot be guaranteed, and folding (or overlapping) of grid lines may occur.

5.2 Control functions

Sometimes it may be desirable to depart from the degree of smoothness resulting from solving eqns (5.3) and introduce some variation of grid spacing. For example, where we expect large gradients in fluid flow variables in a boundary-layer region, we may seek a higher grid density there. If we simply use eqns (5.3) to generate the grid, we have no control of grid density in the interior of R, and boundary layers cannot be properly resolved. (Note that we can, however, select grid points on the boundary as we please – this is one of the most desirable aspects of the 'inverse problem' approach to grid generation.) A standard method for controlling grid density is to vary eqns (5.1) by adding user-specified 'inhomogeneous' terms to the right-hand sides, so that the equations become the Poisson equations

$$\nabla^2\xi = P(\xi,\eta) \quad \text{and } \nabla^2\eta = Q(\xi,\eta), \tag{5.6}$$

where $P(\xi,\eta)$, $Q(\xi,\eta)$ are suitably selected *control functions* (or *forcing functions*). We refer to this as the *TTM Method (Thompson, Thames, and Mastin (1974)).*

Exercise 1. Show that the equations of the inverse problem corresponding to eqn (5.6) are

$$g_{22}\frac{\partial^2 x}{\partial\xi^2} - 2g_{12}\frac{\partial^2 x}{\partial\xi\partial\eta} + g_{11}\frac{\partial^2 x}{\partial\eta^2} + g\left(P\frac{\partial x}{\partial\xi} + Q\frac{\partial x}{\partial\eta}\right) = 0$$

$$g_{22}\frac{\partial^2 y}{\partial\xi^2} - 2g_{12}\frac{\partial^2 y}{\partial\xi\partial\eta} + g_{11}\frac{\partial^2 y}{\partial\eta^2} + g\left(P\frac{\partial y}{\partial\xi} + Q\frac{\partial y}{\partial\eta}\right) = 0. \tag{5.7}$$

Typical control functions

A set of possible control functions was proposed by Thompson, Thames, and Mastin (1974) as follows, and we state some properties of these functions without proof:

$$P(\xi,\eta) = -\sum_{n=1}^{N} a_n\frac{(\xi-\xi_n)}{|\xi-\xi_n|}e^{-c_n|\xi-\xi_n|} - \sum_{i=1}^{I} b_i\frac{(\xi-\xi_i)}{|\xi-\xi_i|}e^{-d_i[(\xi-\xi_i)^2+(\eta-\eta_i)^2]^{\frac{1}{2}}}$$

$$Q(\xi,\eta) = -\sum_{n=1}^{N} a_n\frac{(\eta-\eta_n)}{|\eta-\eta_n|}e^{-c_n|\eta-\eta_n|} - \sum_{i=1}^{I} b_i\frac{(\eta-\eta_i)}{|\eta-\eta_i|}e^{-d_i[(\xi-\xi_i)^2+(\eta-\eta_i)^2]^{\frac{1}{2}}}. \tag{5.8}$$

Here N is the number of lines (co-ordinate lines $\xi = \xi_n$ and $\eta = \eta_n$) and I the number of points (with $\xi = \xi_i, \eta = \eta_i$, $0 \leqslant \xi_i, \eta_i \leqslant 1$) to which the grid is to be attracted, and a_n, c_n, b_i, d_i are positive parameters. The first term in the expression for $P(\xi,\eta)$ has the effect (with typical 'amplitude' a_n) of attracting ξ-lines (curves on which ξ is constant) towards curves $\xi = \xi_n$ in the physical domain, while the second term (with amplitude b_i) attracts ξ-lines towards points (and similarly for $Q(\xi,\eta)$). In each case the attractive effect decays with distance in computational space from the line or point in question according to the 'decay' parameters c_n, d_i.

The functions $(\xi-\xi_n)/|\xi-\xi_n|$ and $(\eta-\eta_n)/|\eta-\eta_n|$ are functions which can take only the values ± 1, and are present to ensure that the attraction takes place on both sides of ξ_n-lines and η_n-lines and in the entire neighbourhood of points (ξ_i, η_i). Taking the amplitudes to be negative turns the attractive effects into repulsive ones.

5.3 Univariate stretching functions

One straightforward way of controlling grid density in the physical domain is through the use of *stretching functions* (several examples of which were given in Section 4.4), combined with a differential model of grid generation. Suppose that the transformation $(x, y) \rightarrow (\chi, \sigma)$ takes the physical domain R in two dimensions onto the square $0 \leqslant \chi, \sigma \leqslant 1$ in $\chi\sigma$-space, such that the co-ordinates χ, σ are boundary-conforming. We regard the $\chi\sigma$-space as an intermediate *parameter space*. A further mapping

$$\chi = f_1(\xi), \quad \sigma = f_2(\eta), \tag{5.9}$$

with $f_1(0) = f_2(0) = 0$, $f_1(1) = f_2(1) = 1$, where the functions f_1 and f_2 are one–one and onto, will map a square in $\xi\eta$-computational space onto the square in parameter space. But a regularly-spaced rectangular grid in computational space will in general map onto an irregularly-spaced grid in parameter space, which will in turn map onto a *body-conforming* grid in physical space (Fig. 5.2), with a different distribution of grid lines from that which would be generated by a regular grid in $\chi\sigma$-space. Appropriate choice of the stretching functions f_1, f_2 may yield a grid in physical space with desirable features.

Suppose that the mapping from physical space to parameter space is achieved through eqns (5.1). Then we have

$$\frac{\partial^2\chi}{\partial x^2} + \frac{\partial^2\chi}{\partial y^2} = 0 \quad \text{and} \quad \frac{\partial^2\sigma}{\partial x^2} + \frac{\partial^2\sigma}{\partial y^2} = 0. \tag{5.10}$$

Now

$$\frac{\partial\chi}{\partial x} = \frac{\partial\xi}{\partial x}\frac{d\chi}{d\xi} = f_1'(\xi)\frac{\partial\xi}{\partial x}$$

and

$$\frac{\partial^2\chi}{\partial x^2} = f_1'(\xi)\frac{\partial^2\xi}{\partial x^2} + f_1''(\xi)\left(\frac{\partial\xi}{\partial x}\right)^2.$$

Similarly,

$$\frac{\partial^2\chi}{\partial y^2} = f_1'(\xi)\frac{\partial^2\xi}{\partial y^2} + f_1''(\xi)\left(\frac{\partial\xi}{\partial y}\right)^2.$$

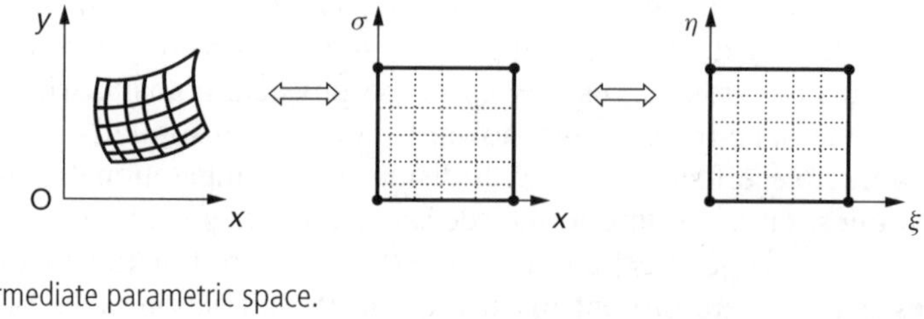

Fig. 5.2 Intermediate parametric space.

Adding and using (5.10), we obtain

$$\frac{\partial^2 \xi}{\partial x^2} + \frac{\partial^2 \xi}{\partial y^2} = -\frac{f_1''}{f_1'}\left[\left(\frac{\partial \xi}{\partial x}\right)^2 + \left(\frac{\partial \xi}{\partial y}\right)^2\right] = -\frac{f_1''}{f_1'} g^{11} = -\left(\frac{f_1''}{f_1'}\right)\left(\frac{g_{22}}{g}\right). \quad (5.11)$$

Similarly,

$$\frac{\partial^2 \eta}{\partial x^2} + \frac{\partial^2 \eta}{\partial y^2} = -\left(\frac{f_2''}{f_2'}\right)\left(\frac{g_{11}}{g}\right). \quad (5.12)$$

The effect of introducing the stretching functions is thus equivalent to using the Poisson system (5.6) with

$$P(\xi, \eta) = -\left(\frac{f_1''}{f_1'}\right)\left(\frac{g_{22}}{g}\right) \quad \text{and } Q(\xi, \eta) = -\left(\frac{f_2''}{f_2'}\right)\left(\frac{g_{11}}{g}\right), \quad (5.13)$$

although of course g_{11}, g_{22}, and g have to be determined as part of the solution.

Using eqn (5.7), the inverse problem is associated with the equations

$$\begin{aligned} g_{22}\frac{\partial^2 x}{\partial \xi^2} - 2g_{12}\frac{\partial^2 x}{\partial \xi \partial \eta} + g_{11}\frac{\partial^2 x}{\partial \eta^2} - \left(\frac{f_1''}{f_1'}\right) g_{22}\frac{\partial x}{\partial \xi} - \left(\frac{f_2''}{f_2'}\right) g_{11}\frac{\partial x}{\partial \eta} = 0, \\ g_{22}\frac{\partial^2 y}{\partial \xi^2} - 2g_{12}\frac{\partial^2 y}{\partial \xi \partial \eta} + g_{11}\frac{\partial^2 y}{\partial \eta^2} - \left(\frac{f_1''}{f_1'}\right) g_{22}\frac{\partial y}{\partial \xi} - \left(\frac{f_2''}{f_2'}\right) g_{11}\frac{\partial y}{\partial \eta} = 0, \end{aligned} \quad (5.14)$$

where g_{11}, g_{12}, g_{22} are given by eqn (1.158).

5.3.1 Orthogonality considerations

Orthogonality is a desirable feature of grids, since the nearer a grid approaches to orthogonality, the more accurate we generally expect numerical solutions to be. In three dimensions it is clear from the discussion of surfaces in Chapter 3 that a co-ordinate line can only be orthogonal at a point to two mutually orthogonal co-ordinate lines lying in a surface if those lines are in the directions of principal curvature of the surface at that point. This is a rather restrictive requirement, which makes orthogonality difficult or impossible to achieve in general. Here we shall confine our discussion to planar regions.

It is convenient to consider the parameter space above, in which the orthogonality condition for the χ, σ co-ordinate curves at any point would be given by

$$\frac{\partial \mathbf{r}}{\partial \chi} \cdot \frac{\partial \mathbf{r}}{\partial \sigma} = \frac{\partial x}{\partial \chi}\frac{\partial x}{\partial \sigma} + \frac{\partial y}{\partial \chi}\frac{\partial y}{\partial \sigma} = 0. \quad (5.15)$$

We can now put

$$\frac{\partial y}{\partial \sigma} = a_r \frac{\partial x}{\partial \chi}, \quad \text{and} \quad \frac{\partial x}{\partial \sigma} = -a_r \frac{\partial y}{\partial \chi}, \quad (5.16)$$

where a_r may be called the *aspect ratio* at a point, since we can write

$$a_r = \frac{\sqrt{\left(\frac{\partial x}{\partial \sigma}\right)^2 + \left(\frac{\partial y}{\partial \sigma}\right)^2}}{\sqrt{\left(\frac{\partial x}{\partial \chi}\right)^2 + \left(\frac{\partial y}{\partial \chi}\right)^2}},$$

which by eqns (1.42) and (1.158) implies that it is the ratio of the lengths of sides of an infinitesimal grid element in physical space corresponding to an infinitesimal element of a uniform grid in parameter space.

If we now apply stretching functions of the form (5.9), we easily obtain

$$\frac{\partial y}{\partial \eta} = a_r \frac{f_2'(\eta)}{f_1'(\xi)} \frac{\partial x}{\partial \xi}, \quad \text{and} \quad \frac{\partial x}{\partial \eta} = -a_r \frac{f_2'(\eta)}{f_1'(\xi)} \frac{\partial y}{\partial \xi}. \tag{5.17}$$

It follows that

$$\frac{\partial x}{\partial \xi}\frac{\partial x}{\partial \eta} + \frac{\partial y}{\partial \xi}\frac{\partial y}{\partial \eta} = 0, \tag{5.18}$$

and hence orthogonality still holds for the ξ, η co-ordinate curves, i.e. it has not been affected by the stretching transformation.

A differential model of grid generation which aims at orthogonality can be written down immediately by regarding the g_{12} terms in eqns (5.14) as zero everywhere. Thus we have

$$g_{22}\frac{\partial^2 x}{\partial \xi^2} + g_{11}\frac{\partial^2 x}{\partial \eta^2} = \left(\frac{f_1''}{f_1'}\right) g_{22}\frac{\partial x}{\partial \xi} + \left(\frac{f_2''}{f_2'}\right) g_{11}\frac{\partial x}{\partial \eta},$$

$$g_{22}\frac{\partial^2 y}{\partial \xi^2} + g_{11}\frac{\partial^2 y}{\partial \eta^2} = \left(\frac{f_1''}{f_1'}\right) g_{22}\frac{\partial y}{\partial \xi} + \left(\frac{f_2''}{f_2'}\right) g_{11}\frac{\partial y}{\partial \eta}. \tag{5.19}$$

These equations can be solved numerically in the computational domain subject to Dirichlet boundary conditions on x and y. However, we can also seek to satisfy the orthogonality condition (5.18) on the boundary. The numerical techniques are pursued in Section 5.6.2.

5.4 Conformal and quasi-conformal mapping

In considering mappings $\xi(x, y)$, $\eta(x, y)$ from plane physical domains to computational domains, it is sometimes mathematically advantageous to introduce complex variables $w = \xi + i\eta$ and $z = x + iy$, the domains then becoming part of complex w and z planes. Since

$$x = (z + \bar{z})/2, \quad y = (z - \bar{z})/2i, \tag{5.20}$$

where the conjugate of z is given by $\bar{z} = x - iy$, a mapping can be expressed as

$$w = \xi(x, y) + i\eta(x, y) = f(z, \bar{z}) \tag{5.21}$$

for some complex function f. In the case where f has no dependence on $\bar{z}$ and, moreover, a derivative $f'(z)$ can be defined (so that $f(z)$ is an *analytic* function), the mapping is *conformal*. In this case the functions $\xi(x, y)$, $\eta(x, y)$ satisfy the Cauchy-Riemann equations

$$\frac{\partial \xi}{\partial x} = \frac{\partial \eta}{\partial y} \quad \text{and} \quad \frac{\partial \xi}{\partial y} = -\frac{\partial \eta}{\partial x}. \tag{5.22}$$

It follows from eqns (1.158), (1.160), and (1.162) that for a grid generated by a conformal transformation we have $g_{11} = g_{22}$, and $g_{12} = 0$, which means that the grid is orthogonal. Furthermore, not only do the functions $\xi(x, y)$, $\eta(x, y)$ satisfy Laplace's eqns (5.1), but it follows from eqns (5.3) that the inverse functions $x(\xi, \eta)$, $y(\xi, \eta)$ also satisfy Laplace's equations in the form (5.5). However, although conformal mapping techniques are well-established for solving Laplace's equations in two dimensions in numerous areas of science and engineering (see for example Nehari (1975)), they have the disadvantages in the area of grid generation that, firstly, they are essentially applicable in two dimensions only, and, secondly, that grid density in the physical domain is not controlled.

The same disadvantages apply to *quasiconformal mapping*, which can be used to generate a wider range of grids. A positive feature of conformal and quasiconformal mapping, however, is that the Jacobian of the transformation is always positive, so that the grids generated are non-overlapping (not folded).

In quasiconformal mapping the complex function $f(z, \bar{z})$ in eqn (5.21) is taken to satisfy the *Beltrami Equation*

$$\frac{\partial f}{\partial \bar{z}} - H(z, \bar{z}) \frac{\partial f}{\partial z} = 0, \tag{5.23}$$

where the function $H(z, \bar{z})$ has real and imaginary parts $\mu(x, y)$, $\nu(x, y)$, so that

$$H(z, \bar{z}) = \mu(x, y) + i\nu(x, y). \tag{5.24}$$

The case of conformal mapping occurs when we take $\mu(x, y) = \nu(x, y) = 0$.

To substitute for f from eqn (5.21) into eqn (5.23), we need

$$\begin{aligned} \frac{\partial f}{\partial \bar{z}} &= \frac{\partial}{\partial \bar{z}}(\xi + i\eta) = \frac{\partial x}{\partial \bar{z}} \frac{\partial}{\partial x}(\xi + i\eta) + \frac{\partial y}{\partial \bar{z}} \frac{\partial}{\partial y}(\xi + i\eta) \\ &= \frac{1}{2}\left(\frac{\partial \xi}{\partial x} + i\frac{\partial \eta}{\partial x}\right) - \frac{1}{2i}\left(\frac{\partial \xi}{\partial y} + i\frac{\partial \eta}{\partial y}\right) \\ &= \frac{1}{2}\left(\frac{\partial \xi}{\partial x} - \frac{\partial \eta}{\partial y}\right) + \frac{1}{2}i\left(\frac{\partial \xi}{\partial y} + \frac{\partial \eta}{\partial x}\right). \end{aligned} \tag{5.25}$$

Similarly,

$$\frac{\partial f}{\partial z} = \frac{1}{2}\left(\frac{\partial \xi}{\partial x} + \frac{\partial \eta}{\partial y}\right) + \frac{1}{2}i\left(-\frac{\partial \xi}{\partial y} + \frac{\partial \eta}{\partial x}\right). \tag{5.26}$$

Exercise 2. Show that equating real and imaginary parts in eqn (5.23) gives the pair of equations

$$(1 - \mu)\frac{\partial \xi}{\partial x} + \nu\left(-\frac{\partial \xi}{\partial y} + \frac{\partial \eta}{\partial x}\right) - (1 + \mu)\frac{\partial \eta}{\partial y} = 0, \tag{5.27}$$

$$\nu\left(\frac{\partial\xi}{\partial x}+\frac{\partial\eta}{\partial y}\right)-(1+\mu)\frac{\partial\xi}{\partial y}-(1-\mu)\frac{\partial\eta}{\partial x}=0, \tag{5.28}$$

which are equivalent to

$$\beta\frac{\partial\xi}{\partial x}+\gamma\frac{\partial\xi}{\partial y}=-\frac{\partial\eta}{\partial x}, \tag{5.29}$$

$$\alpha\frac{\partial\xi}{\partial x}+\beta\frac{\partial\xi}{\partial y}=\frac{\partial\eta}{\partial y}, \tag{5.30}$$

where $\alpha=\frac{(1-\mu)^2+\nu^2}{1-\mu^2-\nu^2}$, $\beta=\frac{-2\nu}{1-\mu^2-\nu^2}$, $\gamma=\frac{(1+\mu)^2+\nu^2}{1-\mu^2-\nu^2}$. Show that

$$\alpha\gamma-\beta^2=1. \tag{5.31}$$

We also have, since $H(z,\bar{z})=\frac{\partial f}{\partial\bar{z}}\Big/\frac{\partial f}{\partial z}$,

$$|H(z,\bar{z})|^2=\mu^2+\nu^2=\frac{\left|\dfrac{\partial f}{\partial\bar{z}}\right|^2}{\left|\dfrac{\partial f}{\partial z}\right|^2}$$

$$=\frac{\left(\dfrac{\partial\xi}{\partial x}-\dfrac{\partial\eta}{\partial y}\right)^2+\left(\dfrac{\partial\xi}{\partial y}+\dfrac{\partial\eta}{\partial x}\right)^2}{\left(\dfrac{\partial\xi}{\partial x}+\dfrac{\partial\eta}{\partial y}\right)^2+\left(-\dfrac{\partial\xi}{\partial y}+\dfrac{\partial\eta}{\partial x}\right)^2}=\frac{g_{11}+g_{22}-2\sqrt{g}}{g_{11}+g_{22}+2\sqrt{g}},$$

using eqns (5.25), (5.26), (1.158), (1.160), and (1.162). The identity

$$\alpha+\gamma=\frac{2(1+\mu^2+\nu^2)}{1-\mu^2-\nu^2}=\frac{g_{11}+g_{22}}{\sqrt{g}} \tag{5.32}$$

then follows.

In the case of conformal mapping, we have $\alpha=\gamma=1$, and $\beta=0$, and eqns (5.29), (5.30) reduce to the Cauchy-Riemann equations. More generally, these equations form a pair of first-order partial differential equations which could be used to generate a grid for some choice of μ, ν. From eqns (1.162) the inverse equations can immediately be written as

$$\frac{\partial x}{\partial\xi}=\alpha\frac{\partial y}{\partial\eta}-\beta\frac{\partial x}{\partial\eta}, \tag{5.33}$$

$$\frac{\partial y}{\partial\xi}=\beta\frac{\partial y}{\partial\eta}-\gamma\frac{\partial x}{\partial\eta}. \tag{5.34}$$

Exercise 3. Show that the Jacobian of the transformation from (ξ,η) to (x,y) can be expressed as

$$J=\sqrt{g}=\frac{\partial x}{\partial\xi}\frac{\partial y}{\partial\eta}-\frac{\partial x}{\partial\eta}\frac{\partial y}{\partial\xi}=\frac{1}{\alpha}\left\{\left(\frac{\partial x}{\partial\xi}\right)^2+\left(\frac{\partial x}{\partial\eta}\right)^2\right\}=\frac{1}{\gamma}\left\{\left(\frac{\partial y}{\partial\xi}\right)^2+\left(\frac{\partial y}{\partial\eta}\right)^2\right\}. \tag{5.35}$$

This result shows that the Jacobian is generally positive as long as α is, and so grids generated will be non-overlapping. In general the grids will not be orthogonal.

Eliminating η from eqns (5.29), (5.30) leads to the following second-order partial differential equation for ξ, which is elliptic because of eqn (5.31):

$$\alpha\frac{\partial^2\xi}{\partial x^2}+2\beta\frac{\partial^2\xi}{\partial x\partial y}+\gamma\frac{\partial^2\xi}{\partial y^2}+\left(\frac{\partial\alpha}{\partial x}+\frac{\partial\beta}{\partial y}\right)\frac{\partial\xi}{\partial x}+\left(\frac{\partial\beta}{\partial x}+\frac{\partial\gamma}{\partial y}\right)\frac{\partial\xi}{\partial y}=0. \tag{5.36}$$

It is straightforward to show that the same equation must be satisfied by η.

By eliminating $\partial y/\partial\eta$ from eqns (5.33) and (5.34) and using eqn (5.31), we obtain the relation

$$\frac{\partial y}{\partial\xi}=\frac{\beta}{\alpha}\frac{\partial x}{\partial\xi}-\frac{1}{\alpha}\frac{\partial x}{\partial\eta}. \tag{5.37}$$

Equating the mixed derivatives $\partial^2 y/\partial\xi\partial\eta$ and $\partial^2 y/\partial\eta\partial\xi$ obtained from eqns (5.33) and (5.37) now yields the second order p.d.e. for $x(\xi,\eta)$:

$$\frac{\partial}{\partial\eta}\left(\frac{\beta}{\alpha}\frac{\partial x}{\partial\xi}-\frac{1}{\alpha}\frac{\partial x}{\partial\eta}\right)=\frac{\partial}{\partial\xi}\left(\frac{1}{\alpha}\frac{\partial x}{\partial\xi}+\frac{\beta}{\alpha}\frac{\partial x}{\partial\eta}\right).$$

This becomes, on performing the differentiations, writing $\frac{\partial\alpha}{\partial\xi}=\frac{\partial\alpha}{\partial x}\frac{\partial x}{\partial\xi}+\frac{\partial\alpha}{\partial y}\frac{\partial y}{\partial\xi}$, etc., and simplifying,

$$\frac{\partial^2 x}{\partial\xi^2}+\frac{\partial^2 x}{\partial\eta^2}-\left(\frac{\partial\beta}{\partial y}+\frac{\partial\alpha}{\partial x}\right)\sqrt{g}=0, \tag{5.38}$$

where $\sqrt{g}$ is given by eqn (5.35). A similar calculation gives

$$\frac{\partial^2 y}{\partial\xi^2}+\frac{\partial^2 y}{\partial\eta^2}-\left(\frac{\partial\beta}{\partial x}+\frac{\partial\gamma}{\partial y}\right)\sqrt{g}=0. \tag{5.39}$$

If we take $\beta=0$ and hence $\gamma=\alpha^{-1}$, eqns (5.33) and (5.34) reduce to the *generalized Cauchy-Riemann equations*

$$\begin{cases}\dfrac{\partial x}{\partial\xi}=\alpha\dfrac{\partial y}{\partial\eta}\\[2ex]\dfrac{\partial y}{\partial\xi}=-\dfrac{1}{\alpha}\dfrac{\partial x}{\partial\eta}.\end{cases} \tag{5.40}$$

As in conformal mapping, grid density in the interior of the physical domain is not controlled by the quasiconformal mapping method. However, stretching functions could be used to overcome this deficiency. More information can be found in Mastin and Thompson (1984) and in Mastin (1991). Further details on quasiconformal mapping are contained in Ahlfors (1996).

5.5 Numerical techniques

5.5.1 The Thomas Algorithm

In the numerical solution of the partial differential equations serving as differential models of grid generation, finite-differencing frequently leads to a set of linear equations involving a tridiagonal matrix (one in which the only non-zero entries appear along the main diagonal and the two adjacent parallel diagonals on either side). These can usually be solved efficiently using the Thomas Algorithm (otherwise known as the Tridiagonal Matrix Algorithm), which is based on Gaussian Elimination. A typical set of equations for the $(N-1)$ unknowns $u_1, u_2, \ldots, u_{N-1}$, is

$$-a_i u_{i-1} + b_i u_i - c_i u_{i+1} = d_i, \quad i = 1, 2, \ldots, (N-1), \tag{5.41}$$

where the values of u_0 and u_N are supposed known due to the given boundary conditions.

The equivalent matrix equation is

$$\mathbf{A}\mathbf{u} = \mathbf{d} \tag{5.42}$$

where

$$A = \begin{pmatrix} b_1 & -c_1 & 0 & 0 & - & - & 0 \\ -a_2 & b_2 & -c_2 & 0 & - & - & - \\ 0 & -a_3 & b_3 & -c_3 & - & - & - \\ 0 & 0 & - & - & - & - & - \\ - & - & - & - & - & - & 0 \\ - & -- & - & 0 & -a_{N-2} & b_{N-2} & -c_{N-2} \\ 0 & - & - & 0 & 0 & -a_{N-1} & b_{N-1} \end{pmatrix},$$

$$\mathbf{u} = \begin{pmatrix} u_1 \\ u_2 \\ - \\ - \\ - \\ - \\ u_{N-1} \end{pmatrix}, \quad \mathbf{d} = \begin{pmatrix} d_1 \\ d_2 \\ d_3 \\ - \\ - \\ d_{N-2} \\ d_{N-1} \end{pmatrix}.$$

Here we have transferred the terms $-a_1 u_0$ and $-c_{N-1} u_N$ to the right-hand side, where they are assumed to have been incorporated into the d_1 and d_{N-1} terms, respectively. So now we can assume $a_1 = c_{N-1} = 0$ in eqn (5.41).

A is *diagonally dominant* if

$$b_i > 0 \quad \text{and } b_i > |a_i| + |c_i|, \quad i = 1, 2, \ldots, (N-1) \tag{5.43}$$

The Thomas Algorithm replaces eqn (5.41) with a simpler recurrence relation from which u_{i-1} has been eliminated, thereby reducing A to an upper triangular matrix. The equations can then be solved by simple back substitution.

Suppose that we seek a recurrence relation of the form

$$u_i = P_i u_{i+1} + Q_i, \quad i = 1, 2, \ldots, (N-1) \tag{5.44}$$

This is consistent with eqn (5.41) if

$$-a_i(P_{i-1}u_i + Q_{i-1}) + b_i u_i - c_i u_{i+1} = d_i, \quad i = 2, 3, \ldots, (N-1)$$

or

$$u_i(b_i - a_i P_{i-1}) = c_i u_{i+1} + d_i + a_i Q_{i-1},$$

which is itself consistent with eqn (5.44) if

$$P_i = \frac{c_i}{b_i - a_i P_{i-1}} \quad \text{and } Q_i = \frac{d_i + a_i Q_{i-1}}{b_i - a_i P_{i-1}}, \quad i = 2, 3, \ldots, (N-1). \tag{5.45}$$

Since we now have $a_1 = 0$,

$$P_1 = \frac{c_1}{b_1} \quad \text{and } Q_1 = \frac{d_1}{b_1}. \tag{5.46}$$

We are now able to generate recursively the values of $P_2, P_3, \ldots, P_{N-2}$, and $Q_2, Q_3, \ldots, Q_{N-1}$. Since we have taken $c_{N-1} = 0$, we also have $P_{N-1} = 0$.

Equation (5.44) now gives the value of u_{N-1} as

$$u_{N-1} = Q_{N-1}. \tag{5.47}$$

We can now use eqn (5.44) to solve successively for $u_{N-2}, u_{N-3}, \ldots, u_1$.

For the Thomas Algorithm to be well-conditioned, it is sufficient that

$$|P_i| \leqslant 1, \quad i = 1, 2, \ldots, N-1, \tag{5.48}$$

since by eqn (5.44) the error in the computed value of u_{i+1} is multiplied by P_i in calculating u_i. It may be shown that the conditions for diagonal dominance (5.43) guarantee that (5.48) is satisfied.

The Thomas Algorithm is a powerful and convenient solution method for a set of linear equations of the form (5.41), requiring computer storage and computer time of the order of N rather than N^2 or N^3. It can be extended to 'block-tridiagonal' systems in which a_i, b_i, c_i are square matrices and the unknowns u_i are vectors.

5.5.2 Jacobi, Gauss-Seidel, SOR methods

In solving matrix equations $A\mathbf{u} = \mathbf{d}$ for the $n \times 1$ column vector $\mathbf{u}$ when the $n \times n$ matrix A is large but very sparse (i.e. the entries are mostly zeros), iterative methods are usually employed. Suppose we write the equations as

$$u_1 = \frac{1}{a_{11}}[d_1 - (a_{12}u_2 + a_{13}u_3 + \cdots + a_{1n}u_n)]$$

$$u_2 = \frac{1}{a_{22}}[d_2 - (a_{21}u_1 + a_{23}u_3 + \cdots + a_{2n}u_n)]$$

$$\ldots$$

$$u_n = \frac{1}{a_{nn}}[d_n - (a_{n1}u_1 + a_{n2}u_2 + \cdots + a_{n(n-1)}u_{n-1})],$$

assuming that none of the diagonal entries of A are zero. Then possible choices for iterative schemes, given an initial guess $u_i^{(1)}$, $i = 1, 2, \dots, n$, are

$$
\begin{aligned}
u_1^{(k+1)} &= \frac{1}{a_{11}}[d_1 - (a_{12}u_2^{(k)} + a_{13}u_3^{(k)} + \cdots + a_{1n}u_n^{(k)})] \\
u_2^{(k+1)} &= \frac{1}{a_{22}}[d_2 - (a_{21}u_1^{(k)} + a_{23}u_3^{(k)} + \cdots + a_{2n}u_n^{(k)})] \\
&\cdots \\
u_n^{(k+1)} &= \frac{1}{a_{nn}}[d_n - (a_{n1}u_1^{(k)} + a_{n2}u_2^{(k)} + \cdots + a_{n(n-1)}u_{n-1}^{(k)})],
\end{aligned} \tag{5.49}
$$

which is the *Jacobi* method, and

$$
\begin{aligned}
u_1^{(k+1)} &= \frac{1}{a_{11}}[d_1 - (a_{12}u_2^{(k)} + a_{13}u_3^{(k)} + \cdots + a_{1n}u_n^{(k)})] \\
u_2^{(k+1)} &= \frac{1}{a_{22}}[d_2 - (a_{21}u_1^{(k+1)} + a_{23}u_3^{(k)} + \cdots + a_{2n}u_n^{(k)})] \\
&\cdots \\
u_n^{(k+1)} &= \frac{1}{a_{nn}}[d_n - (a_{n1}u_1^{(k+1)} + a_{n2}u_2^{(k+1)} + \cdots + a_{n(n-1)}u_{n-1}^{(k+1)})],
\end{aligned} \tag{5.50}
$$

called the *Gauss-Seidel* method, in which the new value $u_1^{(k+1)}$ is used in the second equation to calculate $u_2^{(k+1)}$, both of these values are used in the third equation to calculate $u_3^{(k+1)}$, and the continual updating proceeds until the last equation. Thus Gauss-Seidel involves immediate replacement of old u_i values in the appropriate location and therefore requires less storage space than the Jacobi method. It is also generally faster than the Jacobi method.

Equations (5.50) can be written as

$$
u_i^{(k+1)} = \frac{1}{a_{ii}}\left[d_i - \sum_{j=1}^{i-1} a_{ij}u_j^{(k+1)} - \sum_{j=i+1}^{n} a_{ij}u_j^{(k)}\right],
$$

$$
i = 1, 2, \dots, n \quad (i \text{ not summed}). \tag{5.51}
$$

The SOR (Successive Over-Relaxation) method introduces an extra parameter ω, called the *acceleration parameter*, which can speed up convergence of the iteration. The scheme is given by

$$
u_i^{(k+1)} = u_i^{(k)} + \frac{\omega}{a_{ii}}\left[d_i - \sum_{j=1}^{i-1} a_{ij}u_j^{(k+1)} - a_{ii}u_i^{(k)} - \sum_{j=i+1}^{n} a_{ij}u_j^{(k)}\right] \tag{5.52}
$$

$$
= \frac{\omega}{a_{ii}}\left[d_i - \sum_{j=1}^{i-1} a_{ij}u_j^{(k+1)} - \sum_{j=i+1}^{n} a_{ij}u_j^{(k)}\right] + (1-\omega)u_i^{(k)}, \quad i = 1, 2, \dots, n, \tag{5.53}
$$

which gives a weighted average of the old value of u_i and the value given by eqn (5.51). When $\omega = 1$ the SOR method is the same as Gauss-Seidel. 'Over-relaxation' refers to choosing a value of ω between 1 and 2, while 'under-relaxation' would involve taking $0 < \omega < 1$.

5.5.3 The conjugate gradient method

The conjugate gradient method involves an iterative scheme for solving the matrix equation $A\mathbf{u} = \mathbf{d}$ for the $n \times 1$ column vector $\mathbf{u}$, given the $n \times n$ matrix A and column vector $\mathbf{d}$, which is of general use, but it is particularly appropriate for matrices A which are symmetric and *positive definite*, satisfying $\mathbf{u}^T A\mathbf{u} > 0$ for any $n \times 1$ column vector $\mathbf{u}$ not identically zero. Under these circumstances this method solves an equivalent problem, that of minimizing the function

$$E(u_1, u_2, \ldots, u_n) = \frac{1}{2}\sum_{i=1}^{n}\sum_{j=1}^{n} a_{ij}u_i u_j - \sum_{i=1}^{n} d_i u_i, \tag{5.54}$$

where a_{ij} and d_i are the elements of A and $\mathbf{d}$ respectively.

Exercise 4. Establish this equivalence, by demonstrating that E has one critical point $\bar{\mathbf{u}}$ (where all partial derivatives vanish) satisfying $A\bar{\mathbf{u}} = \mathbf{d}$, and then showing that it is a minimum (by representing E in the neighbourhood of the critical point in terms of $\mathbf{u} = \bar{\mathbf{u}} + \mathbf{h}$, or directly evaluating the second-order terms in the Taylor Series expansion of E about $\bar{\mathbf{u}}$.

The matrix formulation of E is

$$E = \tfrac{1}{2}\mathbf{u}^T A\mathbf{u} - \mathbf{d}^T\mathbf{u}. \tag{5.55}$$

A simple numerical approach to the problem of minimization is provided by the *method of steepest descents*, in which, starting from some given point $\mathbf{u}_i$ (regarded as a point in an n-dimensional space), an iterative step involves a search for a minimum in the value of E along the direction of steepest decrease of E at $\mathbf{u}_0$. This direction $\mathbf{r}$ is that opposite to the direction of the gradient vector (the vector of partial derivatives) of E, which gives the direction of steepest increase. At a point $\mathbf{u}$ the gradient vector is given by $A\mathbf{u} - \mathbf{d}$ (since A is symmetric), so that we take

$$\mathbf{r} = -(A\mathbf{u} - \mathbf{d}) = \mathbf{d} - A\mathbf{u}. \tag{5.56}$$

Thus when $\mathbf{u} = \mathbf{u}_i$ we have $\mathbf{r} = \mathbf{r}_i = \mathbf{d} - A\mathbf{u}_i$, which we may also regard as the *residual* at that point, bearing in mind the equivalent problem $A\mathbf{u} = \mathbf{d}$.

Putting $\mathbf{u} = \mathbf{u}_i + \lambda_i\mathbf{r}_i = \mathbf{u}_i + \lambda_i(\mathbf{d} - A\mathbf{u}_i)$ into eqn (5.55) (with no summation over repeated indices) gives

$$\begin{aligned}
E &= \tfrac{1}{2}\mathbf{u}_i^T A\mathbf{u}_i - \mathbf{d}^T\mathbf{u}_i + \lambda_i\left(\tfrac{1}{2}\mathbf{u}_i^T A\mathbf{r}_i + \tfrac{1}{2}\mathbf{r}_i^T A\mathbf{u}_i - \mathbf{d}^T\mathbf{r}_i\right) + \tfrac{1}{2}(\lambda_i)^2\mathbf{r}_i^T A\mathbf{r}_i \\
&= \tfrac{1}{2}\mathbf{u}_i^T A\mathbf{u}_i - \mathbf{d}^T\mathbf{u}_i + \lambda_i(\mathbf{u}_i^T A\mathbf{r}_i - \mathbf{d}^T\mathbf{r}_i) + \tfrac{1}{2}(\lambda_i)^2\mathbf{r}_i^T A\mathbf{r}_i \\
&= \tfrac{1}{2}\mathbf{u}_i^T A\mathbf{u}_i - \mathbf{d}^T\mathbf{u}_i + \lambda_i(\mathbf{u}_i^T A - \mathbf{d}^T)\mathbf{r}_i + \tfrac{1}{2}(\lambda_i)^2\mathbf{r}_i^T A\mathbf{r}_i \\
&= \tfrac{1}{2}\mathbf{u}_i^T A\mathbf{u}_i - \mathbf{d}^T\mathbf{u}_i - \lambda_i\mathbf{r}_i^T\mathbf{r}_i + \tfrac{1}{2}(\lambda_i)^2\mathbf{r}_i^T A\mathbf{r}_i,
\end{aligned} \tag{5.57}$$

where, for example, the symmetry of A has been used in the identity

$$\tfrac{1}{2}\mathbf{u}_i^T A\mathbf{r}_i = \left(\tfrac{1}{2}\mathbf{u}_i^T A\mathbf{r}_i\right)^T = \tfrac{1}{2}\mathbf{r}_i^T A^T\mathbf{u}_i = \tfrac{1}{2}\mathbf{r}_i^T A\mathbf{u}_i.$$

It is easy to see that the expression (5.57) is minimized when λ_i takes the (positive) value

$$\lambda_i = \frac{\mathbf{r}_i^T\mathbf{r}_i}{\mathbf{r}_i^T A\mathbf{r}_i},$$

giving the iterative scheme

$$\mathbf{u}_{i+1} = \mathbf{u}_i + \left(\frac{\mathbf{r}_i^T\mathbf{r}_i}{\mathbf{r}_i^T A\mathbf{r}_i}\right)\mathbf{r}_i. \tag{5.58}$$

This scheme has the feature that the direction of search $\mathbf{r}_{i+1}$ is always orthogonal to the direction $\mathbf{r}_i$ at the previous iterative step, since the scalar product

$$\begin{aligned}\mathbf{r}_{i+1}^T\mathbf{r}_i &= (\mathbf{d}^T - \mathbf{u}_{i+1}^T A)\mathbf{r}_i = \mathbf{d}^T\mathbf{r}_i - \left[\mathbf{u}_i^T + \left(\frac{\mathbf{r}_i^T\mathbf{r}_i}{\mathbf{r}_i^T A\mathbf{r}_i}\right)\mathbf{r}_i^T\right]A\mathbf{r}_i \\ &= (\mathbf{d}^T - \mathbf{u}_i^T A)\mathbf{r}_i - \mathbf{r}_i^T\mathbf{r}_i = \mathbf{r}_i^T\mathbf{r}_i - \mathbf{r}_i^T\mathbf{r}_i = 0.\end{aligned}$$

However, this can often lead to an inefficient solution procedure, and the method of steepest descents can be quite slow. Various methods have been proposed to accelerate the procedure, and the conjugate gradient method may be regarded as one of these. Here the direction of search becomes $\mathbf{p}_i$, so that at each iterative step (when A is symmetric and positive definite) we search for the minimum value of E with $\mathbf{u} = \mathbf{u}_i + \lambda_i\mathbf{p}_i$ and varying λ_i. Thus

$$\begin{aligned}E &= \tfrac{1}{2}(\mathbf{u}_i^T + \lambda_i\mathbf{p}_i^T)A(\mathbf{u}_i + \lambda_i\mathbf{p}_i) - \mathbf{d}^T(\mathbf{u}_i + \lambda_i\mathbf{p}_i) \\ &= \tfrac{1}{2}\mathbf{u}_i^T A\mathbf{u}_i - \mathbf{d}^T\mathbf{u}_i + \lambda_i(\mathbf{u}_i^T A - \mathbf{d}^T)\mathbf{p}_i + \tfrac{1}{2}(\lambda_i)^2\mathbf{p}_i^T A\mathbf{p}_i \\ &= \tfrac{1}{2}\mathbf{u}_i^T A\mathbf{u}_i - \mathbf{d}^T\mathbf{u}_i - \lambda_i\mathbf{r}_i^T\mathbf{p}_i + \tfrac{1}{2}(\lambda_i)^2\mathbf{p}_i^T A\mathbf{p}_i,\end{aligned}$$

where the residual $\mathbf{r}_i$ is given by

$$\mathbf{r}_i = \mathbf{d} - A\mathbf{u}_i. \tag{5.59}$$

We minimize E as a function of λ_i by taking

$$\lambda_i = \frac{\mathbf{r}_i^T\mathbf{p}_i}{\mathbf{p}_i^T A\mathbf{p}_i}. \tag{5.60}$$

It remains to set up an iterative scheme for the selection of $\mathbf{p}_i$, and we choose

$$\mathbf{p}_{i+1} = \mathbf{r}_{i+1} + \alpha_i\mathbf{p}_i \tag{5.61}$$

for some scalar α_i. The iteration will begin with $\mathbf{u} = \mathbf{u}_0$ and $\mathbf{p}_0 = \mathbf{r}_0$, as in the method of steepest descents.

Note that from eqn (5.59) we have

$$\mathbf{r}_{i+1} = \mathbf{r}_i - A(\mathbf{u}_{i+1} - \mathbf{u}_i) = \mathbf{r}_i - \lambda_i A\mathbf{p}_i. \tag{5.62}$$

It follows that

$$\mathbf{p}_i^T \mathbf{r}_{i+1} = \mathbf{p}_i^T \mathbf{r}_i - \lambda_i \mathbf{p}_i^T A \mathbf{p}_i = 0 \tag{5.63}$$

when λ_i satisfies eqn (5.60).

Suppose that, following a search for a minimum along the direction $\mathbf{p}_i$ from the 'point' $\mathbf{u}_i$, we have reached the point $\mathbf{u}_{i+1}$. We can calculate $\mathbf{r}_{i+1}$ from eqn (5.62), and now we seek the value of α_i in eqn (5.61) which will yield the new direction of search $\mathbf{p}_{i+1}$.

Substituting $\mathbf{u} = \mathbf{u}_{i+1} + \lambda_{i+1}\mathbf{p}_{i+1} = \mathbf{u}_{i+1} + \lambda_{i+1}(\mathbf{r}_{i+1} + \alpha_i \mathbf{p}_i)$ into eqn (5.55) gives, in a similar manner to the derivation of eqn (5.57),

$$\begin{aligned} E &= \tfrac{1}{2}\mathbf{u}_{i+1}^T A \mathbf{u}_{i+1} - \mathbf{d}^T \mathbf{u}_{i+1} - \lambda_{i+1}\mathbf{r}_{i+1}^T \mathbf{r}_{i+1} - \lambda_{i+1}\alpha_i \mathbf{r}_{i+1}^T \mathbf{p}_i \\ &\quad + \tfrac{1}{2}(\lambda_{i+1})^2 \{\mathbf{r}_{i+1}^T A \mathbf{r}_{i+1} + 2\alpha_i(\mathbf{p}_i^T A \mathbf{r}_{i+1}) + (\alpha_i)^2 \mathbf{p}_i^T A \mathbf{p}_i\} \\ &= E_{i+1} - \lambda_{i+1}\mathbf{r}_{i+1}^T \mathbf{r}_{i+1} + \tfrac{1}{2}(\lambda_{i+1})^2 \{\mathbf{r}_{i+1}^T A \mathbf{r}_{i+1} + 2\alpha_i(\mathbf{p}_i^T A \mathbf{r}_{i+1}) + (\alpha_i)^2 \mathbf{p}_i^T A \mathbf{p}_i\}, \end{aligned}$$

using eqn (5.63), where E_{i+1} is the value of E at $\mathbf{u}_{i+1}$. To minimize E, we clearly must choose α_i so that the final term in brackets is minimized, and thus we obtain

$$\alpha_i = -\frac{\mathbf{p}_i^T A \mathbf{r}_{i+1}}{\mathbf{p}_i^T A \mathbf{p}_i}. \tag{5.64}$$

The iterative scheme is now complete, but we make the observation that

$$\mathbf{p}_{i+1}^T A \mathbf{p}_i = \mathbf{r}_{i+1}^T A \mathbf{p}_i + \alpha_i \mathbf{p}_i^T A \mathbf{p}_i = 0 \tag{5.65}$$

when α_i satisfies eqn (5.64). Thus the direction $\mathbf{p}_{i+1}$ is orthogonal to the direction of the vector $A\mathbf{p}_i$. Since A is symmetric, it is also the case that $\mathbf{p}_i$ is orthogonal to the vector $A\mathbf{p}_{i+1}$. The two directions $\mathbf{p}_i$, $\mathbf{p}_{i+1}$ are said to be *conjugate* with respect to the symmetric matrix A.

To summarize, given a starting guessed solution $\mathbf{u}_0$, an initial residual $\mathbf{r}_0 = \mathbf{d} - A\mathbf{u}_0$, and an initial direction $\mathbf{p}_0 = \mathbf{r}_0$, the conjugate gradient method is defined by the iterated cycle of steps given by

- $\lambda_i = (\mathbf{r}_i^T \mathbf{p}_i)/(\mathbf{p}_i^T A \mathbf{p}_i)$;
- $\mathbf{u}_{i+1} = \mathbf{u}_i + \lambda_i \mathbf{p}_i$;
- $\mathbf{r}_{i+1} = \mathbf{r}_i - \lambda_i A \mathbf{p}_i$;
- $\alpha_i = -(\mathbf{p}_i^T A \mathbf{r}_{i+1})/(\mathbf{p}_i^T A \mathbf{p}_i)$;
- $\mathbf{p}_{i+1} = \mathbf{r}_{i+1} + \alpha_i \mathbf{p}_i$.

5.6 Numerical solutions of Winslow equations

Obtaining a grid in two dimensions typically involves the following steps:

1. Set up a rectangular grid in the computational domain and obtain an initial guess for the values of x and y at the grid points. This may be achieved by Transfinite Interpolation, or Hermite interpolation, or simply by a univariate interpolation between two opposite boundaries.

2. Obtain the generating system of equations for the Inverse Problem, with x, y as the dependent variables and ξ, η as independent variables.
3. Discretize the generating equations using second-order accurate finite-difference approximations.
4. Solve the resulting system of algebraic equations iteratively in the computational plane subject to the given boundary conditions.

5.6.1 Thomas Algorithm

Here we show how to solve eqns (5.3) using the Thomas Algorithm. Because we have a two-dimensional problem, the method has to be used in a 'line-by-line' fashion. Suppose that we have a grid in the square (or rectangular) computational $\xi\eta$ domain with equal increments $\Delta\xi, \Delta\eta$ in ξ and η and with grid points labelled by integer values of i and j, for example as shown in Fig. 5.3. Finite differences applied to the terms in eqn (5.3) give

$$(g_{11})_{i,j} = \left[\left(\frac{\partial x}{\partial \xi}\right)^2 + \left(\frac{\partial y}{\partial \xi}\right)^2\right]_{i,j} = \left(\frac{x_{i+1,j} - x_{i-1,j}}{2\Delta\xi}\right)^2 + \left(\frac{y_{i+1,j} - y_{i-1,j}}{2\Delta\xi}\right)^2 \tag{5.66}$$

$$(g_{22})_{i,j} = \left[\left(\frac{\partial x}{\partial \eta}\right)^2 + \left(\frac{\partial y}{\partial \eta}\right)^2\right]_{i,j} = \left(\frac{x_{i,j+1} - x_{i,j-1}}{2\Delta\eta}\right)^2 + \left(\frac{y_{i,j+1} - y_{i,j-1}}{2\Delta\eta}\right)^2 \tag{5.67}$$

$$\begin{aligned}(g_{12})_{i,j} &= \left[\left(\frac{\partial x}{\partial \xi}\right)\left(\frac{\partial x}{\partial \eta}\right) + \left(\frac{\partial y}{\partial \xi}\right)\left(\frac{\partial y}{\partial \eta}\right)\right]_{i,j} \\ &= \left(\frac{x_{i+1,j} - x_{i-1,j}}{2\Delta\xi}\right)\left(\frac{x_{i,j+1} - x_{i,j-1}}{2\Delta\eta}\right) \\ &\quad + \left(\frac{y_{i+1,j} - y_{i-1,j}}{2\Delta\xi}\right)\left(\frac{y_{i,j+1} - y_{i,j-1}}{2\Delta\eta}\right)\end{aligned} \tag{5.68}$$

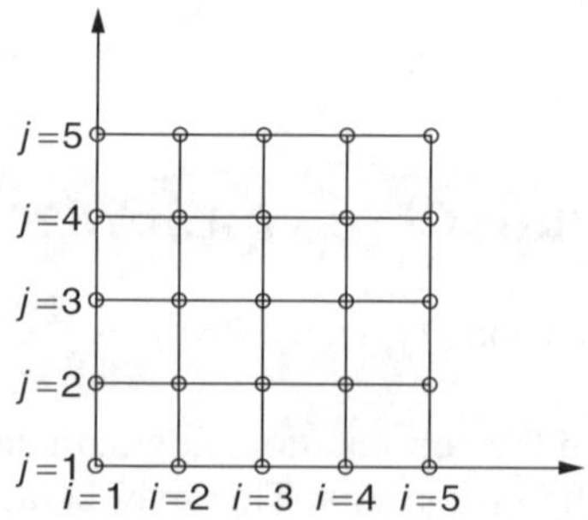

Fig. 5.3 Rectangular array of points.

$$\left(\frac{\partial^2 x}{\partial \xi^2}\right)_{i,j} = \frac{x_{i+1,j} - 2x_{i,j} + x_{i-1,j}}{(\Delta\xi)^2} \tag{5.69}$$

$$\left(\frac{\partial^2 x}{\partial \eta^2}\right)_{i,j} = \frac{x_{i,j+1} - 2x_{i,j} + x_{i,j-1}}{(\Delta\eta)^2} \tag{5.70}$$

$$\left(\frac{\partial^2 x}{\partial \xi \partial \eta}\right)_{i,j} = \frac{x_{i+1,j+1} + x_{i-1,j-1} - x_{i-1,j+1} - x_{i+1,j-1}}{4(\Delta\xi)(\Delta\eta)}, \tag{5.71}$$

with similar expressions for the second derivatives $\partial^2 y/\partial\xi^2, \partial^2 y/\partial\eta^2, \partial^2 y/\partial\xi\partial\eta$.

Equations (5.3) can now be written in the approximate form

$$\begin{aligned}(g_{22})_{i,j}\frac{(x_{i+1,j} - 2x_{i,j} + x_{i-1,j})}{(\Delta\xi)^2} &+ (g_{11})_{i,j}\frac{(x_{i,j+1} - 2x_{i,j} + x_{i,j-1})}{(\Delta\eta)^2} \\ -2(g_{12})_{i,j}&\frac{(x_{i+1,j+1} + x_{i-1,j-1} - x_{i-1,j+1} - x_{i+1,j-1})}{4(\Delta\xi)(\Delta\eta)} = 0,\end{aligned} \tag{5.72}$$

with a similar expression for y instead of x. Re-arranging, we get

$$\begin{aligned}\left[\frac{2(g_{22})_{i,j}}{(\Delta\xi)^2} + \frac{2(g_{11})_{i,j}}{(\Delta\eta)^2}\right]x_{i,j} = &\frac{(g_{22})_{i,j}}{(\Delta\xi)^2}(x_{i+1,j} + x_{i-1,j}) + \frac{(g_{11})_{i,j}}{(\Delta\eta)^2}(x_{i,j+1} + x_{i,j-1}) \\ &-2(g_{12})_{i,j}\frac{(x_{i+1,j+1} + x_{i-1,j-1} - x_{i-1,j+1} - x_{i+1,j-1})}{4(\Delta\xi)(\Delta\eta)}\end{aligned} \tag{5.73}$$

plus a similar equation for y.

If we now put

$$b_{i,j} = \left[\frac{2(g_{22})_{i,j}}{(\Delta\xi)^2} + \frac{2(g_{11})_{i,j}}{(\Delta\eta)^2}\right], \quad a_{i,j} = c_{i,j} = \frac{(g_{22})_{i,j}}{(\Delta\xi)^2},$$

$$\begin{aligned}d_{i,j} &= \frac{(g_{11})_{i,j}}{(\Delta\eta)^2}(x_{i,j+1} + x_{i,j-1}) \\ &\quad -2(g_{12})_{i,j} \times \frac{(x_{i+1,j+1} + x_{i-1,j-1} - x_{i-1,j+1} - x_{i+1,j-1})}{4(\Delta\xi)(\Delta\eta)} \\ e_{i,j} &= \frac{(g_{11})_{i,j}}{(\Delta\eta)^2}(y_{i,j+1} + y_{i,j-1}) \\ &\quad -2(g_{12})_{i,j} \times \frac{(y_{i+1,j+1} + y_{i-1,j-1} - y_{i-1,j+1} - y_{i+1,j-1})}{4(\Delta\xi)(\Delta\eta)},\end{aligned}$$

we get an equation of the form (5.41), and a similar one for y:

$$\begin{aligned}-a_{i,j}x_{i-1,j} + b_{i,j}x_{i,j} - c_{i,j}x_{i+1,j} &= d_{i,j} \\ -a_{i,j}y_{i-1,j} + b_{i,j}y_{i,j} - c_{i,j}y_{i+1,j} &= e_{i,j}.\end{aligned} \tag{5.74}$$

These equations may be solved 'line-by-line' at fixed j by the Thomas Algorithm, as shown in Fig. 5.3. The method is iterative, and an initial guess for the values of x and y at the interior grid nodes, given the values of x and y at the boundary nodes,

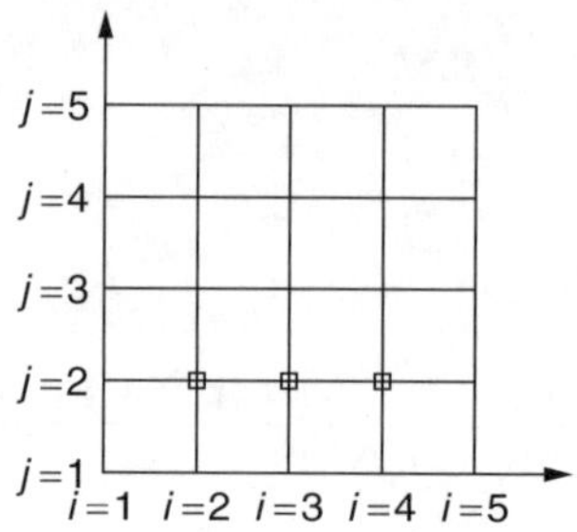

Fig. 5.4 Interior grid nodes for $j = 2$.

could be obtained, for example, by Transfinite Interpolation. This would give starting values for the coefficients $a_{i,j}$, etc., in eqn (5.74). As indicated in Fig. 5.4, the next step would be to use the Thomas Algorithm to evaluate x and y at interior grid points on the line $j = 2$. Having updated x and y values on the line $j = 2$, and re-calculated the coefficients, we would proceed to the line $j = 3$ and again use the Thomas Algorithm to evaluate x and y there. Thus the method involves a *sweep* from 'South' to 'North' (with *traverse*, using the Thomas Algorithm, from 'West' to 'East'). Clearly we could re-formulate the method so that we exploit the Thomas Algorithm along lines of fixed i, so that we sweep from West to East, while traversing from South to North. It can be seen from eqn (5.73) that in this case we have to solve

$$\begin{aligned} -a^*_{i,j}x_{i,j-1} + b_{i,j}x_{i,j} - c^*_{i,j}x_{i,j+1} &= d^*_{i,j} \\ -a^*_{i,j}y_{i,j-1} + b_{i,j}y_{i,j} - c^*_{i,j}y_{i,j+1} &= e^*_{i,j}, \end{aligned} \tag{5.75}$$

where $b_{i,j}$ take the same values as above, $a^*_{i,j} = c^*_{i,j} = (g_{11})_{i,j}/(\Delta\eta)^2$, and $d^*_{i,j}$ and $e^*_{i,j}$ are given by different expressions which can easily be found.

The ADI (Alternating Direction Implicit) method is commonly used to organize the sequence of traverses and sweeps. This procedure involves first carrying out a sweep from South to North, say, with traverses from West to East according to eqn (5.74), immediately followed by a sweep from West to East with traverses from South to North according to eqn (5.75).

The accompanying disk contains five programs for solving the Winslow equations in various situations using the Thomas Algorithm or SOR. See Section 5.13.

5.6.2 Orthogonality

Equations (5.19) may be discretized, given the univariate stretching functions $f_1(\xi)$, $f_2(\eta)$, and solved using a line-by-line iterative procedure (Thomas Algorithm) with ADI as described above. During any one iteration (involving one complete solution sweep) boundary values for x and y are temporarily held constant. However, after the iteration step has been completed, we can adjust the boundary values of x and y so as to satisfy the orthogonality condition (5.18) at the boundary.

The procedure is illustrated in Fig. 5.5. We focus on a grid point P in the physical domain which is adjacent to a boundary curve with equation $y = y_B(x)$ in cartesians

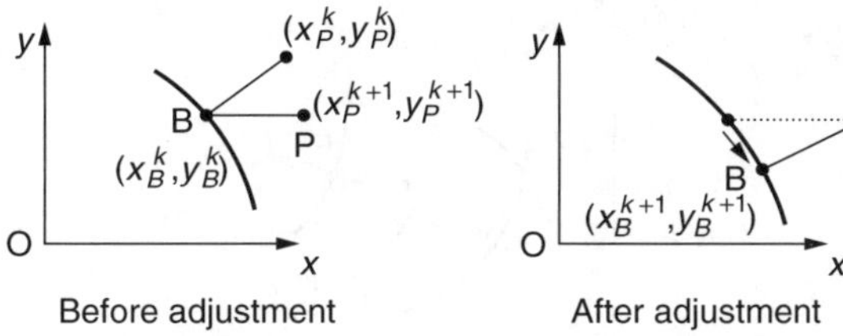

Fig. 5.5 Adjusting position of nodes on boundary to achieve orthogonality.

and has been located with co-ordinates (x_P^{k+1}, y_P^{k+1}) at the $(k+1)$th iteration step. Point B, with co-ordinates (x_B^k, y_B^k), is on the boundary, and these co-ordinates have been used in the iteration to determine the position of P. The aim is to move B along the boundary to a new position (x_B^{k+1}, y_B^{k+1}) such that the line PB will cut the boundary curve at right angles.

The slope $y_B'(x_B^k)$ of the boundary curve at the old position of B is used to give an approximate value $-(y_B')^{-1}$ for the required slope of PB, so that for the co-ordinates (x_B^{k+1}, y_B^{k+1}) of the new position of B we have

$$y_B^{k+1} - y_P^{k+1} = -(y_B')^{-1}(x_B^{k+1} - x_P^{k+1}). \tag{5.76}$$

Moreover, with first-order accuracy,

$$y_B^{k+1} - y_B^k = (y_B')(x_B^{k+1} - x_B^k). \tag{5.77}$$

These two simultaneous equations can be solved for x_B^{k+1}, and then, instead of solving for y_B^{k+1} as well, a value can be obtained by substituting into the equation of the boundary curve, i.e.

$$y_B^{k+1} = y_B(x_B^{k+1}). \tag{5.78}$$

In the case where y_B' is zero, a large value, say 10^{30}, may be used for $(y_B')^{-1}$. This effectively sets $x_B^{k+1} = x_P^{k+1}$. In the case where y_B' is infinite, we can switch the roles of x_B^{k+1} and y_B^{k+1} by solving the simultaneous equations for y_B^{k+1} and substituting into the boundary curve in the inverse form $x = x_B(y)$ to obtain x_B^{k+1}.

The solution procedure proposed here may be summarized as follows:

1. Guess values of x and y in the interior of the computational domain for the given boundary data.
2. Calculate the coefficients g_{11}, g_{22} and the 'source' terms on the right-hand side in the discretized form of eqns (5.19), and solve for new interior values of x and then y using the Thomas Algorithm applied in an ADI manner.
3. Adjust the boundary values of x and y so that orthogonality at the boundaries is achieved.
4. Check whether the convergence criterion is satisfied. If not, return to step 2 until the criterion is satisfied.

Finally, we can check how close the grid is to orthogonality by estimating the angle of intersection between ξ and η lines. At a grid-point P this may be done by fitting

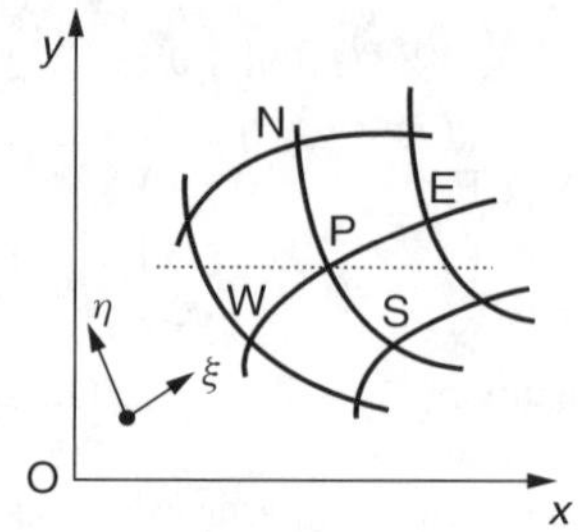

Fig. 5.6 Assessing closeness to orthogonality at P.

a parabola to the nodes S, P, N and another to W, P, E as shown in Fig. 5.6. The gradients of the tangents to these parabolas at P and the angles that they make with Ox can then be found; summing these angles gives the required estimate.

The accompanying disk contains a program for carrying out this procedure with specified stretching functions. It is listed in Section 5.13.

5.7 One-dimensional grids

5.7.1 Grid control

Some simple concepts of grid control, including the idea of a *weight function*, may be illustrated in the construction of a one-dimensional grid (here just a set of grid points) on the interval $a < x < b$ in 'physical' space. This interval will be mapped onto the one-dimensional interval $0 < \xi < 1$ in 'computational' space.

The one-dimensional version of Laplace's equation is

$$\frac{d^2\xi}{dx^2} = 0,$$

with solution $\xi = (x-a)/(b-a)$, the inverse being $x = a+(b-a)\xi$, a linear map which takes a uniformly-spaced set of points in computational space to a uniformly-spaced set of points in physical space.

If we introduce a control function $P(\xi)$, continuous in ξ, the grid generating equation, in place of eqn (5.6), is the ordinary differential equation

$$\frac{d^2\xi}{dx^2} = P(\xi), \tag{5.79}$$

where the mapping $\xi = \xi(x)$ satisfies the end-conditions $\xi(a) = 0$, $\xi(b) = 1$. To formulate the inverse problem, we have

$$\frac{d^2x}{d\xi^2} = \frac{d}{d\xi}\left(\frac{dx}{d\xi}\right) = \frac{dx}{d\xi}\frac{d}{dx}\left(\frac{d\xi}{dx}\right)^{-1} = -\frac{dx}{d\xi}\left(\frac{d\xi}{dx}\right)^{-2}\frac{d^2\xi}{dx^2} = -\left(\frac{dx}{d\xi}\right)^3\left(\frac{d^2\xi}{dx^2}\right).$$

Thus eqn (5.79) becomes

$$\frac{d^2x}{d\xi^2} = -\left(\frac{dx}{d\xi}\right)^3 P(\xi), \tag{5.80}$$

with end-conditions $x(0) = a$, $x(1) = b$. Although this is a non-linear differential equation, and the existence of solutions (in particular, single-valued solutions for which x increases monotonically with ξ) is not guaranteed, we may integrate once to give

$$\left(\frac{\mathrm{d}x}{\mathrm{d}\xi}\right)^{-2} = 2\int_c^\xi P(t)\,\mathrm{d}t$$

for some constant c. This indicates that some restrictions on the possible values of P are necessary for solutions to exist (and not to involve folding).

Another approach to grid generation in this situation may be derived by specifying a weight function $\varphi(\xi)$, which has the property that the grid spacing between points x_i, x_{i+1} in the physical plane is proportional to the product of the grid spacing between corresponding points ξ_i, ξ_{i+1} in the computational plane and the weight function evaluated at the mid-point of ξ_i and ξ_{i+1}. Thus

$$x_{i+1} - x_i = K\varphi\left(\frac{\xi_i + \xi_{i+1}}{2}\right)(\xi_{i+1} - \xi_i), \tag{5.81}$$

where K is the constant of proportionality. The weight function is defined in the computational domain $0 \leqslant \xi \leqslant 1$ and takes only positive values.

Revising the definition a little so that it applies in a general way without having to specify particular grid points, we proceed to the limit as the grid spacing tends to zero, and re-write eqn (5.81) as

$$\frac{\mathrm{d}x}{\mathrm{d}\xi} = K\varphi(\xi). \tag{5.82}$$

Eliminating K, we obtain the equation

$$\frac{\mathrm{d}}{\mathrm{d}\xi}\left(\frac{1}{\varphi(\xi)}\frac{\mathrm{d}x}{\mathrm{d}\xi}\right) = 0, \tag{5.83}$$

which is equivalent to

$$\frac{\mathrm{d}^2x}{\mathrm{d}\xi^2} = \frac{1}{\varphi}\frac{\mathrm{d}\varphi}{\mathrm{d}\xi}\frac{\mathrm{d}x}{\mathrm{d}\xi}. \tag{5.84}$$

This is the non-conservative version of eqn (5.83). Thus, given $\varphi(\xi)$, we have a differential model of grid generation, with the same boundary conditions $x(0) = a$, $x(1) = b$.

Comparing eqn (5.84) with (5.80) shows that they are equivalent if we put

$$P(\xi) = -\frac{1}{K^2\varphi^3}\frac{\mathrm{d}\varphi}{\mathrm{d}\xi}. \tag{5.85}$$

Note that integration of eqn (5.82) with respect to ξ over the whole range from $\xi = 0$ to 1 gives

$$K = (b-a)\Big/\int_0^1 \varphi(\xi)\,\mathrm{d}\xi. \tag{5.86}$$

As an example, we take

$$\varphi(\xi) = \begin{cases} 1, & 0 \leqslant \xi \leqslant \xi_0 \\ 2, & \xi_0 < \xi \leqslant 1 \end{cases}, \tag{5.87}$$

where the effect will be to double the spacing between grid points in one part of the physical domain.

Exercise 5. With $\varphi(\xi)$ given by eqn (5.87), show by integrating eqn (5.82) in the two intervals $0 \leqslant \xi < \xi_0$ and $\xi_0 < \xi \leqslant 1$ and eliminating K that

$$x(\xi) = \begin{cases} a + \dfrac{(b-a)}{(2-\xi_0)}\xi, & 0 \leqslant \xi < \xi_0 \\ \dfrac{2a - b\xi_0}{2-\xi_0} + 2\xi\dfrac{(b-a)}{2-\xi_0}, & \xi_0 < \xi \leqslant 1 \end{cases}. \tag{5.88}$$

In this example the grid spacing in the physical plane changes at $x = x_0$, where $x_0 = a + (b-a)\xi_0/(2-\xi_0)$. This gives

$$\xi_0 = \frac{2(x_0 - a)}{b + x_0 - 2a}.$$

For more complicated weight functions, it may be difficult to ensure that changes in grid spacing in the physical plane occur precisely where they are wanted. It may be more convenient to use weight functions which are functions of physical co-ordinates, so that in the one-dimensional case we have $\varphi(x)$, with

$$x_{i+1} - x_i = K\varphi\left(\frac{x_{i+1} + x_i}{2}\right)(\xi_{i+1} - \xi_i) \tag{5.89}$$

instead of eqn (5.81), with the limiting forms

$$\frac{\mathrm{d}x}{\mathrm{d}\xi} = K\varphi(x), \tag{5.90}$$

$$\frac{\mathrm{d}}{\mathrm{d}\xi}\left(\frac{1}{\varphi(x)}\frac{\mathrm{d}x}{\mathrm{d}\xi}\right) = 0, \tag{5.91}$$

in place of eqns (5.82) and (5.83), giving

$$\frac{\mathrm{d}^2x}{\mathrm{d}\xi^2} = \frac{1}{\varphi}\frac{\mathrm{d}\varphi}{\mathrm{d}x}\left(\frac{\mathrm{d}x}{\mathrm{d}\xi}\right)^2. \tag{5.92}$$

This is a non-linear second-order differential equation (with the same end-conditions $x(0) = a$, $x(1) = b$), with no guaranteed solution in general. Integration of eqn (5.90) yields

$$K = \int_a^b \frac{1}{\varphi(x)}\mathrm{d}x. \tag{5.93}$$

Equation (5.92) is equivalent to eqn (5.80) if we express P as a function of x and put

$$P(x) = -\frac{1}{K\varphi^2}\frac{\mathrm{d}\varphi}{\mathrm{d}x}. \tag{5.94}$$

5.7.2 Numerical aspects

Weight function equation

Here we present a standard finite-difference scheme for solving eqn (5.83). The interval $0 \leqslant \xi \leqslant 1$ is divided into m equal intervals, and we can label $(m+1)$ discrete points $\xi_i = i\Delta\xi$, $i = 0, 1, \ldots, m$, with $\Delta\xi = 1/m$. It is useful to be able to evaluate certain quantities at intermediate points, so that we have the finite-difference approximations at the point corresponding to i:

$$\left[\frac{d}{d\xi}\left\{\frac{dx/d\xi}{\varphi}\right\}\right]_i \backsimeq \frac{1}{\Delta\xi}\left\{\left(\frac{dx/d\xi}{\varphi}\right)_{i+\frac{1}{2}} - \left(\frac{dx/d\xi}{\varphi}\right)_{i-\frac{1}{2}}\right\}, \tag{5.95}$$

with

$$\left(\frac{dx}{d\xi}\right)_{i+\frac{1}{2}} \backsimeq \frac{x_{i+1}-x_i}{\Delta\xi}, \quad \left(\frac{dx}{d\xi}\right)_{i-\frac{1}{2}} \backsimeq \frac{x_i - x_{i-1}}{\Delta\xi}. \tag{5.96}$$

Then this finite-difference version of eqn (5.83) becomes

$$\frac{1}{(\Delta\xi)^2}\left\{\frac{1}{\varphi_{i-\frac{1}{2}}}x_{i-1} - \left(\frac{1}{\varphi_{i-\frac{1}{2}}} + \frac{1}{\varphi_{i+\frac{1}{2}}}\right)x_i + \frac{1}{\varphi_{i+\frac{1}{2}}}x_{i+1}\right\} = 0,$$

or

$$-a_i x_{i-1} + b_i x_i - c_i x_{i+1} = 0, \quad i = 1, 2, \ldots, (m-1), \tag{5.97}$$

with $a_i = \frac{1}{\varphi_{i-\frac{1}{2}}}$, $b_i = \left(\frac{1}{\varphi_{i-\frac{1}{2}}} + \frac{1}{\varphi_{i+\frac{1}{2}}}\right)$, $c_i = \frac{1}{\varphi_{i+\frac{1}{2}}}$, $i = 1, 2, \ldots, (m-1)$. Note that $b_i = a_i + c_i$.

Incorporating the end-conditions $x_0 = a$, $x_m = b$, we obtain the tridiagonal matrix equation

$$\begin{pmatrix} b_1 & -c_1 & 0 & 0 & - & - & 0 \\ -a_2 & b_2 & -c_2 & 0 & - & - & - \\ 0 & -a_3 & b_3 & -c_3 & - & - & - \\ 0 & 0 & - & - & - & - & - \\ - & - & - & - & - & - & 0 \\ - & -- & - & 0 & -a_{m-2} & b_{m-2} & -c_{m-2} \\ 0 & - & - & 0 & 0 & -a_{m-1} & b_{m-1} \end{pmatrix} \begin{pmatrix} x_1 \\ x_2 \\ - \\ - \\ - \\ - \\ x_{m-1} \end{pmatrix} = \begin{pmatrix} a_1 a \\ 0 \\ 0 \\ - \\ - \\ 0 \\ c_{m-1} b \end{pmatrix}. \tag{5.98}$$

The matrix of coefficients is also symmetric, since $c_i = a_{i+1}$. Solutions may be obtained efficiently using the Thomas Algorithm or SOR. On the accompanying disk the subdirectory *Book/one.d.gds* contains the file *line.SOR.f*, which applies SOR to the problem.

Control function equation

A numerical scheme for solving eqn (5.80) is shown here. With the same uniform grid as above along the computational ξ-axis, we have

$$\left(\frac{d^2x}{d\xi^2}\right)_i \backsimeq \frac{x_{i+1} - 2x_i + x_{i-1}}{(\Delta\xi)^2} \quad \text{and} \quad \left(\frac{dx}{d\xi}\right)_i \backsimeq \frac{x_{i+1} - x_{i-1}}{2(\Delta\xi)},$$

with $(P(\xi))_i = P(\xi_i) = P_i$. Hence eqn (5.80) becomes

$$\frac{x_{i+1} - 2x_i + x_{i-1}}{(\Delta\xi)^2} = -P_i \left(\frac{x_{i+1} - x_{i-1}}{2(\Delta\xi)}\right)^3,$$

which we write as

$$-a_i x_{i+1} + b_i x_i - c_i x_{i+1} = h_i, \quad i = 1, 2, \ldots, (m-1), \tag{5.99}$$

where $a_i = c_i = 1$, $b_i = 2$, and $h_i = P_i(x_{i+1} - x_{i-1})^3/8(\Delta\xi)$.

Because of the dependence of the terms h_i on the solution, these equations can be solved in an iterative manner as follows:

(a) Guess a reasonable initial set of values x_i, for example by linear interpolation, given the end-conditions $x_0 = a$ and $x_m = b$.

(b) Evaluate the set of values h_i.

(c) Solve the set of matrix equations for a new set of values x_i; this can be done by Gaussian elimination.

(d) Return to step (b) and continue the iteration until the difference between successive sets of values x_i, as measured by $\max_i |x_i^{new} - x_i^{old}|$, is less than some prescribed tolerance.

5.8 Three-dimensional grid generation

Extending eqn (5.6) to three dimensions leads naturally to the set of Poisson equations for the curvilinear co-ordinates x^i:

$$\nabla^2 x^i = P^i, \quad i = 1, 2, 3, \tag{5.100}$$

where the P^is are three control functions. Writing eqn (1.114) in its full three-dimensional vector form

$$g^{ij}\frac{\partial^2 \mathbf{r}}{\partial x^i \partial x^j} = -(\nabla^2 x^j)\frac{\partial \mathbf{r}}{\partial x^j}, \tag{5.101}$$

leads immediately to the inverted form

$$g g^{ij}\frac{\partial^2 \mathbf{r}}{\partial x^i \partial x^j} + g P^j \frac{\partial \mathbf{r}}{\partial x^j} = \mathbf{0}, \tag{5.102}$$

which may be expressed, using eqn (1.34) and putting $x^1 = \xi$, $x^2 = \eta$, $x^3 = \zeta$, as

$$D\mathbf{r} + g P^j \frac{\partial \mathbf{r}}{\partial x^j} = \mathbf{0},$$

where the second-order differential operator D is given by

$$D = G_1\frac{\partial^2}{\partial\xi^2} + G_2\frac{\partial^2}{\partial\eta^2} + G_3\frac{\partial^2}{\partial\zeta^2} + 2G_4\frac{\partial^2}{\partial\xi\partial\eta} + 2G_5\frac{\partial^2}{\partial\xi\partial\zeta} + 2G_6\frac{\partial^2}{\partial\eta\partial\zeta}. \tag{5.103}$$

5.9 Surface-grid generation model

A natural extension of the differential model given by eqns (5.6) to grid generation on a surface in three dimensions is obtained by replacing the Laplacian operator by the Beltrami operator, as defined in Section 3.9. Thus if the surface is to be covered by a curvilinear co-ordinate system of ξ^α curves, we could force $\xi(=\xi^1)$ and $\eta(=\xi^2)$ to satisfy the equations

$$\Delta_B \xi = P, \quad \Delta_B \eta = Q, \tag{5.104}$$

where P, Q are control functions. Note that the surface may already be effectively covered by parametric curves u^α, $\alpha = 1, 2$, but these may not give rise to a satisfactory grid. We assume here that the surface has four edges (four known space-curve segments) which can be mapped onto the edges of a unit square in the ξ^α computational plane.

Substituting into eqn (3.165) with the ξ^α system instead of u^α immediately gives the 'inverse' equation

$$a^{\alpha\beta}\frac{\partial^2 \mathbf{r}}{\partial \xi^\alpha \partial \xi^\beta} + P\frac{\partial \mathbf{r}}{\partial \xi} + Q\frac{\partial \mathbf{r}}{\partial \eta} = 2\kappa_m \mathbf{N},$$

which after multiplying through by a may be written as

$$D\mathbf{r} + a\left(P\frac{\partial \mathbf{r}}{\partial \xi} + Q\frac{\partial \mathbf{r}}{\partial \eta}\right) = R\mathbf{N}, \tag{5.105}$$

where the operator D is given by

$$D = a_{22}\frac{\partial^2}{\partial \xi^2} - 2a_{12}\frac{\partial^2}{\partial \xi \partial \eta} + a_{11}\frac{\partial^2}{\partial \eta^2}, \tag{5.106}$$

and

$$R = 2a\kappa_m = aa^{\alpha\beta}b_{\alpha\beta} = a_{22}b_{11} - 2a_{12}b_{12} + a_{11}b_{22}. \tag{5.107}$$

Equation (5.105) comprises three scalar equations for the cartesian components (x, y, z) of grid-points in physical space. These can be solved numerically when the equation of the surface is given, either explicitly, implicitly, or in terms of parameters. Specification of the two *constraining* eqns (5.104) provides the *core* of the method.

Here we shall just consider the case where $P = Q = 0$, and take the surface to be defined by the explicit equation $z = f(x, y)$. Then eqn (5.105) reduces to the three equations

$$\begin{aligned}
a_{22}\frac{\partial^2 x}{\partial \xi^2} - 2a_{12}\frac{\partial^2 x}{\partial \xi \partial \eta} + a_{11}\frac{\partial^2 x}{\partial \eta^2} &= RN_x \\
a_{22}\frac{\partial^2 y}{\partial \xi^2} - 2a_{12}\frac{\partial^2 y}{\partial \xi \partial \eta} + a_{11}\frac{\partial^2 y}{\partial \eta^2} &= RN_y \\
a_{22}\frac{\partial^2 z}{\partial \xi^2} - 2a_{12}\frac{\partial^2 z}{\partial \xi \partial \eta} + a_{11}\frac{\partial^2 z}{\partial \eta^2} &= RN_z,
\end{aligned} \tag{5.108}$$

where the cartesian components, from eqn (3.114), of **N** are, denoting partial derivatives by subscripts,

$$N_x = -\frac{f_x}{\sqrt{[1+(f_x)^2+(f_y)^2]}},$$
$$N_y = -\frac{f_y}{\sqrt{[1+(f_x)^2+(f_y)^2]}} \tag{5.109}$$
$$N_z = \frac{1}{\sqrt{[1+(f_x)^2+(f_y)^2]}},$$

and, moreover, making use of eqn (3.113),

$$R = a\left\{\frac{[1+(f_x)^2]f_{yy} - 2f_x f_y f_{xy} + [1+(f_y)^2]f_{xx}}{[1+(f_x)^2+(f_y)^2]^{3/2}}\right\}. \tag{5.110}$$

In addition we have, of course,

$$a_{11} = (x_\xi)^2 + (y_\xi)^2 + (z_\xi)^2, \quad a_{22} = (x_\eta)^2 + (y_\eta)^2 + (z_\eta)^2,$$
$$a_{12} = x_\xi x_\eta + y_\xi y_\eta + z_\xi z_\eta, \quad a = a_{11}a_{22} - (a_{12})^2.$$

Given the values of x, y, z on the edges of the computational square (corresponding to the four known space-curves), the aim now is to solve the partial differential eqns (5.108) for the cartesian co-ordinates of grid-points in the interior of the computational domain. These equations depend on having an explicit representation $z = f(x, y)$ of the surface. In many cases in practice such a representation has to be constructed from surface data involving a finite number of points, using least square fits or 'bicubic spline' methods. Numerical solution of eqns (5.108) may be carried out using any suitable method which has proved to be robust in elliptic grid-generation problems, for example LSOR (line, or line-by-line, successive over-relaxation) based on an initially guessed x, y, z field.

5.10 Hyperbolic grid generation

Grid generation based on the solution of elliptic partial differential equations with Dirichlet boundary conditions can be expensive in terms of computer time. There are situations where, instead of trying to match a curvilinear co-ordinate system to four boundary curves in two dimensions, or six surfaces in three dimensions, it may be more convenient to start with a single boundary and march outwards into the physical domain, using a hyperbolic partial differential equation as the basis of computation. This approach is suggested by the classical method of characteristics for second-order hyperbolic equations, where, starting from some non-characteristic initial curve in two dimensions, one may construct a network of characteristic curves in the solution domain.

Here we present a system of non-linear equations proposed by Steger and Chaussee (1980) for generating a two-dimensional grid with orthogonality and with control of grid-cell area. We start with a single boundary curve on which η is taken to be zero. The two first-order partial differential equations for the cartesian co-ordinates x, y as functions of ξ, η are:

$$g_{12} = \mathbf{g}_1 \cdot \mathbf{g}_2 = x_\xi x_\eta + y_\xi y_\eta = 0,$$
$$|\mathbf{g}_1 \times \mathbf{g}_2| = x_\xi y_\eta - x_\eta y_\xi = V, \tag{5.111}$$

where the first equation imposes orthogonality, and the second implies, by eqn (1.43), that V is a measure of cell-area (assuming that $\delta\xi\delta\eta$ is the same for each grid cell). In general we can let V be a function of ξ, η, so that, for example, we could increase grid-density near the boundary $\eta = 0$ by ensuring that V is small there.

It follows from eqns (5.111) that

$$x_\eta = -\frac{V}{g_{11}} y_\xi, \quad y_\eta = \frac{V}{g_{11}} x_\xi, \tag{5.112}$$

where $g_{11} = (x_\xi)^2 + (y_\xi)^2$.

The subdirectory *Book/hyper.gds* on the accompanying disk contains a program for solving these equations numerically using a marching procedure. The cell-area V is prescribed in the form

$$V = K\sqrt{g_{11}} e^{-\lambda(1-\eta)}, \tag{5.113}$$

where K and λ are constants, and the equations are discretized with first-order accuracy in the η-direction and second-order accuracy in the ξ-direction as

$$\frac{(x_{i,j+1} - x_{i,j})}{\Delta\eta} = -\left(\frac{V}{g_{11}}\right)_{i,j} \left(\frac{y_{i+1,j} - y_{i-1,j}}{2\Delta\xi}\right),$$
$$\frac{(y_{i,j+1} - y_{i,j})}{\Delta\eta} = \left(\frac{V}{g_{11}}\right)_{i,j} \left(\frac{x_{i+1,j} - x_{i-1,j}}{2\Delta\xi}\right), \tag{5.114}$$

where

$$(g_{11})_{i,j} = \left(\frac{x_{i+1,j} - x_{i-1,j}}{2\Delta\xi}\right)^2 + \left(\frac{y_{i+1,j} - y_{i-1,j}}{2\Delta\xi}\right)^2.$$

These equations determine explicitly the values $x_{i,j+1}$ and $y_{i,j+1}$ in terms of values $x_{i-1,j}, x_{i,j}, x_{i+1,j}, y_{i-1,j}, y_{i,j}$, and $y_{i+1,j}$. Hence the method itself may be called explicit. Only data on values of x and y on the initial boundary $\eta = 0$ are required to start the process. The numerical solution starts from this boundary and marches outward in the direction of increasing η, eventually constructing new co-ordinate lines $\xi = 0, 1$ and $\eta = 1$.

5.11 Solving the hosted equations

5.11.1 An example

In this section we illustrate the task of solving a partial differential equation, given that the solution domain has been discretized by the generation of a suitable grid. We take the two-dimensional form of Laplace's Equation as an example. The subdirectory *Book/p.d.Equation* on the accompanying disk, as listed in the last section of this chapter, contains two files, the first of which, *Conjugate.Laplace.f*, simply shows how to solve Laplace's Equation

$$\frac{\partial^2\varphi}{\partial x^2}+\frac{\partial^2\varphi}{\partial y^2}=0$$

in a square solution domain using the Conjugate Gradient method. Exact solutions, such as $u = e^x \sin y$, with appropriate boundary conditions, may be used to assess the numerical accuracy of the method.

For non-rectangular domains we need to be able to transform the equation to general curvilinear co-ordinates ξ, η. A convenient expression for $\nabla^2\varphi$ was obtained in eqn (1.173), with the addition of eqns (1.174) and (1.175). The file *laplacian.general.f* contains a program which solves the discretized form of these equations using SOR. First we set

$$\begin{aligned}
&g_{11}/\sqrt{g} = g_{oo}, \quad -2g_{12}/\sqrt{g} = g_{ot}, \quad g_{22}/\sqrt{g} = g_{tt},\\
&\sqrt{g}\nabla^2\xi = \text{Delzi}, \quad \sqrt{g}\nabla^2\eta = \text{Delet}
\end{aligned}$$

so that eqns (1.174) and (1.175) can be written

$$\begin{aligned}
\text{Delzi} &= g_{tt}\frac{(x_\eta y_{\xi\xi} - y_\eta x_{\xi\xi})}{\sqrt{g}} + g_{ot}\frac{(x_\eta y_{\xi\eta} - y_\eta x_{\xi\eta})}{\sqrt{g}} + g_{oo}\frac{(x_\eta y_{\eta\eta} - y_\eta x_{\eta\eta})}{\sqrt{g}},\\
\text{Delet} &= g_{tt}\frac{(y_\xi x_{\xi\xi} - x_\xi y_{\xi\xi})}{\sqrt{g}} + g_{ot}\frac{(y_\xi x_{\xi\eta} - x_\xi y_{\xi\eta})}{\sqrt{g}} + g_{oo}\frac{(y_\xi x_{\eta\eta} - x_\xi y_{\eta\eta})}{\sqrt{g}}.
\end{aligned}$$

Discretization of eqn (1.173) can now be represented by

$$\begin{aligned}
&\{(g_{tt}*\varphi)_{j+1,k} - 2(g_{tt}*\varphi)_{j,k} + (g_{tt}*\varphi)_{j-1,k}\}\\
&\quad + \{(g_{oo}*\varphi)_{j,k+1} - 2(g_{oo}*\varphi)_{j,k} + (g_{oo}*\varphi)_{j,k-1}\}\\
&\quad + 0.25\,\{(g_{ot}*\varphi)_{j+1,k+1} + (g_{ot}*\varphi)_{j-1,k-1}\\
&\qquad - (g_{ot}*\varphi)_{j+1,k-1} - (g_{ot}*\varphi)_{j-1,k+1}\}\\
&\quad + 0.5\,\{(\text{Delzi}*\varphi)_{j+1,k} - (\text{Delzi}*\varphi)_{j-1,k}\}\\
&\quad + 0.5\,\{(\text{Delet}*\varphi)_{j,k+1} - (\text{Delet}*\varphi)_{j,k-1}\} = 0
\end{aligned} \tag{5.115}$$

in the transformed domain of curvilinear co-ordinates.

Taking terms containing $\varphi_{j,k}$ to one side of the equation, we can generate a 'temporary' value

$$\begin{aligned}
\varphi^t_{j,k} &= \left\{\frac{(g_{tt}*\varphi)_{j+1,k} + (g_{tt}*\varphi)_{j-1,k} + (g_{oo}*\varphi)_{j,k+1} + (g_{oo}*\varphi)_{j,k-1}}{2[(g_{tt})_{j,k} + (g_{oo})_{j,k}]}\right\}\\
&\quad + 0.5\left\{\frac{(\text{Delzi}*\varphi)_{j+1,k} - (\text{Delzi}*\varphi)_{j-1,k}}{2[(g_{tt})_{j,k} + (g_{oo})_{j,k}]}\right\}
\end{aligned}$$

$$+0.5\left\{\frac{(\text{Delet}*\varphi)_{j,k+1}-(\text{Delet}*\varphi)_{j,k-1}}{2[(g_{tt})_{j,k}+(g_{oo})_{j,k}]}\right\}$$
$$+0.25\left\{\frac{\begin{array}{c}(g_{ot}*\varphi)_{j+1,k+1}+(g_{ot}*\varphi)_{j-1,k-1}\\-(g_{ot}*\varphi)_{j+1,k-1}-(g_{ot}*\varphi)_{j-1,k+1}\end{array}}{2[(g_{tt})_{j,k}+(g_{oo})_{j,k}]}\right\}. \tag{5.116}$$

An iterative scheme employing SOR is now given by

$$\varphi^{n+1}_{j,k}=\varphi^{n}_{j,k}+\omega(\varphi^{t}_{j,k}-\varphi^{n}_{j,k}), \tag{5.117}$$

where ω is an over-relaxation factor which we have taken in the program to be 1.5.

The program uses this method to solve Laplace's equation in the region between concentric circular arcs shown in Fig. 4.2 of Chapter 4, subject to the Dirichlet boundary conditions:

$$\varphi=0\quad\text{when }\theta=0,\qquad\varphi=\frac{1}{r}\sin\alpha\quad\text{when }\varphi=\alpha,$$
$$\varphi=\frac{1}{r_1}\sin\theta\quad\text{when }r=r_1,\qquad\varphi=\frac{1}{r_2}\sin\theta\quad\text{when }r=r_2. \tag{5.118}$$

For these boundary conditions the exact solution $\varphi=\frac{1}{r}\sin\theta$ exists, and this enables us to assess the accuracy of the numerical procedure.

5.11.2 More general steady-state equation

Suppose that the basic partial differential equation to be solved in physical space for the field variable φ has the form

$$\nabla\cdot(\mathbf{v}\varphi)+\nabla\cdot(\nu\nabla\varphi)+S=0, \tag{5.119}$$

where S is a source term, ν could be a diffusion coefficient, and $\mathbf{v}$ is a vector field (fluid velocity). In cartesian co-ordinates this takes the form

$$\frac{\partial}{\partial y_i}(v_i\varphi)+\frac{\partial}{\partial y_i}\left(\nu\frac{\partial\varphi}{\partial y_i}\right)+S=0,$$

which transforms to the consistent generalized tensor form

$$(v^i\varphi)_{,i}+(\nu g^{ij}\varphi_{,j})_{,i}+S=0. \tag{5.120}$$

Using eqns (1.128), (1.122), (1.111), and (1.134), this may be expressed as

$$\begin{aligned}&v^i\varphi_{,i}+v^i_{,i}\varphi+g^{ij}(\nu\varphi_{,j})_{,i}+S\\&=v^i\frac{\partial\varphi}{\partial x^i}+\varphi\,\text{div}\mathbf{v}+g^{ij}\left[\frac{\partial}{\partial x^i}\left(\nu\frac{\partial\varphi}{\partial x^j}\right)-\Gamma^k_{ij}\left(\nu\frac{\partial\varphi}{\partial x^k}\right)\right]+S\\&=v^i\frac{\partial\varphi}{\partial x^i}+\varphi\mathbf{g}^i\cdot\frac{\partial\mathbf{v}}{\partial x^i}+g^{ij}\frac{\partial}{\partial x^i}\left(\nu\frac{\partial\varphi}{\partial x^j}\right)+(\nabla^2x^k)\nu\frac{\partial\varphi}{\partial x^k}+S=0,\end{aligned}$$

giving finally

$$[v^i + \nu(\nabla^2 x^i)]\frac{\partial\varphi}{\partial x^i} + g^{ij}\frac{\partial}{\partial x^i}\left(\nu\frac{\partial\varphi}{\partial x^j}\right) + \varphi\mathbf{g}^i \cdot \frac{\partial\mathbf{v}}{\partial x^i} + S = 0, \qquad (5.121)$$

where the $\mathbf{g}^i$s are given by eqns (1.8) or (1.161), $v^i = \mathbf{g}^i \cdot \mathbf{v}$, and i, j are summed from 1 to 2 or from 1 to 3, depending on whether the problem is in one or two dimensions.

Exercise 6. Making use of eqns (1.120), (1.118), and (1.135), show that another (conservative) form of the equation is

$$\frac{\partial}{\partial x^i}\left[\sqrt{g}\left(v^i\varphi + \nu g^{ij}\frac{\partial\varphi}{\partial x^j}\right)\right] + \sqrt{g}S = 0. \qquad (5.122)$$

Now, following the generation of a structured grid for the physical space, with the x^i co-ordinate curves coinciding by definition with grid lines, the hosted equation may be discretized and solved on the square or rectangular domain in x^i computational space, subject to the given boundary conditions.

Typical boundary conditions

$$c_1\varphi + c_2\frac{\partial\varphi}{\partial n} = c_3 \qquad (5.123)$$

in physical space transform, according to eqn (1.193), into the condition

$$c_1\varphi + c_2\frac{1}{\sqrt{g^{ii}}}g^{ij}\frac{\partial\varphi}{\partial x^j} = c_3, \qquad (5.124)$$

with summation over j but not i, on a boundary $x^i = const.$ in computational space.

5.12 Multiblock grid generation

When the geometry of the solution domain is complex, as is generally the case in engineering problems, one approach is to divide it into subdomains (blocks) of simpler geometries. Structured grid generation can be carried out in each of the sub-domains using algebraic methods or differential models, and the resulting grids can then be patched together at the common boundaries. A major drawback of this ‘multiblocking’, or ‘block-structuring’, technique has been the difficulty of automating the process of domain decomposition. Dividing the domain into subdomains while keeping track of data relating to the boundaries between the blocks and to the ‘connectivity’ of the blocks is not difficult when the number of subdomains is relatively small, so that the process can be performed manually. However, the generation of this initial data-structure can be unacceptably time-consuming when the number of subdomains is large (say, into the hundreds).

Approaches to automating this procedure were suggested by Allwright (1988) and Eiseman, Cheng, and Hauser (1994), and there are currently commercially available Computer Aided Design (CAD) codes which can be used. These perform block geometry generation automatically, together with curve-fitting, typically using cubic splines, along the common boundaries of the sub-domains. When the overall solution domain geometry is not too complex, however, it may be convenient to decompose it into a small number of blocks of fairly simple geometry, in each of which a grid can be generated with desirable features, such as orthogonality. For two dimensional problems the following procedure (as implemented on the accompanying disk in the subdirectory *Book/Winslow gds*, file *orthog.g.f*) is proposed here for patching at the common boundaries.

Consider typical (two-dimensional) blocks A and B, whose common boundary can be represented as a cubic spline. Suppose that the length of this boundary, which may be calculated using the integral formula for a plane curve, is L_{AB}. Next suppose that a structured grid is generated independently for each block using an elliptic generator, say, so that block A may be mapped onto the square $0 < \xi < 1$, $0 < \eta < 1$ in computational space, while block B is mapped onto a similar square using a different mapping. The grid in block A is generated using a discrete set of ξ-values, $0 = \xi_1, \xi_2, \ldots, \xi_n = 1$, and a discrete set of η-values. To give some measure of control the sizes of successive intervals $(\xi_i - \xi_{i-1})$ between ξ-values are either taken to follow a geometric progression in the program *orthog.g.f.*, or are obtained using a stretching function.

The *same* set of ξ-values is used to generate the grid in block B, according to the mapping for block B. A point on the boundary in physical space corresponding to ξ_i has known cartesian co-ordinates (x_{Ai}, y_{Ai}), according to the mapping of block A, with respect to some set of cartesian axes. Similarly, these ξ-values produce grid-nodes along the boundary according to the mapping of block B with, in general, different cartesian co-ordinates (x_{Bi}, y_{Bi}). Note that the number of grid-nodes along the boundary here is the same for both A and B.

The following algorithm is used to match the boundary grid-points of A to those in B:

1. The closeness of the matching is tested by calculating a residual defined by

$$\text{ResMAT} = \sum_{i=1}^{n} \left(\frac{|x_{Ai} - x_{Bi}|}{L_{AB}} + \frac{|y_{Ai} - y_{Bi}|}{L_{AB}} \right).$$

2. If ResMAT is greater than some user-specified value, say10^{-4}, then some adjustment of the A nodes is carried out so that they will coincide more closely with those of B.
3. For typical grid-points of block A and block B corresponding to a value ξ_i, for some i, the length L_1 of the boundary curve between them may be calculated. In the program this is done by dividing the difference $(x_{Bi} - x_{Ai})$ into ten equal parts Δx, calculating the corresponding Δys using the cubic spline expression for the boundary curve, and then calculating the approximation

$$L_1 = \sum_{1}^{10} \sqrt{\{(\Delta x)^2 + (\Delta y)^2\}}.$$

4. A similar calculation gives the length L_2 of the boundary curve between the grid-nodes (x_{Bi}, y_{Bi}) and $(x_{A(i+1)}, y_{A(i+1)})$. The sum $(L_1 + L_2)$ then gives an approximation to the distance along the curve between (x_{Ai}, y_{Ai}) and $(x_{A(i+1)}, y_{A(i+1)})$.

5. A linear interpolation formula is now used to generate an improved value for ξ_i, namely
$$\frac{\xi_i^{New} - \xi_i}{L_1} = \frac{\xi_{i+1} - \xi_i}{L_1 + L_2}.$$
which gives $\xi_i^{New} = \xi_i + \frac{L_1}{L_1+L_2}(\xi_{i+1} - \xi_i)$.
6. Having adjusted (in a loop) all the values of ξ_i, we regenerate the two grids and test again the value of ResMAT. The process is iterated until this value is within the specified limit.

5.13 Website programs

5.13.1 Subdirectory: Book/Winslow.gds

This subdirectory contains six files. The first five solve the Winslow eqns (5.3) or the Winslow equations with control functions (5.7) (TTM equations) for various planar two-dimensional geometries, using either SOR or the Thomas Algorithm. They are:

1. W.SOR.f
2. W.adaptive.f
3. W.SOR.circle.f
4. W.adaptive.circle.f
5. W.trid.f
   A number of grids obtained for domains of various shapes using these programs are shown in Figs 5.7–5.12
   The sixth file is
6. orthog.g.f

This program generates a two-dimensional planar orthogonal grid with grid-density control by solving the Winslow eqns (5.19), incorporating univariate stretching functions in both transformed co-ordinates ξ, η. The code has been written with the possibility of employing finite-volume methods in mind, for example for solving the

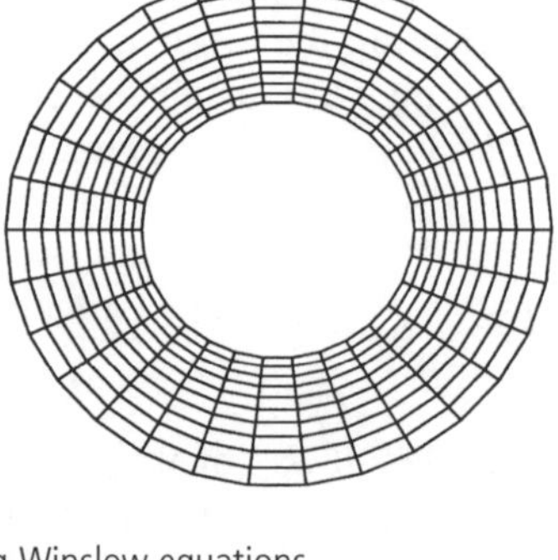

Fig. 5.7 Grid generated employing Winslow equations.

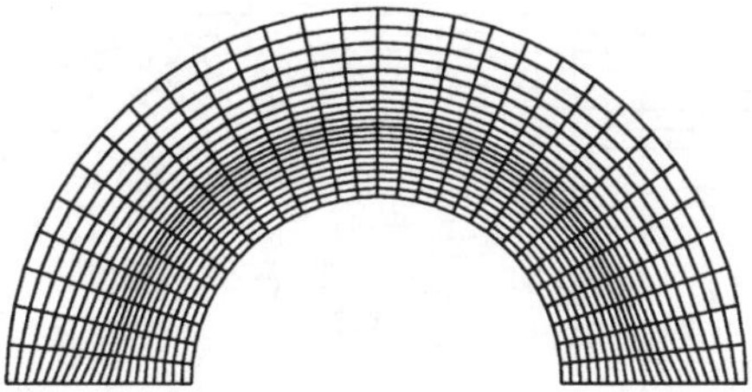

Fig. 5.8 TTM equations.

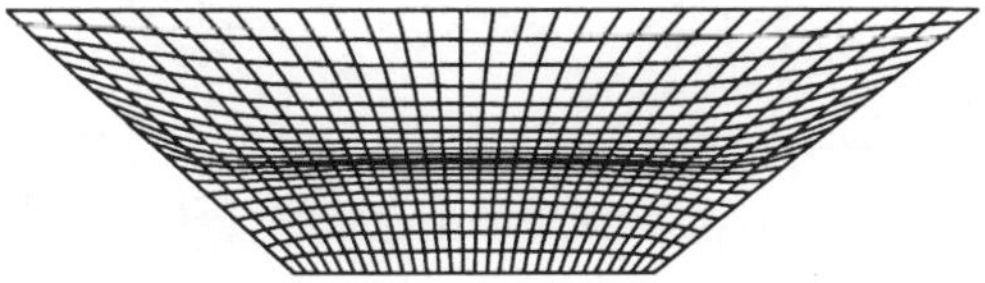

Fig. 5.9 TTM equations.

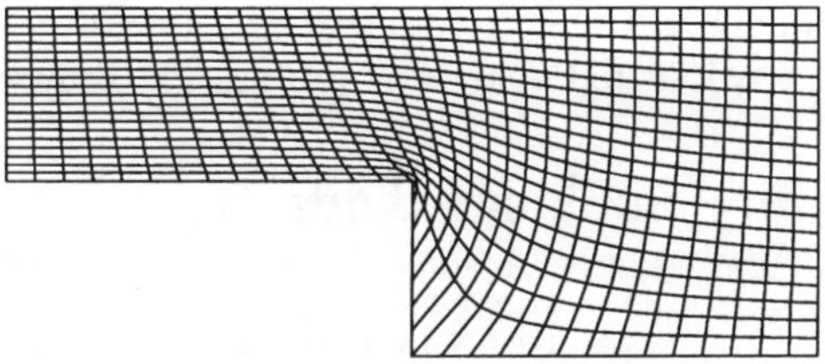

Fig. 5.10 Winslow equations.

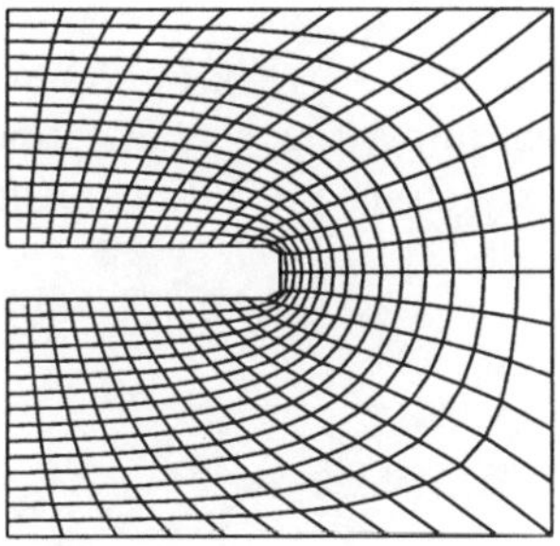

Fig. 5.11 Winslow equations.

Navier-Stokes equations. Thus the code contains provisions for calculating (using numerical integration) internodal distances, grid-cell areas, and, when the underlying physical problem is axi-symmetric, grid-cell volumes.

To run this program the user requires a NAG (Numerical Algorithm Group) Library. If the physical boundary is obtained by direct physical measurement, cubic-spline routines are used to construct a smooth curve. There is a switch in the program that

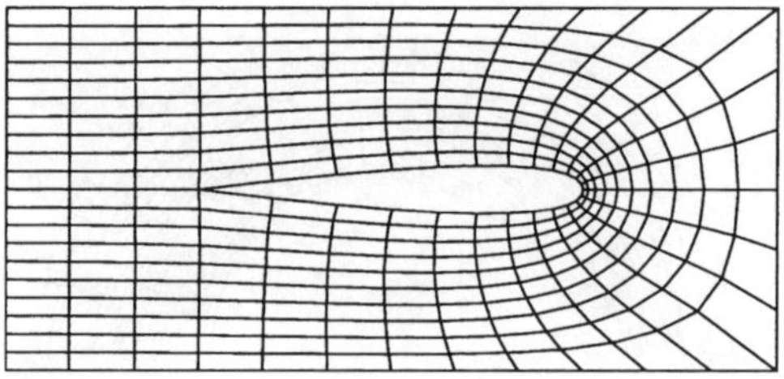

Fig. 5.12 Winslow equations.

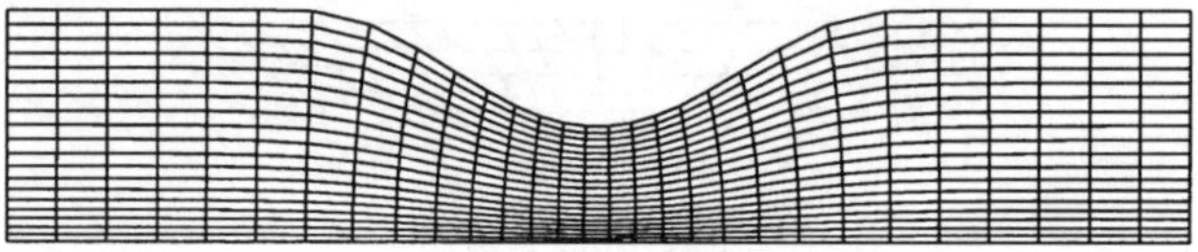

Fig. 5.13 Orthogonal Winslow.

by-passes these routines if the boundary-data is provided through an analytical function. However, the program does not run without the presence of a NAG Library.

An example of a grid calculated using this program is shown in Fig. 5.13.

5.13.2 Subdirectory: Book/one.d.gds

This contains two files, the second of which listed below is relevant to the present chapter. The first is explained in the next chapter.

1. curve.SOR.f
2. line.SOR.f

The program here solves eqn (5.97) by SOR, the weight function being specified by the user. In fact the code prompts the user to choose one of the four following functions, which are, respectively, constant, bilinear, quadratic, and exponential:

(a) $\varphi(\xi) = 1$, a constant;
(b) $\varphi(\xi) = \lambda\left(|\xi - \xi_0| + 1\right);$
(c) $\varphi(\xi) = \lambda\left\{(\xi - \xi_0)^2 + 1\right\};$
(d) $\varphi(\xi) = e^{\lambda(\xi - \xi_0)}$,

where the λs are constants, and 1s have been added to two right-hand sides to ensure that the weight functions remain positive.

5.13.3 Subdirectory: Book/hyper.gds

There is one file in this subdirectory:

1. hyperbolic.f

The operation of the program in this file has been described in Section 5.10.

5.13.4 Subdirectory: Book/p.d.Equations

This contains two files, referred to in Section 5.11. They are:

1. Conjugate.Laplace.f
2. Laplacian.general.f

6

Variational methods and adaptive grid generation

6.1 Introduction

Despite the availability of a variety of algebraic and differential models for grid generation, many practitioners experience substantial difficulties when applying these models to new problems. The grids generated may turn out to be badly skewed (with large departures from orthogonality), compressed, or expanded, folding may occur, and sometimes even convergence fails. Ever since grid generation methods began to be studied seriously in the late 1960s, variational approaches have been used in attempts to attain a deeper understanding of the limitations and strengths of the various differential models. In this chapter we present an introduction to variational methods and show how they may be used to control grid 'quality'.

The quality of the grids used to discretize the physical domain when solving partial differential equations very much affects the accuracy of the numerical solutions obtained. Given that cost constraints allow only a limited number of grid nodes to be introduced into a solution domain, one aspect of quality is related to the density of grid nodes in regions where there are large gradients of field variables. A grid-generating scheme should be able to allocate more grid nodes where large gradients occur (and fewer where field variables vary smoothly). Moreover, in transient problems where the solution develops in time, regions of high gradients may not be known *a priori*, and there is therefore a need for some mechanism which can automatically detect regions of high gradients as they arise, so that grid density can be increased there. In other words, a grid generation procedure may be coupled to the numerical solution, producing grids which may be called *solution-adaptive*, or *dynamically adaptive*. Algorithms which automatically concentrate and disperse grid nodes in this way are called *adaptive*. Adaptive methods may also be used in steady-state problems when regions of high gradients are not known in advance.

As well as the optimum spacing of grid points, the quality of grids depends on factors such as cell-areas (or volumes) and the angles between grid lines (how far a grid departs from orthogonality). An ideal structured grid would be an orthogonal grid with grid-node density able to cope with sharp solution gradients. In many cases, however, the geometrical complexity of the physical domain makes the construction of such a grid difficult or impossible, and a compromise has to be reached. The attraction of the variational approach [see Liseikin (1999) for detailed references] has been that it can facilitate intuitive control over these various factors, while at the same time providing firm theoretical foundations.

In the following section we review some basic ideas of variational calculus.

6.2 Euler-Lagrange equations

We have already considered one application of variational methods in Chapter 3, where an outline of the derivation of the general equations for geodesics on surfaces in E^3 was presented. These eqns (3.67) are *necessary* conditions that geodesics must satisfy in order to make the integral in eqn (3.66) stationary. Such integrals, which depend for their value on the particular functions, $u^1(t)$ and $u^2(t)$ in that case, being substituted into the integrand, are called *functionals* in the classical Calculus of Variations. *Sufficient* conditions for maxima or minima are, however, usually more difficult to establish than in the corresponding calculus of functions. Thus we may be able to solve the Euler-Lagrange equations (the necessary conditions) to obtain solution functions, which are called *extremals*, but it is not usually easy to establish their nature. It may be pointed out, in addition, that variational problems may have no solution. For example, the functional may be bounded below while at the same time a function that actually minimizes it may not exist.

Here we give a brief account of the main types of variational problems that we shall encounter, with a discussion of the corresponding Euler-Lagrange equations.

In Chapter 3 we met the functional of the form, writing x_1, x_2 in place of u^1, u^2,

$$I = \int_a^b F(x_1, x_2, \dot{x}_1, \dot{x}_2)\,\mathrm{d}t,$$

where F is regarded formally as a function of four independent variables, and the functions x_1 and x_2 are required to take prescribed values at the ends $t = a$ and b of the interval of integration. The Euler-Lagrange equations are the two ordinary differential equations

$$\begin{aligned}\frac{\mathrm{d}}{\mathrm{d}t}\left(\frac{\partial F}{\partial \dot{x}_1}\right) - \frac{\partial F}{\partial x_1} &= 0,\\ \frac{\mathrm{d}}{\mathrm{d}t}\left(\frac{\partial F}{\partial \dot{x}_2}\right) - \frac{\partial F}{\partial x_2} &= 0.\end{aligned} \tag{6.1}$$

Since in general the terms $\partial F/\partial \dot{x}_1$ and $\partial F/\partial \dot{x}_2$ contain terms in $\dot{x}_1$ and $\dot{x}_2$, these are two second-order differential equations, and there are four prescribed boundary conditions to be satisfied.

The argument sketched in Chapter 3 also applies to the case

$$I = \int_a^b F(x_1, x_2, \dot{x}_1, \dot{x}_2, t)\,\mathrm{d}t,$$

and the conclusions, eqns (6.1), are the same.

It is also straightforward to generalize the argument to cover the case

$$I = \int_a^b F(x_1, x_2, \ldots, x_n, \dot{x}_1, \dot{x}_2, \ldots, \dot{x}_n, t)\,\mathrm{d}t \tag{6.2}$$

for n extremal functions $x_1(t), x_2(t), \ldots, x_n(t)$, to be determined subject to the conditions that they make I stationary and that they satisfy certain prescribed end conditions. The Euler-Lagrange equations are

$$\frac{\mathrm{d}}{\mathrm{d}t}\left(\frac{\partial F}{\partial \dot{x}_i}\right) - \frac{\partial F}{\partial x_i} = 0, \quad i = 1, 2, \ldots, n, \tag{6.3}$$

n second-order differential equations to be satisfied subject to $2n$ boundary conditions. Of course, this also holds for the case $n = 1$.

The differentiation operators appearing in eqn (6.3) have to be interpreted carefully. The partial derivatives involve formal differentiation of F as a function of $(2n + 1)$ variables, while the ordinary (or *total*) derivative operator $\mathrm{d}/\mathrm{d}t$ acting in the first term regards everything as a function of the single variable t.

In the case where F does not depend explicitly on t, we can infer from the 'total derivative' Chain Rule and eqn (6.3) that extremals satisfy the equation

$$\begin{aligned}\frac{\mathrm{d}F}{\mathrm{d}t} &= \sum_{i=1}^{n} \frac{\partial F}{\partial x_i}\frac{\mathrm{d}x_i}{\mathrm{d}t} + \sum_{i=1}^{n} \frac{\partial F}{\partial \dot{x}_i}\frac{\mathrm{d}\dot{x}_i}{\mathrm{d}t} \\ &= \sum_{i=1}^{n}\left[\frac{\mathrm{d}}{\mathrm{d}t}\left(\frac{\partial F}{\partial \dot{x}_i}\right)\frac{\mathrm{d}x_i}{\mathrm{d}t} + \frac{\partial F}{\partial \dot{x}_i}\frac{\mathrm{d}\dot{x}_i}{\mathrm{d}t}\right] = \sum_{i=1}^{n}\frac{\mathrm{d}}{\mathrm{d}t}\left(\dot{x}_i\frac{\partial F}{\partial \dot{x}_i}\right).\end{aligned}$$

Hence we obtain

$$\frac{\mathrm{d}}{\mathrm{d}t}\left[F - \sum_{i=1}^{n} \dot{x}_i \frac{\partial F}{\partial \dot{x}_i}\right] = 0. \tag{6.4}$$

A further generalization of (6.2) is the functional

$$I = \int_a^b F(x_1, x_2, \ldots, x_n, \dot{x}_1, \dot{x}_2, \ldots, \dot{x}_n, \ddot{x}_1, \ddot{x}_2, \ldots, \ddot{x}_n, t)\,\mathrm{d}t, \tag{6.5}$$

where the unknown functions $x_1(t), \ldots, x_n(t)$ have their values and the values of their derivatives prescribed at the end-points. Variations $\delta x_1, \ldots, \delta x_n$ (all functions of t) in these functions from some supposed extremal functions should produce a variation in I which is zero to first order. Thus, using first-order increment formulas,

$$\begin{aligned}0 = \delta I &= \int_a^b \delta F\,\mathrm{d}t = \int_a^b \left(\sum_{i=1}^{n}\frac{\partial F}{\partial x_i}\delta x_i + \sum_{i=1}^{n}\frac{\partial F}{\partial \dot{x}_i}\delta \dot{x}_i + \sum_{i=1}^{n}\frac{\partial F}{\partial \ddot{x}_i}\delta \ddot{x}_i\right)\mathrm{d}t \\ &= \int_a^b \sum_{i=1}^{n}\left[\left(\frac{\partial F}{\partial x_i} - \frac{\mathrm{d}}{\mathrm{d}t}\left(\frac{\partial F}{\partial \dot{x}_i}\right)\right)\delta x_i - \frac{\mathrm{d}}{\mathrm{d}t}\left(\frac{\partial F}{\partial \ddot{x}}\right)\delta \dot{x}_i\right]\mathrm{d}t\end{aligned}$$

after integration by parts, the integrated parts vanishing because all δx_i and $\delta \dot{x}_i$, $i = 1, 2, \ldots, n$, vanish at the ends because of the boundary conditions. Note that for the integration by parts we assume that $\delta(\dot{x}_i) = \mathrm{d}(\delta x_i)/\mathrm{d}t$ and $\delta(\ddot{x}) = \mathrm{d}(\delta \dot{x})/\mathrm{d}t$.

One more integration by parts gives

$$0 = \delta I = \int_a^b \sum_{i=1}^{n} \left[\frac{\partial F}{\partial x_i} - \frac{\mathrm{d}}{\mathrm{d}t}\left(\frac{\partial F}{\partial \dot{x}_i}\right) + \frac{\mathrm{d}^2}{\mathrm{d}t^2}\left(\frac{\partial F}{\partial \ddot{x}_i}\right)\right] \delta x_i \,\mathrm{d}t$$

with another integrated part vanishing.

Since this must hold for arbitrary independent variations δx_i, it follows straightforwardly (using a proof by contradiction) that the following equations must hold for all t:

$$\frac{\partial F}{\partial x_i} - \frac{\mathrm{d}}{\mathrm{d}t}\left(\frac{\partial F}{\partial \dot{x}_i}\right) + \frac{\mathrm{d}^2}{\mathrm{d}t^2}\left(\frac{\partial F}{\partial \ddot{x}_i}\right) = 0, \quad i = 1, 2, \ldots, n. \tag{6.6}$$

These are the n fourth-order ordinary differential equations (the Euler-Lagrange equations for this case), with $4n$ end conditions, whose solutions are the extremals for the variational problem $\delta I = 0$ with I given by eqn (6.5).

Another type of generalization arises from considering functionals

$$I = \iint_R F(x, y, u, u_x, u_y)\,\mathrm{d}x\,\mathrm{d}y, \tag{6.7}$$

which involve a double integral over a domain R of the xy-plane, where the integrand depends explicitly on a function $u(x, y)$ and its partial derivatives u_x, u_y. We assume that the value of u is prescribed on the boundary C of R. To make I stationary we equate to zero the first-order expression for the increment δI

$$0 = \delta I = \iint_R \delta F\,\mathrm{d}x\,\mathrm{d}y = \iint_R \left(\frac{\partial F}{\partial u}\delta u + \frac{\partial F}{\partial u_x}\delta u_x + \frac{\partial F}{\partial u_y}\delta u_y\right)\mathrm{d}x\,\mathrm{d}y$$

corresponding to a variation δu (a function of x and y).

We also assume that $\delta u_x = \partial(\delta u)/\partial x$, so that integration of the second term by parts with respect to x with y fixed gives

$$\begin{aligned}\iint_R \frac{\partial F}{\partial u_x}\delta u_x\,\mathrm{d}x\,\mathrm{d}y &= \int \left(\int \frac{\partial F}{\partial u_x}\delta u_x\,\mathrm{d}x\right)\mathrm{d}y \\ &= \int \left(\left[\frac{\partial F}{\partial u_x}\delta u\right]_{x_1}^{x_2} - \int_{x_1}^{x_2} \frac{\partial}{\partial x}\left(\frac{\partial F}{\partial u_x}\right)\delta u\,\mathrm{d}x\right)\mathrm{d}y,\end{aligned}$$

where x_1 and x_2 are values of x on C at given values of y. Since u is prescribed on C, we must have $\delta u = 0$ on C, and hence

$$\iint_R \frac{\partial F}{\partial u_x}\delta u_x\,\mathrm{d}x\,\mathrm{d}y = -\iint_R \frac{\partial}{\partial x}\left(\frac{\partial F}{\partial u_x}\right)\delta u\,\mathrm{d}x\,\mathrm{d}y.$$

Similarly we get

$$\iint_R \frac{\partial F}{\partial u_y}\delta u_y\,\mathrm{d}x\,\mathrm{d}y = -\iint_R \frac{\partial}{\partial y}\left(\frac{\partial F}{\partial u_y}\right)\delta u\,\mathrm{d}x\,\mathrm{d}y.$$

Consequently

$$\delta I = \iint_R \left(\frac{\partial F}{\partial u} - \frac{\partial}{\partial x}\left(\frac{\partial F}{\partial u_x}\right) - \frac{\partial}{\partial y}\left(\frac{\partial F}{\partial u_y}\right)\right)\delta u\,\mathrm{d}x\,\mathrm{d}y = 0,$$

and this vanishes for arbitrary variations δu in R only if the extremal function u satisfies the second-order partial differential equation (the Euler-Lagrange equation for this problem)

$$\frac{\partial F}{\partial u} - \frac{\partial}{\partial x}\left(\frac{\partial F}{\partial u_x}\right) - \frac{\partial}{\partial y}\left(\frac{\partial F}{\partial u_y}\right) = 0 \tag{6.8}$$

everywhere in R. Some care is again required in interpreting this equation, in regard to the partial derivatives in the second and third terms. In the case of the second term, for example, the partial derivative $\partial F/\partial u_x$ differentiates F formally (as a function of the five variables x, y, u, u_x, u_y) with respect to u_x. In the subsequent differentiation $\partial/\partial x$, however, $\partial F/\partial u_x$ must be regarded as a function of the two variables x and y only.

As an example, consider

$$I = \frac{1}{2}\iint_R \{(u_x)^2 + (u_y)^2\}\,\mathrm{d}x\,\mathrm{d}y, \tag{6.9}$$

where u takes prescribed values on the boundary of R. Then

$$F = \frac{1}{2}(u_x)^2 + \frac{1}{2}(u_y)^2, \quad \frac{\partial F}{\partial u_x} = u_x, \quad \frac{\partial F}{\partial u_y} = u_y,$$

and

$$\frac{\partial F}{\partial u} - \frac{\partial}{\partial x}\left(\frac{\partial F}{\partial u_x}\right) - \frac{\partial}{\partial y}\left(\frac{\partial F}{\partial u_y}\right) = 0 - \frac{\partial}{\partial x}(u_x) - \frac{\partial}{\partial y}(u_y) = -(u_{xx} + u_{yy}).$$

Thus the extremal for the variational problem $\delta I = 0$ must satisfy Laplace's equation

$$\frac{\partial^2 u}{\partial x^2} + \frac{\partial^2 u}{\partial y^2} = 0 \tag{6.10}$$

in R. In this problem I can take only non-negative values, and so is bounded below by zero. This suggests that the extremal must *minimize* I, and in this particular case this is indeed quite straightforward to prove, due to the quadratic nature of F.

The more general case

$$I = \iint_R F(x, y, u, v, u_x, u_y, v_x, v_y)\,\mathrm{d}x\,\mathrm{d}y \tag{6.11}$$

with two functions $u(x, y)$, $v(x, y)$, leads to two Euler-Lagrange equations

$$\frac{\partial F}{\partial u} - \frac{\partial}{\partial x}\left(\frac{\partial F}{\partial u_x}\right) - \frac{\partial}{\partial y}\left(\frac{\partial F}{\partial u_y}\right) = 0,$$
$$\frac{\partial F}{\partial v} - \frac{\partial}{\partial x}\left(\frac{\partial F}{\partial v_x}\right) - \frac{\partial}{\partial y}\left(\frac{\partial F}{\partial v_y}\right) = 0, \tag{6.12}$$

which are two simultaneous second-order partial differential equations for u and v. In Section 6.4 of this chapter these equations will be the focus of interest.

6.3 One-dimensional grid generation

6.3.1 Variational approach

According to the analysis of Section 5.7, the transformation $x = x(\xi)$ between the one-dimensional computational domain $0 \leqslant \xi \leqslant 1$ and the physical domain $a \leqslant x \leqslant b$, where the grid-point density is governed by a weight function $\varphi(\xi)$, must satisfy eqn (5.83). This equation is actually the Euler-Lagrange equation of the variational problem $\delta I = 0$, where the functional I is given by

$$I = \frac{1}{2}\int_0^1 \frac{1}{\varphi(\xi)}\left(\frac{\mathrm{d}x}{\mathrm{d}\xi}\right)^2 \mathrm{d}\xi. \tag{6.13}$$

In fact, because the integrand is reasonably simple, we can reach without difficulty the stronger conclusion that the solution of (5.83), subject to the end conditions $x(0) = a$, $x(1) = b$, *minimizes* I. For suppose that $x = \hat{x}(\xi)$ satisfies eqn (5.83) and the end conditions. If we consider a varied function $x = \hat{x}(\xi) + v(\xi)$, also satisfying the end conditions, so that $v(0) = v(1) = 0$, we obtain, substituting into eqn (6.13),

$$I = \frac{1}{2}\int_a^b \frac{1}{\varphi(\xi)}\left(\frac{\mathrm{d}\hat{x}}{\mathrm{d}\xi} + \frac{\mathrm{d}v}{\mathrm{d}\xi}\right)^2 \mathrm{d}\xi = \hat{I} + \int_0^1 \frac{1}{\varphi(\xi)}\frac{\mathrm{d}\hat{x}}{\mathrm{d}\xi}\frac{\mathrm{d}v}{\mathrm{d}\xi}\,\mathrm{d}\xi + \frac{1}{2}\int_0^1 \frac{1}{\varphi(\xi)}\left(\frac{\mathrm{d}v}{\mathrm{d}\xi}\right)^2 \mathrm{d}\xi,$$

where $\hat{I}$ is the value of I when $x = \hat{x}$. The middle term vanishes since, on integration by parts, we get

$$\left[\frac{1}{\varphi(\xi)}\frac{\mathrm{d}\hat{x}}{\mathrm{d}\xi}v\right]_0^1 - \int_0^1 \frac{\mathrm{d}}{\mathrm{d}\xi}\left(\frac{1}{\varphi(\xi)}\frac{\mathrm{d}\hat{x}}{\mathrm{d}\xi}\right) v\,\mathrm{d}\xi,$$

where the integrated part is zero because of the end conditions satisfied by $v(\xi)$ and the integral is zero because $\hat{x}(\xi)$ satisfies eqn (5.83).

Hence we have the exact equation

$$I = \hat{I} + \frac{1}{2}\int_0^1 \frac{1}{\varphi(\xi)}\left(\frac{\mathrm{d}v}{\mathrm{d}\xi}\right)^2 \mathrm{d}\xi \geqslant \hat{I},$$

since $\varphi(\xi)$ takes positive values. So the minimizing property of $\hat{x}(\xi)$ is established.

On the other hand, if we consider the functional

$$\tilde{I} = \frac{1}{2}\int_0^1 \frac{1}{[\varphi(x)]^2}\left(\frac{\mathrm{d}x}{\mathrm{d}\xi}\right)^2 \mathrm{d}\xi, \tag{6.14}$$

the extremal which makes I stationary must satisfy the Euler-Lagrange equation

$$\begin{aligned}
\frac{\mathrm{d}}{\mathrm{d}\xi}\left(\frac{\partial \tilde{I}}{\partial x_\xi}\right) - \frac{\partial \tilde{I}}{\partial x} &= \frac{\mathrm{d}}{\mathrm{d}\xi}\left(\frac{x_\xi}{[\varphi(x)]^2}\right) + \frac{1}{[\varphi(x)]^3}\frac{\mathrm{d}\varphi}{\mathrm{d}x}\left(\frac{\mathrm{d}x}{\mathrm{d}\xi}\right)^2 \\
&= \frac{1}{[\varphi(x)]^2}\frac{\mathrm{d}^2x}{\mathrm{d}\xi^2} - \frac{2}{[\varphi(x)]^3}\left(\frac{\mathrm{d}x}{\mathrm{d}\xi}\right)^2\frac{\mathrm{d}\varphi}{\mathrm{d}x} + \frac{1}{[\varphi(x)]^3}\frac{\mathrm{d}\varphi}{\mathrm{d}x}\left(\frac{\mathrm{d}x}{\mathrm{d}\xi}\right)^2 \\
&= \frac{1}{[\varphi(x)]^2}\frac{\mathrm{d}^2x}{\mathrm{d}\xi^2} - \frac{1}{[\varphi(x)]^3}\frac{\mathrm{d}\varphi}{\mathrm{d}x}\left(\frac{\mathrm{d}x}{\mathrm{d}\xi}\right)^2 = 0,
\end{aligned}$$

giving finally

$$\frac{d^2x}{d\xi^2} - \frac{1}{\varphi(x)}\frac{d\varphi}{dx}\left(\frac{dx}{d\xi}\right)^2 = 0. \tag{6.15}$$

This is the same equation as (5.92), so the transformation $x = x(\xi)$ which gives a grid according to the weight function $\varphi(x)$ makes I stationary (although minimizing properties are now not so easy to prove).

Exercise 1. As an alternative derivation of the Euler-Lagrange equation, use the fact that the integrand in eqn (6.14) is independent of ξ to show directly from eqn (6.4) that

$$\frac{1}{[\varphi(x)]^2}\left(\frac{dx}{d\xi}\right)^2 = const.$$

(differentiation of this equation yielding eqn (6.15)), and hence $dx/d\xi$ is proportional to $\varphi(x)$.

Returning to eqn (6.13), the variational approach suggests a way of discretizing the problem. Taking $x_0 = a$ and $x_m = b$ as given end points and $x_1, x_2, \ldots, x_{m-1}$ as the unknown interior points of the grid, a representation of the integral I, with a uniform division of the range of integration into points $\xi_i = i/m$, $i = 0, 1, 2, \ldots, m$, is

$$I \backsimeq \sum_{i=1}^{m} \frac{1}{2\varphi_{i-\frac{1}{2}}}\left(\frac{x_i - x_{i-1}}{\Delta\xi}\right)^2 \Delta\xi, \tag{6.16}$$

where $\varphi(\xi)$ is evaluated at mid-points of the sub-intervals; $\varphi_{i-\frac{1}{2}} = \varphi(\xi_i - 1/2m)$.

Hence we have a problem of ordinary calculus in which, instead of minimizing a functional, we want to minimize the function (since $\Delta\xi$ is a constant)

$$\begin{aligned} f(x_1, x_2, \ldots, x_{m-1}) &= \sum_{i=1}^{m} \frac{(x_i - x_{i-1})^2}{2\varphi_{i-\frac{1}{2}}} \\ &= \frac{(x_1 - x_0)^2}{2\varphi_{\frac{1}{2}}} + \frac{(x_2 - x_1)^2}{2\varphi_{\frac{3}{2}}} + \cdots + \frac{(x_m - x_{m-1})^2}{2\varphi_{m-\frac{1}{2}}}. \end{aligned} \tag{6.17}$$

Equating the partial derivatives $\partial f/\partial x_j$ to zero gives the $(m-1)$ equations

$$\frac{(x_j - x_{j-1})}{\varphi_{j-\frac{1}{2}}} - \frac{(x_{j+1} - x_j)}{\varphi_{j+\frac{1}{2}}} = 0, \quad j = 1, 2, \ldots, (m-1). \tag{6.18}$$

This is a system of $(m-1)$ linear equations for $(m-1)$ unknowns $x_1, x_2, \ldots, x_{m-1}$, which we encountered in Chapter 5 at eqn (5.97). An equivalent system is clearly

$$\frac{(x_1 - x_0)}{\varphi_{\frac{1}{2}}} = \frac{(x_2 - x_1)}{\varphi_{\frac{3}{2}}} = \cdots = \frac{(x_m - x_{m-1})}{\varphi_{m-\frac{1}{2}}} = K', \tag{6.19}$$

for some constant K'. These equations are consistent with the condition (5.81) that the distance between grid points is proportional to the value of the weight function at the mid-point of the corresponding ξ-interval.

6.3.2 Dynamic adaptation

Ideally the numerical solution of the hosted equations should be dynamically coupled to the process of grid generation, so that, in regions of the physical domain where large gradients (or higher-order spatial derivatives) of the dependent variables are found to occur, grid density may be increased to allow more accurate resolution. In two or three dimensions re-distribution of grid points can result in distortion of grid cells, but in one dimension the task is simpler.

Here we present a grid generation technique which is equivalent to that defined by the transformation from computational space to physical space given by eqn (5.90). This is itself equivalent to the 'Equidistribution Principle' proposed by Boor (1974), which requires that the errors in the numerical solution of a problem should be uniformly distributed throughout the solution domain. Given a set of grid points $a = x_0$, $x_1, \ldots, x_{n-1}, x_n = b$, this may be expressed as

$$\int_{x_i}^{x_{i+1}} W(x)\,\mathrm{d}x = K, \quad \text{constant}, \; i = 0, 1, \ldots, (n-1), \tag{6.20}$$

where the weight function $W(x)$ varies with the error at a point x. Clearly $W(x)$ is equivalent to the *reciprocal* of the weight function $\varphi(x)$ as given in eqn (5.90) through the modified form of eqn (6.20) given by

$$W(x)\frac{\mathrm{d}x}{\mathrm{d}\xi} = K, \quad \text{with } x(0) = a, \quad x(1) = b, \tag{6.21}$$

so that with an evenly distributed grid in the ξ interval $0 < \xi < 1$ the distance between corresponding points in the x interval $a < x < b$ is *inversely* proportional to the local value of $W(x)$.

Integrating eqn (6.21) gives

$$\int_a^b W(x)\,\mathrm{d}x = K\int_0^1 \mathrm{d}\xi = K, \tag{6.22}$$

and

$$\int_a^x W(x)\,\mathrm{d}x = K\int_0^\xi \mathrm{d}\xi = K\xi,$$

showing that the transformation from the physical domain to the computational domain is given by

$$\xi(x) = \frac{\displaystyle\int_a^x W(x)\,\mathrm{d}x}{\displaystyle\int_a^b W(x)\,\mathrm{d}x}. \tag{6.23}$$

In one-dimensional adaptive grid generation the weight function will in practice be taken to depend directly on the derivatives of the solution to the hosted equation, say $u(x)$ in the physical domain, rather than on some explicit representation of the error in the solution. Given that the largest numerical errors tend to occur where the lowest derivatives have high values, we want $W(x)$ to have a high value in regions in particular where the first derivative $u_x (= du/dx)$ has a high value so that grid-points locally are close together. The most commonly used weight functions are

$$W(x) = \begin{cases} u_x(x) \\ \sqrt{1 + \alpha^2 (u_x)^2} \\ (1 + \beta\kappa)\sqrt{1 + \alpha^2 (u_x)^2}, \end{cases} \tag{6.24}$$

similar to the expressions given in eqns (2.58), (2.59), and (2.60), where here κ is the curvature of the solution, given by

$$\kappa = \frac{|u_{xx}|}{[1 + (u_x)^2]^{3/2}} \tag{6.25}$$

as in eqn (2.17), and α, β are parameters to be chosen by the user.

Note that with $W(x) = u_x$, eqn (6.21) becomes

$$\frac{du}{dx}\frac{dx}{d\xi} = \frac{du}{d\xi} = K, \tag{6.26}$$

which has the consequence that the increments in the values of u between the equally-spaced points on the ξ interval *and* between the corresponding unequally-spaced grid points in the x interval remain the same. Figure 6.1 shows an example where a decreasing gradient of u with x results in a higher density of grid points at low values of x where u_x is higher.

This choice of weight function may have the disadvantage that the grid-point spacing is too large when u_x is small. An alternative is

$$W(x) = \sqrt{1 + (u_x)^2} \tag{6.27}$$

with the particular choice of $\alpha = 1$ in (6.24), which approximates to u_x when u_x is large. This choice can be neatly interpreted in the ux plane in terms of the distance s between points on the solution curve, since

$$ds = \sqrt{(du)^2 + (dx)^2} = \sqrt{1 + (u_x)^2}\, dx = W(x)\, dx,$$

so that eqn (6.21) becomes

$$\frac{ds}{dx}\frac{dx}{d\xi} = \frac{ds}{d\xi} = K. \tag{6.28}$$

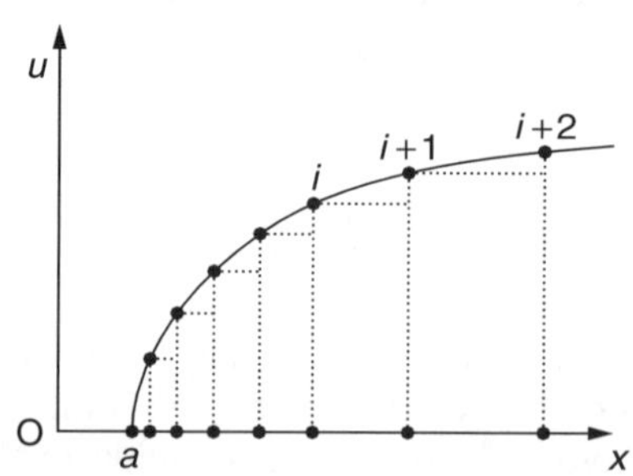

Fig. 6.1 Adaptation in one dimension.

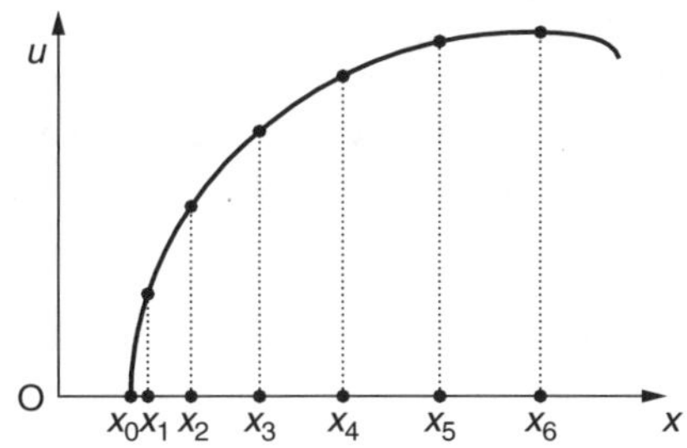

Fig. 6.2 Equidistant grid.

This has the consequence that the increments in *arc-length* on the solution curve $u(x)$ between points corresponding to the equally-spaced ξ points and the unequally-spaced x points remain constant. See Fig. 6.2 for an illustration of a typical grid. Such a grid in the physical domain which produces an equidistribution of arc-length along the solution curve is sometimes referred to as *equidistant*. Clearly in parts of the physical domain where u_x is small the grid points will tend to be uniformly distributed.

When the parameter α is incorporated to give the second choice in eqn (6.24), it is not difficult to see that the grid still has an equidistant property, provided that the solution curve is plotted with u as a function of x/α. For then we can write arc-length as

$$\mathrm{d}\tilde{s} = \sqrt{1 + \left(\frac{\mathrm{d}u}{\mathrm{d}(x/\alpha)}\right)^2}\,\mathrm{d}(x/\alpha) = \sqrt{1 + \alpha^2 (u_x)^2}\,\mathrm{d}(x/\alpha) = W(x)\,\mathrm{d}(x/\alpha),$$

giving from eqn (6.21)

$$\frac{\mathrm{d}\tilde{s}}{\mathrm{d}(x/\alpha)}\frac{\mathrm{d}x}{\mathrm{d}\xi} = \alpha\frac{\mathrm{d}\tilde{s}}{\mathrm{d}\xi} = K.$$

With $\alpha > 1$, the solution curve is compressed, and the effect on regions where solution gradients are high is to increase the density of grid-points compared with the choice $\alpha = 1$ above. In flatter regions of the solution curve the effect is to widen the grid-spacing.

The weight functions presented so far may not be satisfactory close to extrema in the solution curve where $u_x = 0$. Such a neighbourhood may be treated as if it were flat locally, and the density of grid points may be insufficient to resolve the behaviour of the solution adequately. Incorporating the curvature term as in the third possible weight function of eqn (6.24) tends to reduce the spacing between grid points where the curvature of the solution curve is high. Appropriate adjustment of the parameters α, β can help to achieve a successful balance between increasing grid-density in regions of high solution gradient and of high curvature.

6.3.3 Space-curves

A fundamental aspect of generating a grid on a space-curve $\mathbf{r} = \mathbf{r}(\chi)$ was presented in Section 2.4. in terms of obtaining a re-parameterization $\chi \to \xi$, whereby a uniformly-spaced set of points in ξ-space (the computational space $0 \leqslant \xi \leqslant 1$) maps to a non-uniformly spaced set of points in χ-space (the parametric space $0 \leqslant \chi \leqslant 1$), which generate a suitable set of points $\mathbf{r}(\chi)$ on the curve. Here we show that mappings $\chi(\xi)$ may be obtained as the solutions of Euler-Lagrange equations for some variational problems.

An appropriate functional incorporating a weight function $\varphi(\mathbf{r})$ in physical space suggested by eqn (6.14) would be

$$I = \frac{1}{2}\int_0^1 \frac{g_{11}}{[\varphi(\mathbf{r})]^2}\,\mathrm{d}\xi,$$

where $g_{11} = (x_\xi)^2 + (y_\xi)^2 + (z_\xi)^2$. We may regard $\mathbf{r}$ and hence φ as a function of the parameter χ, and consider the slightly more general functional

$$I = \frac{1}{2}\int_0^1 \frac{H(g_{11})}{[\varphi(\chi)]^2}\mathrm{d}\xi, \tag{6.29}$$

for some function H. By eqn (2.54), we can write

$$I = \frac{1}{2}\int_0^1 \frac{H(\tilde{g}_{11}[\chi_\xi]^2)}{[\varphi(\chi)]^2}\mathrm{d}\xi. \tag{6.30}$$

where $\tilde{g}_{11} = (x_\chi)^2 + (y_\chi)^2 + (z_\chi)^2$, which may be regarded as a function of χ.

Thus the integrand is not explicitly dependent on ξ, and we can call on eqn (6.4) to write the Euler-Lagrange equation as

$$\frac{\mathrm{d}}{\mathrm{d}\xi}\left[\frac{H}{\varphi^2} - \chi_\xi \frac{1}{\varphi^2}\frac{\partial}{\partial \chi_\xi} H(\tilde{g}_{11}[\chi_\xi]^2)\right] = 0,$$

from which we obtain

$$\frac{\mathrm{d}}{\mathrm{d}\xi}\left[\frac{H - 2(\chi_\xi)^2\tilde{g}_{11}H'(\tilde{g}_{11}[\chi_\xi]^2)}{\varphi^2}\right] = 0. \tag{6.31}$$

In the particular case where $H = g_{11}$, this equation reduces to

$$\frac{\mathrm{d}}{\mathrm{d}\xi}\left[\frac{\tilde{g}_{11}[\chi_\xi]^2 - 2\tilde{g}_{11}[\chi_\xi]^2}{\varphi^2}\right] = -\frac{\mathrm{d}}{\mathrm{d}\xi}\left[\frac{\tilde{g}_{11}[\chi_\xi]^2}{\varphi^2}\right] = 0, \tag{6.32}$$

which yields

$$\frac{\tilde{g}_{11}[\chi_\xi]^2}{\varphi^2} = \frac{g_{11}}{\varphi^2} = const. \tag{6.33}$$

The end conditions are of course $\chi(0) = 0$, $\chi(1) = 1$.

For the special case $\varphi = 1$, we obtain $g_{11} = const.$ and hence also $\sqrt{g_{11}} = const.$ By eqn (2.47), equally spaced points on the ξ interval will produce a set of points on the curve with equal arc-length between points.

On the accompanying disk the subdirectory *Book/one.d.gds* contains a file *curve.SOR.f* which solves eqn (6.32) in the case of a plane parabolic curve defined by

$$\begin{cases} x = \chi \\ y = \chi(1-\chi) \end{cases}. \tag{6.34}$$

Here we have

$$\tilde{g}_{11} = 1 + (1 - 2\chi)^2 = 2 - 4\chi + 4\chi^2. \tag{6.35}$$

More generally, the program discretizes the non-linear eqn (6.32) as

$$\frac{\left(\dfrac{\tilde{g}_{11}\chi_\xi^2}{\varphi^2}\right)_{i+\frac{1}{2}} - \left(\dfrac{\tilde{g}_{11}\chi_\xi^2}{\varphi^2}\right)_{i-\frac{1}{2}}}{\Delta\xi} = 0,$$

and takes

$$\left(\frac{\tilde{g}_{11}\chi_\xi^2}{\varphi^2}\right)_{i+\frac{1}{2}} = \left(\frac{\tilde{g}_{11}\chi_\xi}{\varphi^2}\right)_{i+\frac{1}{2}} (\chi_\xi)_{i+\frac{1}{2}} = \left(\frac{\tilde{g}_{11}\chi_\xi}{\varphi^2}\right)_{i+\frac{1}{2}} \left(\frac{\chi_{i+1} - \chi_i}{\Delta\xi}\right)$$

with, similarly,

$$\left(\frac{\tilde{g}_{11}\chi_\xi^2}{\varphi^2}\right)_{i-\frac{1}{2}} = \left(\frac{\tilde{g}_{11}\chi_\xi}{\varphi^2}\right)_{i-\frac{1}{2}} \left(\frac{\chi_i - \chi_{i-1}}{\Delta\xi}\right).$$

Thus, re-arranging terms, we have

$$\frac{1}{\Delta\xi}\left\{\left(\frac{\tilde{g}_{11}\chi_\xi}{\varphi^2}\right)_{i-\frac{1}{2}} \chi_{i-1} - \left[\left(\frac{\tilde{g}_{11}\chi_\xi}{\varphi^2}\right)_{i+\frac{1}{2}} + \left(\frac{\tilde{g}_{11}\chi_\xi}{\varphi^2}\right)_{i-\frac{1}{2}}\right]\chi_i \right.$$
$$\left. + \left(\frac{\tilde{g}_{11}\chi_\xi}{\varphi^2}\right)_{i+\frac{1}{2}} \chi_{i+1}\right\} = 0. \tag{6.36}$$

This gives a tridiagonal matrix equation for χ_i, $i = 1, 2, \ldots, (m-1)$, subject to the boundary conditions $\chi_0 = 0$, $\chi_1 = 1$, which may be solved by the Thomas Algorithm or by SOR. The elements of the matrix 'lag' the solution at each iteration. The idea here is that we guess an initial distribution for $\chi(\xi)$, for example a uniform distribution $\chi = \xi$, with corresponding values of the matrix elements to start the iteration. Then, having solved the matrix equation, we evaluate the updated values of the matrix elements according to the new values of χ and repeat the process.

The program on the disk evaluates the lagged elements as

$$\left(\frac{\tilde{g}_{11}\chi_\xi}{\varphi^2}\right)_{i+\frac{1}{2}} = \left(\frac{\tilde{g}_{11}}{\varphi^2}\right)_{i+\frac{1}{2}} (\chi_\xi)_{i+\frac{1}{2}} = \left(\frac{\tilde{g}_{11}}{\varphi^2}\right)_{i+\frac{1}{2}} \left(\frac{\chi_{i+1} - \chi_i}{\Delta\xi}\right),$$
$$\left(\frac{\tilde{g}_{11}\chi_\xi}{\varphi^2}\right)_{i-\frac{1}{2}} = \left(\frac{\tilde{g}_{11}}{\varphi^2}\right)_{i-\frac{1}{2}} (\chi_\xi)_{i-\frac{1}{2}} = \left(\frac{\tilde{g}_{11}}{\varphi^2}\right)_{i-\frac{1}{2}} \left(\frac{\chi_i - \chi_{i-1}}{\Delta\xi}\right),$$

with simple averaging to give

$$\left(\frac{\tilde{g}_{11}}{\varphi^2}\right)_{i+\frac{1}{2}} = \frac{1}{2}\left[\left(\frac{\tilde{g}_{11}}{\varphi^2}\right)_{i+1} + \left(\frac{\tilde{g}_{11}}{\varphi^2}\right)_i\right],$$
$$\left(\frac{\tilde{g}_{11}}{\varphi^2}\right)_{i-\frac{1}{2}} = \frac{1}{2}\left[\left(\frac{\tilde{g}_{11}}{\varphi^2}\right)_i + \left(\frac{\tilde{g}_{11}}{\varphi^2}\right)_{i-1}\right].$$

For example, in the case considered here, from eqns (6.34) and (6.35), we would have

$$\left(\frac{\tilde{g}_{11}}{\varphi^2}\right)_i = \frac{1}{\varphi_i^2}(2 - 4\chi_i + 4\chi_i^2). \tag{6.37}$$

The presence of the user-specified weight function φ implies that the grid can be generated adaptively, so that grid generation can be coupled dynamically to the solution of the hosted equations. For example, the gradient of field variables can be used as a weight function to ensure that the grid-density increases where there are sharp changes in these variables.

The program on the file employs SOR, and takes constant $\varphi = 1$. Having obtained the solution $\chi(\xi)$, we can find the set of χ points corresponding to a uniformly spaced set of ξ points, and then from eqns (6.34) calculate the co-ordinates of the corresponding points on the curve. According to the observations above for constant φ these points should be equally spaced (in terms of arc-length) along the curve.

It may be pointed out that for this case ($\varphi = 1$) eqns (6.32) admit an analytic solution which gives χ implicitly in terms of ξ, and in principle this may be compared with the numerical one to assess accuracy.

6.4 Two-dimensional grids

In the variational approach grid properties such as grid-node spacing, cell area, and orthogonality can be controlled by minimizing an appropriate functional. In fact these properties can be controlled simultaneously by choosing a functional which is a weighted combination of specific functionals which control individual properties. The appropriate choice of weight functions may, however, not be clear-cut, since the question of which grid generator is best is not as readily answered in two and three-dimensional problems as it is in one dimension. But variational methods allow intuition to be used in selecting functionals appropriate to a particular problem.

Here we shall be mainly concerned with 'length' (L) functionals, 'area' (A) functionals, and 'orthogonality' (O) functionals, and various combinations of these. These functionals will be expressed in terms of the transformation $x = x(\xi, \eta)$, $y = y(\xi, \eta)$ from a computational domain to a two-dimensional physical domain (or its inverse), with a uniform grid in computational space transforming to a curvilinear co-ordinate grid in physical space. By eqn (1.42), the length of elements of grid lines in physical space, given a fixed uniform grid in computational space, is governed by the terms g_{11} and g_{22} of the covariant metric tensor, while the angle between intersecting grid lines is related to g_{12}. In a completely orthogonal grid g_{12} would be zero everywhere. The area of a grid cell, using eqns (1.44), (1.159), and (1.160), depends on the local value of the Jacobian $\sqrt{g}$ of the transformation. So here we discuss variational problems where the functionals depend on g_{11}, g_{22} g_{12}, and $\sqrt{g}$. The associated Euler-Lagrange equations then provide the corresponding differential model of grid generation. (On the floppy disk included with this book we present a number of numerical examples with various geometries and varying degrees of complexity to show how to solve the Euler-Lagrange equations associated with certain functionals, and it is hoped that this will help readers to write their own programs to generate grids for different geometries to match particular requirements.)

6.4.1 The *L*-functional and the Winslow model

We begin with the observation that according to eqns (6.9), (6.10), and (6.12), the Euler-Lagrange equations for the functional

$$I = \frac{1}{2}\iint_{R'} \{(x_\xi)^2 + (x_\eta)^2 + (y_\xi)^2 + (y_\eta)^2\}\,\mathrm{d}\xi\,\mathrm{d}\eta, \tag{6.38}$$

where R' is the computational domain $0 \leqslant \xi \leqslant 1$, $0 \leqslant \eta \leqslant 1$, are the pair of Laplace's equations

$$\begin{aligned}\frac{\partial^2 x}{\partial \xi^2} + \frac{\partial^2 x}{\partial \eta^2} &= 0\\ \frac{\partial^2 y}{\partial \xi^2} + \frac{\partial^2 y}{\partial \eta^2} &= 0.\end{aligned} \tag{6.39}$$

This is the differential model presented in eqns (5.5). So we see that the associated transformation minimizes the functional I, which by eqns (1.158) may be written as

$$I = \frac{1}{2}\iint_{R'} (g_{11} + g_{22})\,\mathrm{d}\xi\,\mathrm{d}\eta. \tag{6.40}$$

In view of the remarks above on the properties of g_{11} and g_{22}, we call this the L(length)-*functional*. Equations (6.39) are to be solved subject to boundary conditions such that x and y are specified on the four (b,t,l,r) boundaries of the physical domain. Thus

$$\begin{aligned}x(\xi,0) &= x_b(\xi); & x(\xi,1) &= x_t(\xi)\\ x(0,\eta) &= x_l(\eta); & x(1,\eta) &= x_r(\eta)\\ y(\xi,0) &= y_b(\xi); & y(\xi,1) &= y_t(\xi)\\ y(0,\eta) &= y_l(\eta); & y(1,\eta) &= y_r(\eta).\end{aligned}$$

As mentioned in Section 5.1, this transformation can generate satisfactory grids, but the Jacobian is not guaranteed to be non-zero, and folding of the grid can occur, particularly when the physical domain is not convex.

If, by contrast, we formulate the variational problem

$$\delta\left\{\frac{1}{2}\iint_R \left[\left(\frac{\partial \xi}{\partial x}\right)^2 + \left(\frac{\partial \xi}{\partial y}\right)^2 + \left(\frac{\partial \eta}{\partial x}\right)^2 + \left(\frac{\partial \eta}{\partial y}\right)^2\right]\mathrm{d}x\,\mathrm{d}y\right\} = 0, \tag{6.41}$$

where the integration is taken over the physical domain R and we seek the transformation $\xi = \xi(x, y)$, $\eta = \eta(x, y)$ which minimizes the integral, then the Euler-Lagrange equations must be the pair of Laplace's equations

$$\begin{aligned}\frac{\partial^2 \xi}{\partial x^2} + \frac{\partial^2 \xi}{\partial y^2} &= 0\\ \frac{\partial^2 \eta}{\partial x^2} + \frac{\partial^2 \eta}{\partial y^2} &= 0,\end{aligned} \tag{6.42}$$

and this is the Winslow model, eqns (5.1). Thus transformations satisfying the Winslow model are extremals of the functional (6.41), which may be written, using (1.158), (1.162), and (1.163), as

$$\frac{1}{2}\iint_R (g^{11} + g^{22})\,\mathrm{d}x\,\mathrm{d}y. \tag{6.43}$$

Transforming the double integral to the computational domain gives

$$\frac{1}{2}\iint_{R'} \left(\frac{g_{22} + g_{11}}{g}\right)(\sqrt{g}\,\mathrm{d}\xi\,\mathrm{d}\eta) = \frac{1}{2}\iint_{R'} \left(\frac{g_{11} + g_{22}}{\sqrt{g}}\right)\mathrm{d}\xi\,\mathrm{d}\eta. \tag{6.44}$$

The Winslow transformation, mapping a square region to R, always gives an unfolded grid, except for problems which can arise due to the nature of the discretization process (i.e. truncation errors). In particular, this can happen when R is not convex.

6.4.2 The weighted *L*-functional

The concept of a weight function in one dimension can be generalized to higher dimensions. Suppose that we wish the lengths of grid cells in the ξ and η directions at a point corresponding to (ξ, η) in the computational plane to be proportional to the positive weight functions $\varphi(\xi, \eta)$, $\psi(\xi, \eta)$, respectively. Then a functional analogous to (6.13) is given by

$$\begin{aligned} I_L &= \frac{1}{2}\iint_{R'} \left(\frac{(x_\xi)^2 + (y_\xi)^2}{\varphi(\xi, \eta)} + \frac{(x_\eta)^2 + (y_\eta)^2}{\psi(\xi, \eta)}\right)\mathrm{d}\xi\,\mathrm{d}\eta \\ &= \frac{1}{2}\iint_{R'} \left(\frac{g_{11}}{\varphi(\xi, \eta)} + \frac{g_{22}}{\psi(\xi, \eta)}\right)\mathrm{d}\xi\,\mathrm{d}\eta. \end{aligned} \tag{6.45}$$

The two Euler-Lagrange equations for the variational problem $\delta I = 0$ are

$$\begin{aligned} \frac{\partial}{\partial \xi}\left(\frac{1}{\varphi}\frac{\partial x}{\partial \xi}\right) + \frac{\partial}{\partial \eta}\left(\frac{1}{\psi}\frac{\partial x}{\partial \eta}\right) &= 0 \\ \frac{\partial}{\partial \xi}\left(\frac{1}{\varphi}\frac{\partial y}{\partial \xi}\right) + \frac{\partial}{\partial \eta}\left(\frac{1}{\psi}\frac{\partial y}{\partial \eta}\right) &= 0, \end{aligned} \tag{6.46}$$

which we can write as the single vector equation

$$\frac{\partial}{\partial \xi}\left(\frac{1}{\varphi}\frac{\partial \mathbf{r}}{\partial \xi}\right) + \frac{\partial}{\partial \eta}\left(\frac{1}{\psi}\frac{\partial \mathbf{r}}{\partial \eta}\right) = \mathbf{0}, \tag{6.47}$$

where $\mathbf{r} = \binom{x}{y}$.

Carrying out the partial differentiations gives the set of equations

$$A_{11}\mathbf{r}_{\xi\xi} + A_{12}\mathbf{r}_{\xi\eta} + A_{22}\mathbf{r}_{\eta\eta} + \mathbf{S} = \mathbf{0}, \tag{6.48}$$

where A_{11}, A_{12}, A_{22} are the symmetric 2×2 matrices

$$A_{11} = \frac{1}{\varphi}\begin{pmatrix} 1 & 0 \\ 0 & 1 \end{pmatrix}, \quad A_{12} = \begin{pmatrix} 0 & 0 \\ 0 & 0 \end{pmatrix}, \quad A_{22} = \frac{1}{\psi}\begin{pmatrix} 1 & 0 \\ 0 & 1 \end{pmatrix}, \tag{6.49}$$

and

$$\mathbf{S} = -\begin{pmatrix} \frac{1}{\varphi^2}\varphi_\xi x_\xi + \frac{1}{\psi^2}\psi_\eta x_\eta \\ \frac{1}{\varphi^2}\varphi_\xi y_\xi + \frac{1}{\psi^2}\psi_\eta y_\eta \end{pmatrix} = -\begin{pmatrix} \frac{1}{\varphi^2}x_\xi & \frac{1}{\psi^2}x_\eta \\ \frac{1}{\varphi^2}y_\xi & \frac{1}{\psi^2}y_\eta \end{pmatrix}\begin{pmatrix} \varphi_\xi \\ \psi_\eta \end{pmatrix} \tag{6.50}$$

and partial differentiation (including that of the vector $\mathbf{r}$) is denoted by suffixes.

6.4.3 The weighted area-functional

The area-functional generally provides a more satisfactory grid than the length-functional. The objective here is to generate a grid such that the area of grid-cells is proportional to some given positive weight function $\varphi(\xi, \eta)$ in computational space. Given that the area of grid-cells in the physical domain corresponding to uniform grid-cells of fixed area in the computational domain is proportional to $\sqrt{g}$, a natural way to define a functional by analogy with the length functionals above would be to put

$$I_A = \frac{1}{2}\iint_{R'} \frac{g}{\varphi(\xi, \eta)} \mathrm{d}\xi\, \mathrm{d}\eta = \frac{1}{2}\iint_{R'} \frac{J^2}{\varphi(\xi, \eta)} \mathrm{d}\xi\, \mathrm{d}\eta, \tag{6.51}$$

where the Jacobian $J = (x_\xi y_\eta - x_\eta y_\xi)$.

Exercise 2. Show that the Euler-Lagrange equations in this case can be written as

$$\left(\frac{J\mathbf{r}_\eta}{\varphi}\right)_\xi - \left(\frac{J\mathbf{r}_\xi}{\varphi}\right)_\eta = \mathbf{0}.$$

It may be seen that on further differentiation these equations reduce to

$$(J/\varphi)_\xi \mathbf{r}_\eta - (J/\varphi)_\eta \mathbf{r}_\xi = \mathbf{0}. \tag{6.52}$$

Exercise 3. Show that this is equivalent to

$$A_{11}\mathbf{r}_{\xi\xi} + A_{12}\mathbf{r}_{\xi\eta} + A_{22}\mathbf{r}_{\eta\eta} + \mathbf{S} = \mathbf{0}$$

as in (6.48), with

$$A_{11} = \begin{pmatrix} (y_\eta)^2 & -x_\eta y_\eta \\ -x_\eta y_\eta & (x_\eta)^2 \end{pmatrix}, \quad A_{12} = \begin{pmatrix} -2y_\xi y_\eta & x_\xi y_\eta + x_\eta y_\xi \\ x_\xi y_\eta + x_\eta y_\xi & -2x_\xi x_\eta \end{pmatrix},$$

$$A_{22} = \begin{pmatrix} (y_\xi)^2 & -x_\xi y_\xi \\ -x_\xi y_\xi & (x_\xi)^2 \end{pmatrix}, \quad \mathbf{S} = -\frac{J}{\varphi}\begin{pmatrix} y_\eta & -y_\xi \\ -x_\eta & x_\xi \end{pmatrix}\begin{pmatrix} \varphi_\xi \\ \varphi_\eta \end{pmatrix}. \tag{6.53}$$

6.4.4 Orthogonality-functional

An orthogonality (O)-functional without weight function is

$$I_O = \frac{1}{2}\iint_{R'} (g_{12})^2\, \mathrm{d}\xi\, \mathrm{d}\eta, \tag{6.54}$$

where $g_{12} = (x_\xi x_\eta + y_\xi y_\eta)$. Like the previous functionals, this takes only non-negative values. It would take a minimum value of zero only for a completely orthogonal grid (if one exists), with $g_{12} = 0$ everywhere.

Exercise 4. Show that the Euler-Lagrange equations may be expressed as

$$(g_{12}\mathbf{r}_\eta)_\xi + (g_{12}\mathbf{r}_\xi)_\eta = \mathbf{0} \tag{6.55}$$

and also in the form (6.48), with

$$A_{11} = \begin{pmatrix} (x_\eta)^2 & x_\eta y_\eta \\ x_\eta y_\eta & (y_\eta)^2 \end{pmatrix}, \quad A_{12} = \begin{pmatrix} (4x_\xi x_\eta + 2y_\xi y_\eta) & (x_\xi y_\eta + x_\eta y_\xi) \\ (x_\xi y_\eta + x_\eta y_\xi) & (4y_\xi y_\eta + 2x_\xi x_\eta) \end{pmatrix},$$

$$A_{22} = \begin{pmatrix} (x_\xi)^2 & x_\xi y_\xi \\ x_\xi y_\xi & (y_\xi)^2 \end{pmatrix}, \quad \mathbf{S} = \begin{pmatrix} 0 \\ 0 \end{pmatrix}. \tag{6.56}$$

If the resulting minimum (or lower bound) for I_O is greater than zero, then clearly we are as close as we can get to an orthogonal grid in a 'least-squares' sense. With no weight function involved, the 'metric' g_{12} may be said to be *equidistributed*. However, in practice it is found that even equidistribution may fail to produce a satisfactory grid, particularly in non-convex physical domains where a perfectly orthogonal grid may fail to exist, and the functional (6.54) on its own is not recommended.

6.4.5 Combination of functionals

A weighted linear combination of length-, area-, and orthogonality-functional may be formulated to achieve a compromise between the various properties controlled by these functionals separately. So we may consider

$$I = w_L I_L + w_A I_A + w_O I_O, \tag{6.57}$$

where the weights w_L, w_A, w_O, are non-negative constants satisfying $w_L + w_A + w_O = 1$. These weights can be used in numerical experiments to obtain the combination of functionals that produces the most satisfactory grid. For example, the choice $w_L = 0.1$, $w_A = 0.9$, $w_O = 0$ of length and area functionals generally gives smooth, unfolded grids, this combination usually overcoming the separate limitations of lack of smoothness in the area-functional and a tendency for folding in the length-functional. However, in certain geometries the resulting grid may suffer from severe skewness and/or a failure to converge. In such cases there is a need to experiment so as to 'tune' the weight parameters to obtain a satisfactory grid.

The area-orthogonality-functional

The choice of weight parameters $w_L = 0$, $w_A = 0.5$, $w_O = 0.5$ of area and orthogonality functionals leads to a robust automatic grid generator known as the Area-Orthogonality(AO)-functional. A weight function $\varphi(\xi, \eta)$ can also be employed, so that we get a functional

$$I_{AO} = \frac{1}{2} \iint_{R'} \frac{1}{2} \left(\frac{g}{\varphi} + \frac{(g_{12})^2}{\varphi} \right) \mathrm{d}\xi \, \mathrm{d}\eta. \tag{6.58}$$

Since $g = g_{11}g_{22} - (g_{12})^2$, this is equal to

$$I_{AO} = \frac{1}{4}\iint_{R'} \frac{g_{11}g_{22}}{\varphi} \mathrm{d}\xi\, \mathrm{d}\eta. \tag{6.59}$$

Exercise 5. Show that the Euler-Lagrange equations may be expressed as

$$\left(\frac{g_{22}\mathbf{r}_\xi}{\varphi}\right)_\xi + \left(\frac{g_{11}\mathbf{r}_\eta}{\varphi}\right)_\eta = 0, \tag{6.60}$$

and also in the form (6.48), with

$$A_{11} = \begin{pmatrix} (x_\eta)^2 + (y_\eta)^2 & 0 \\ 0 & (x_\xi)^2 + (y_\xi)^2 \end{pmatrix},$$

$$A_{12} = \begin{pmatrix} 4x_\xi x_\eta & 2(x_\xi y_\eta + x_\eta y_\xi) \\ 2(x_\xi y_\eta + x_\eta y_\xi) & 4x_\xi x_\eta \end{pmatrix}$$

$$A_{22} = \begin{pmatrix} (x_\xi)^2 + (y_\xi)^2 & 0 \\ 0 & (x_\eta)^2 + (y_\eta)^2 \end{pmatrix},$$

$$\mathbf{S} = -\frac{1}{\varphi}\begin{pmatrix} x_\xi & x_\eta \\ y_\xi & y_\eta \end{pmatrix}\begin{pmatrix} \varphi_\xi((x_\eta)^2 + (y_\eta)^2) \\ \varphi_\eta((x_\xi)^2 + (y_\xi)^2) \end{pmatrix}. \tag{6.61}$$

The AO-eqns (6.48) with (6.61) are quasilinear and coupled, and can be solved iteratively. Unweighted AO grids (with $\varphi(\xi, \eta) = 1$) are generally smooth and unfolded, being near to orthogonality and having near-uniform areas. The smoothness of the grids is rather unexpected, since the generating equations are not elliptic.

A variation on the AO-functional is the 'AO-squared functional', which is given in unweighted form by

$$I_{AO\text{-}squared} = \iint_{R'} (g_{11}g_{22})^2\, \mathrm{d}\xi\, \mathrm{d}\eta. \tag{6.62}$$

The Euler-Lagrange equations whose solutions minimize this functional generally produce very good grids.

6.4.6 Other orthogonality functionals

Orthogonality two functional

This functional is given by

$$I_{O,2} = \iint_{R'} \frac{(g_{12})^2}{g_{11}g_{22}} \mathrm{d}\xi\, \mathrm{d}\eta. \tag{6.63}$$

Since

$$\frac{(g_{12})^2}{g_{11}g_{22}} = \frac{(\mathbf{r}_\xi \cdot \mathbf{r}_\eta)^2}{(\mathbf{r}_\xi \cdot \mathbf{r}_\xi)(\mathbf{r}_\eta \cdot \mathbf{r}_\eta)} = \left(\frac{\mathbf{r}_\xi}{|\mathbf{r}_\xi|} \cdot \frac{\mathbf{r}_\eta}{|\mathbf{r}_\eta|}\right)^2,$$

the integrand represents the square of the scalar product of unit tangent vectors to the ξ and η co-ordinate curves. Thus this functional controls the angle between intersecting co-ordinate curves.

Orthogonality three functional

Another functional related to orthogonality is

$$I_{O,3} = \iint_{R'} \sqrt{g_{11}g_{22}}\, d\xi\, d\eta. \tag{6.64}$$

Exercise 6. Show that the Euler-Lagrange equations for the functional (6.64) may be expressed as

$$\left(\sqrt{\frac{g_{22}}{g_{11}}}\mathbf{r}_\xi\right)_\xi + \left(\sqrt{\frac{g_{11}}{g_{22}}}\mathbf{r}_\eta\right)_\eta = \mathbf{0}. \tag{6.65}$$

The connection between these equations and orthogonality may be seen by considering eqns (1.164). For an orthogonal grid it is clear that we must have $g_{12} = 0$ everywhere, $g = \sqrt{g_{11}g_{22}}$, $\mathbf{g}^1 = \mathbf{g}_1/g_{11}$, and $\mathbf{g}^2 = \mathbf{g}_2/g_{22}$.

Thus it is necessary (but not sufficient) for a grid to be orthogonal that, by (1.164),

$$\frac{\partial}{\partial \xi}(\sqrt{g}\,\mathbf{g}^1) + \frac{\partial}{\partial \eta}(\sqrt{g}\,\mathbf{g}^2) = \frac{\partial}{\partial \xi}\left(\sqrt{\frac{g_{22}}{g_{11}}}\mathbf{g}_1\right) + \frac{\partial}{\partial \eta}\left(\sqrt{\frac{g_{11}}{g_{22}}}\mathbf{g}_2\right) = \mathbf{0},$$

which is equivalent to eqn (6.65). Equation (6.65) has been much used in the context of generating orthogonal grids – see, for example, Warsi and Thompson (1980) and Ryskin and Leal (1983).

6.4.7 The Liao functionals

Liao and Liu (1993) proposed the grid generation functional

$$I_{li} = \iint_{R'} [(g_{11} + g_{22})^2 - 2g]\, d\xi\, d\eta = \iint_{R'} [(g_{11})^2 + (g_{22})^2 + 2(g_{12})^2]\, d\xi\, d\eta, \tag{6.66}$$

but this has a tendency to produce folded grids. Experience has shown that this tendency is apparently diminished by taking the same functional with the integrand divided by g. This gives, after discarding a constant, the *modified Liao functional*

$$I_{ml} = \iint_{R'} \left(\frac{g_{11} + g_{22}}{\sqrt{g}}\right)^2 d\xi\, d\eta \tag{6.67}$$

which is similar to the Winslow functional (6.44).

The following table shows a list of unweighted functionals that are in use.

Functional	Symbol	Integrand
Length	I_L	$g_{11} + g_{22}$
Orthogonality-1	I_O	$(g_{12})^2$
Area	I_A	$(\sqrt{g})^2$
Orthogonality-2	$I_{O,2}$	$(g_{12})^2/g_{11}g_{22}$

Functional	Symbol	Integrand
Orthogonality-3	$I_{O,3}$	$\sqrt{g_{11}g_{22}}$
Area-Orthogonality	I_{AO}	$g_{11}g_{22}$
AO-squared	$I_{AO\text{-}squared}$	$(g_{11}g_{22})^2$
Winslow	I_{Win}	$(g_{11} + g_{22})/\sqrt{g}$
Liao	I_{li}	$(g_{11})^2 + (g_{22})^2 + 2(g_{12})^2$
Modified Liao	I_{ml}	$([g_{11} + g_{22}]/\sqrt{g})^2$

In all these variational problems the Euler-Lagrange equations lead to matrix equations of the form (6.48), in which the matrices A_{11}, A_{12}, A_{22} are symmetric, with coefficients depending at most on the first partial derivatives of the cartesian co-ordinates x, y and the weight functions. The subdirectory *Book/var.gds* on the accompanying disk contains five files which solve these equations. Each file offers two choices of geometry, and both SOR and the Thomas Algorithm have been used in the numerical solution. The files are listed at the end of this chapter.

6.4.8 Surface grids

A similar approach may be taken to the generation of a grid on a surface, such as one given in the parametric form (3.3) with parameters u, v. Surface metric tensors $a_{\alpha\beta}$, as defined in eqn (3.17), and their determinants a, given by eqn (3.24), can be used in place of g_{ij} and g above. We assume that a surface can be mapped onto the computational space $0 \leqslant \xi \leqslant 1$, $0 \leqslant \eta \leqslant 1$, and, indeed, we may seek the particular mapping

$$u = u(\xi, \eta), \qquad v = v(\xi, \eta)$$

which minimizes some selected functional

$$I = \iint \frac{H(a_{11}, a_{22}, a_{12}, \sqrt{a})}{\{\varphi(u, v)\}^2} \mathrm{d}\xi \, \mathrm{d}\eta, \tag{6.68}$$

where the weight function φ may be used for the purposes of adaptivity.

The metric tensor components here are defined according to the re-parameterization $\mathbf{r} = \mathbf{r}(\xi, \eta)$, so that

$$a_{11} = (x_\xi)^2 + (y_\xi)^2 + (z_\xi)^2, \quad a_{22} = (x_\eta)^2 + (y_\eta)^2 + (z_\eta)^2,$$
$$a_{12} = x_\xi x_\eta + y_\xi y_\eta + z_\xi z_\eta.$$

According to eqn (3.18), these are related to the components $\tilde{a}_{\alpha\beta}$ with respect to the parameters (u, v) by

$$\begin{aligned} a_{11} &= (u_\xi)^2\tilde{a}_{11} + 2u_\xi v_\xi \tilde{a}_{12} + (v_\xi)^2\tilde{a}_{22} \\ a_{22} &= (u_\eta)^2\tilde{a}_{11} + 2u_\eta v_\eta \tilde{a}_{12} + (v_\eta)^2\tilde{a}_{22} \\ a_{12} &= u_\xi u_\eta \tilde{a}_{11} + (u_\eta v_\xi + u_\xi v_\eta)\tilde{a}_{12} + v_\xi v_\eta \tilde{a}_{22}. \end{aligned} \tag{6.69}$$

As an example, we present an orthogonality-functional

$$I = \frac{1}{2} \iint (a_{12})^2 \, d\xi \, d\eta. \tag{6.70}$$

The two corresponding Euler-Lagrange equations are:

$$\begin{aligned} &\frac{\partial}{\partial u}(a_{12})^2 - \frac{\partial}{\partial \xi}\left(\frac{\partial}{\partial u_\xi}(a_{12})^2\right) - \frac{\partial}{\partial \eta}\left(\frac{\partial}{\partial u_\eta}(a_{12})^2\right) = 0, \\ &\frac{\partial}{\partial v}(a_{12})^2 - \frac{\partial}{\partial \xi}\left(\frac{\partial}{\partial v_\xi}(a_{12})^2\right) - \frac{\partial}{\partial \eta}\left(\frac{\partial}{\partial v_\eta}(a_{12})^2\right) = 0. \end{aligned} \tag{6.71}$$

If we put

$$\begin{aligned} A &= u_\eta \tilde{a}_{11} + v_\eta \tilde{a}_{12}, \quad & B &= u_\eta \tilde{a}_{12} + v_\eta \tilde{a}_{22}, \\ C &= u_\xi \tilde{a}_{11} + v_\xi \tilde{a}_{12}, \quad & D &= u_\xi \tilde{a}_{12} + v_\xi \tilde{a}_{22}, \end{aligned} \tag{6.72}$$

we get

$$a_{12} = A u_\xi + B v_\xi = C u_\eta + D v_\eta, \tag{6.73}$$

and eqns (6.71) may be expressed in the form:

$$\begin{aligned} &\frac{\partial}{\partial \xi}(A^2 u_\xi) + \frac{\partial}{\partial \eta}(C^2 u_\eta) + \frac{\partial}{\partial \xi}(AB v_\xi) + \frac{\partial}{\partial \eta}(CD v_\eta) = a_{12}\frac{\partial a_{12}}{\partial u}, \\ &\frac{\partial}{\partial \xi}(AB u_\xi) + \frac{\partial}{\partial \eta}(CD u_\eta) + \frac{\partial}{\partial \xi}(B^2 v_\xi) + \frac{\partial}{\partial \eta}(D^2 v_\eta) = a_{12}\frac{\partial a_{12}}{\partial v}. \end{aligned} \tag{6.74}$$

Note that by eqn (6.69)

$$\frac{\partial a_{12}}{\partial u} = u_\xi u_\eta \frac{\partial \tilde{a}_{11}}{\partial u} + (u_\eta v_\xi + u_\xi v_\eta)\frac{\partial \tilde{a}_{12}}{\partial u} + v_\xi v_\eta \frac{\partial \tilde{a}_{22}}{\partial u},$$

and similarly for $\partial a_{12}/\partial v$.

6.5 Harmonic maps

The Winslow method, with variational formulation based on the functionals (6.43) and (6.44), together with its extension to surface grid generation discussed in Section 5.9, can be regarded as an application of *harmonic maps*. These may be defined in the general context of differentiable mappings from an n-dimensional 'manifold' M with co-ordinates x^i and contravariant metric tensor γ^{ij}, which we may regard here simply as the n-dimensional physical domain, to an n-dimensional manifold N with co-ordinates ξ^i and covariant metric tensor G_{ij}, which will be the computational domain. Thus for practical purposes here N is a square or rectangle in E^2 or a cube in E^3, and M is a two-dimensional region in E^2 or a two-dimensional surface or three-dimensional region in E^3.

Associated with any mapping is a so-called *energy density*

$$e(\xi^i(\mathbf{x})) = \frac{1}{2}\gamma^{ij}(\mathbf{x}) G_{kl}(\xi^i(\mathbf{x})) \frac{\partial \xi^k}{\partial x^i}\frac{\partial \xi^l}{\partial x^j}, \tag{6.75}$$

where the indices i, j, k, l are summed from 1 to n, and $\mathbf{x}$ represents the column vector of curvilinear co-ordinates x^i. The *energy* E of the map is defined as the integral of the energy density over M:

$$E(\xi^i(\mathbf{x})) = \int_M \frac{1}{2}\gamma^{ij}(\mathbf{x})G_{kl}(\xi^i(\mathbf{x}))\frac{\partial\xi^k}{\partial x^i}\frac{\partial\xi^l}{\partial x^j}\mathrm{d}M, \tag{6.76}$$

where $\mathrm{d}M$ represents an element of area or volume of M. By eqns (1.44), (1.160), (1.45), and (3.42), we can put $\mathrm{d}M = \sqrt{\gamma}\,\mathrm{d}x^1\ldots\mathrm{d}x^n$, where $\gamma = \det(\gamma_{ij})$ and γ_{ij} is the covariant metric tensor of M.

A *harmonic map* is then a twice-differentiable map $\xi^i(\mathbf{x})$ which is an extremal of the functional (6.76). Since E can take only non-negative values due to the positive-definiteness of G_{kl}, we would expect the extremal to minimize the functional. It has been shown that, provided that there exists a smooth one–one mapping from M to N which also maps the boundary of M to the boundary of N, and that N is convex with negative or zero curvature, then there exists a harmonic map from M to N which is also unique. The geometric conditions on N are satisfied if N is a unit square in E^2 or a unit cube in E^3.

In the special case where M is a region of E^2 or E^3, we may take x^i to be cartesian co-ordinates y_i and take Euclidean metrics $\gamma_{ij} = \delta_{ij}$, $G_{kl} = \delta_{kl}$ in M and N. The energy functional then becomes, with $\gamma = 1$,

$$E = \int_M \frac{1}{2}\frac{\partial\xi^k}{\partial y_j}\frac{\partial\xi^k}{\partial y_j}\,\mathrm{d}M, \tag{6.77}$$

which in two dimensions becomes

$$E = \frac{1}{2}\iint_M \left[\left(\frac{\partial\xi}{\partial x}\right)^2 + \left(\frac{\partial\xi}{\partial y}\right)^2 + \left(\frac{\partial\eta}{\partial x}\right)^2 + \left(\frac{\partial\eta}{\partial y}\right)^2\right]\mathrm{d}x\,\mathrm{d}y$$

in the obvious notation. We have already encountered this functional at eqn (6.41) and shown that the corresponding Euler-Lagrange equations are the Laplace eqns (6.42), which give rise directly to the Winslow model. Thus the harmonic map in this case involves harmonic functions (which satisfy Laplace's equations). We also have

$$E = \frac{1}{2}\iint_M (g^{11} + g^{22})\,\mathrm{d}x\,\mathrm{d}y,$$

where g^{ij} may be regarded as the contravariant metric tensor induced in M by the mapping $\mathbf{r} \to \xi^i$.

In three dimensions we have, by eqn (1.20),

$$E = \frac{1}{2}\iiint_M (g^{11} + g^{22} + g^{33})\,\mathrm{d}x\,\mathrm{d}y\,\mathrm{d}z = \frac{1}{2}\iiint_M g^{kk}\,\mathrm{d}x\,\mathrm{d}y\,\mathrm{d}z, \tag{6.78}$$

with implied summation over k. The variational problem now involves three co-ordinate functions, say ξ, η, ζ, of three variables x, y, z, and is a natural extension of the two-dimensional problem defined by eqn (6.11). There are three Euler-Lagrange equations, which reduce to

$$\nabla^2\xi = \nabla^2\eta = \nabla^2\zeta = 0, \tag{6.79}$$

where ∇^2 is the three-dimensional Laplacian $(\partial^2/\partial x^2 + \partial^2/\partial y^2 + \partial^2/\partial z^2)$.

When M is a two-dimensional surface in E^3, we can use surface co-ordinates u^α in place of x^i, and the surface metric $a^{\alpha\beta}$ in place of γ^{ij}. With N again a unit square with $G_{ij} = \delta_{ij}$, we get

$$E = \frac{1}{2}\iint_M a^{\alpha\beta}(u^\gamma)\frac{\partial\xi^k}{\partial u^\alpha}\frac{\partial\xi^k}{\partial u^\beta}\sqrt{a}\,\mathrm{d}u^1\,\mathrm{d}u^2, \tag{6.80}$$

where all indices are summed from 1 to 2. Thus we have a variational problem of the form (6.11). The two Euler-Lagrange eqns (6.12) may be expressed here as

$$\frac{\partial(e\sqrt{a})}{\partial\xi^i} - \frac{\partial}{\partial u^\gamma}\left(\frac{\partial(e\sqrt{a})}{\partial\left(\partial\xi^i/\partial u^\gamma\right)}\right) = 0, \quad i = 1, 2, \tag{6.81}$$

with summation over γ. Since no terms in e depend explicitly on ξ^i, and

$$\frac{\partial}{\partial\left(\partial\xi^i/\partial u^\gamma\right)}\left(\frac{\partial\xi^k}{\partial u^\alpha}\frac{\partial\xi^k}{\partial u^\beta}\right) = \delta_{ik}\delta^\gamma_\alpha\frac{\partial\xi^k}{\partial u^\beta} + \frac{\partial\xi^k}{\partial u^\alpha}\delta_{ik}\delta^\gamma_\beta = \frac{\partial\xi^i}{\partial u^\beta}\delta^\gamma_\alpha + \frac{\partial\xi^i}{\partial u^\alpha}\delta^\gamma_\beta,$$

we obtain the equations

$$\frac{1}{2}\frac{\partial}{\partial u^\gamma}\left(\sqrt{a}\,a^{\alpha\beta}\left(\frac{\partial\xi^i}{\partial u^\beta}\delta^\gamma_\alpha + \frac{\partial\xi^i}{\partial u^\alpha}\delta^\gamma_\beta\right)\right) = \frac{1}{2}\frac{\partial}{\partial u^\gamma}\left(\sqrt{a}\left(a^{\gamma\beta}\frac{\partial\xi^i}{\partial u^\beta} + a^{\alpha\gamma}\frac{\partial\xi^i}{\partial u^\alpha}\right)\right) = 0.$$

Using the symmetry of $a^{\alpha\beta}$, we have, finally,

$$\frac{\partial}{\partial u^\gamma}\left(\sqrt{a}\,a^{\gamma\beta}\frac{\partial\xi^i}{\partial u^\beta}\right) = 0, \quad i = 1, 2. \tag{6.82}$$

Pre-multiplying by $1/\sqrt{a}$ shows by eqn (3.160) that the harmonic mapping in this case must satisfy the pair of Beltramian equations

$$\Delta_B\xi^1 = \Delta_B\xi^2 = 0 \tag{6.83}$$

which are the same as eqns (5.104) in the absence of the control functions P, Q.

Returning to the more general form (6.76) with N a unit square or cube and $G_{kl} = \delta_{kl}$, a similar procedure shows that the Euler-Lagrange equations are

$$\frac{1}{\sqrt{\gamma}}\frac{\partial}{\partial x^j}\left(\sqrt{\gamma}\gamma^{jk}\frac{\partial\xi^i}{\partial x^k}\right) = 0, \quad i = 1, \ldots, n, \tag{6.84}$$

which can be expressed as

$$\gamma^{jk}\frac{\partial^2\xi^i}{\partial x^j\partial x^k} + \frac{1}{\sqrt{\gamma}}\frac{\partial\xi^i}{\partial x^k}\frac{\partial}{\partial x^j}(\sqrt{\gamma}\,\gamma^{jk}) = 0. \tag{6.85}$$

This equation can be inverted to produce a different partial differential equation satisfied by $x^i(\xi^k)$, by making use of the identity

$$\frac{\partial^2\xi^i}{\partial x^j\partial x^k}\frac{\partial x^l}{\partial\xi^i} = -\frac{\partial^2 x^l}{\partial\xi^m\partial\xi^i}\frac{\partial\xi^i}{\partial x^j}\frac{\partial\xi^m}{\partial x^k},$$

which follows, similarly to what was presented in Exercise 10, Chapter 1, in the derivation of eqn (1.113), from differentiating with respect to x^k the Chain Rule

$$\frac{\partial \xi^i}{\partial x^j}\frac{\partial x^l}{\partial \xi^i} = \delta^l_j.$$

Thus, multiplying eqn (6.85) by $\partial x^l/\partial \xi^i$ (implying summation over i) gives

$$\gamma^{jk}\frac{\partial \xi^i}{\partial x^j}\frac{\partial \xi^m}{\partial x^k}\frac{\partial^2 x^l}{\partial \xi^m \partial \xi^i} = \frac{1}{\sqrt{\gamma}}\frac{\partial \xi^i}{\partial x^k}\frac{\partial x^l}{\partial \xi^i}\frac{\partial}{\partial x^j}(\sqrt{\gamma}\,\gamma^{jk}),$$

which, with further use of the Chain Rule, gives

$$\gamma^{jk}\frac{\partial \xi^i}{\partial x^j}\frac{\partial \xi^m}{\partial x^k}\frac{\partial^2 x^l}{\partial \xi^m \partial \xi^i} = \frac{1}{\sqrt{\gamma}}\frac{\partial}{\partial x^j}(\sqrt{\gamma}\gamma^{jl}). \tag{6.86}$$

Note that $\gamma^{jk}(\partial \xi^i/\partial x^j)(\partial \xi^m/\partial x^k)$ is a tensor transformation rule for contravariant second-order tensor components applied to the contravariant metric tensor γ^{jk}. The result is a set of contravariant components, say g^{im}, representing the contravariant metric tensor in the new curvilinear co-ordinate system ξ^i in M. Hence we can write

$$g^{im}\frac{\partial^2 x^l}{\partial \xi^m \partial \xi^i} = \frac{1}{\sqrt{\gamma}}\frac{\partial}{\partial x^j}(\sqrt{\gamma}\gamma^{jl}). \tag{6.87}$$

This is a quasi-linear system of elliptic partial differential equations which may be used as the basis of an algorithm for generating structured two-dimensional adaptive grids, grids on surfaces, and three-dimensional grids.

6.5.1 Surface grids

As an example, we consider M to be a two-dimensional surface in E^3 defined in terms of cartesian co-ordinates by $z = f(x, y)$. In fact we take x, y to be parametric co-ordinates for the surface, so that $x^1 = x$, $x^2 = y$. The covariant metric tensor components of the surface with these co-ordinates are given by eqn (3.23), so that

$$\gamma_{11} = 1 + (f_x)^2, \quad \gamma_{12} = (f_x)(f_y), \quad \gamma_{22} = 1 + (f_y)^2, \tag{6.88}$$

and

$$\gamma = \det(\gamma_{ij}) = 1 + (f_x)^2 + (f_y)^2, \tag{6.89}$$

where subscripts denote partial derivatives.

From eqn (3.30) we immediately have the contravariant components

$$\gamma^{11} = \frac{1 + (f_y)^2}{1 + (f_x)^2 + (f_y)^2}, \quad \gamma^{12} = -\frac{(f_x)(f_y)}{1 + (f_x)^2 + (f_y)^2},$$

$$\gamma^{22} = \frac{1 + (f_x)^2}{1 + (f_x)^2 + (f_y)^2}. \tag{6.90}$$

A harmonic map will take M onto a unit square N in the (ξ, η) plane, with $\xi^1 = \xi$, $\xi^2 = \eta$. Straight lines $\xi = const.$, $\eta = const.$ in N are what results from mapping ξ, η curvilinear co-ordinate curves in M. The 'induced' covariant metric components are given by

$$g_{11} = (x_\xi)^2 + (y_\xi)^2 + (z_\xi)^2$$
$$g_{12} = x_\xi x_\eta + y_\xi y_\eta + z_\xi z_\eta$$
$$g_{22} = (x_\eta)^2 + (y_\eta)^2 + (z_\eta)^2.$$

However, since $z = f(x, y)$, we have, using Chain Rules,

$$z_\xi = f_x x_\xi + f_y y_\xi \text{ and } z_\eta = f_x x_\eta + f_y y_\eta, \tag{6.91}$$

giving

$$\begin{aligned} g_{11} &= [1 + (f_x)^2](x_\xi)^2 + 2 f_x f_y x_\xi y_\xi + [1 + (f_y)^2](y_\xi)^2 \\ g_{12} &= [1 + (f_x)^2] x_\xi x_\eta + f_x f_y (x_\xi y_\eta + x_\eta y_\xi) + [1 + (f_y)^2] y_\xi y_\eta \\ g_{22} &= [1 + (f_x)^2](x_\eta)^2 + 2 f_x f_y x_\eta y_\eta + [1 + (f_y)^2](y_\eta)^2. \end{aligned} \tag{6.92}$$

Exercise 7. Verify that

$$g = \det(g_{ij}) = [1 + (f_x)^2 + (f_y)^2](x_\xi y_\eta - x_\eta y_\xi)^2 = \gamma J^2, \tag{6.93}$$

where J is the Jacobian $(x_\xi y_\eta - x_\eta y_\xi)$ of the mapping from the ξ^is to the x^is.

We now have

$$g^{11} = g_{22}/g, \quad g^{12} = -g_{12}/g, \quad g^{22} = g_{11}/g.$$

Substituting into eqns (6.87) gives

$$\begin{aligned} L(x) &= g_{22} x_{\xi\xi} - 2 g_{12} x_{\xi\eta} + g_{22} x_{\eta\eta} \\ &\quad - \frac{g}{\sqrt{\gamma}} \left[\frac{\partial}{\partial x} \left(\frac{1 + (f_y)^2}{\sqrt{\gamma}} \right) - \frac{\partial}{\partial y} \left(\frac{f_x f_y}{\sqrt{\gamma}} \right) \right] = 0, \\ L(y) &= g_{22} y_{\xi\xi} - 2 g_{12} y_{\xi\eta} + g_{22} y_{\eta\eta} \\ &\quad - \frac{g}{\sqrt{\gamma}} \left[-\frac{\partial}{\partial x} \left(\frac{f_x f_y}{\sqrt{\gamma}} \right) + \frac{\partial}{\partial y} \left(\frac{1 + (f_x)^2}{\sqrt{\gamma}} \right) \right] = 0 \end{aligned} \tag{6.94}$$

with γ, g_{11}, g_{12}, g_{22}, and g given by eqns (6.89), (6.92), and (6.93).

To solve these equations numerically, the second-order accurate finite difference approximations (4.3) for first derivatives and (4.4), (4.5), and (4.6) for second derivatives may be used, taking uniform differences h, k in ξ and η, respectively, in the computational domain N. To calculate f_x and f_y we need to invert eqns (6.91), so that

$$f_x = \frac{1}{J}(z_\xi y_\eta - z_\eta y_\xi) \quad \text{and} \quad f_y = \frac{1}{J}(z_\eta x_\xi - z_\xi x_\eta). \tag{6.95}$$

If we re-scale the computational domain so that $h = k = 1$, these equations yield values of f_x and f_y at the i, j grid point given by the approximations

$$[f_x]_{ij} = \frac{(f_{i+1,j} - f_{i-1,j})(y_{i,j+1} - y_{i,j-1}) - (f_{i,j+1} - f_{i,j-1})(y_{i+1,j} - y_{i-1,j})}{(x_{i+1,j} - x_{i-1,j})(y_{i,j+1} - y_{i,j-1}) - (x_{i,j+1} - x_{i,j-1})(y_{i+1,j} - y_{i-1,j})}, \quad (6.96)$$

$$[f_y]_{ij} = \frac{(f_{i,j+1} - f_{i,j-1})(x_{i+1,j} - x_{i-1,j}) - (f_{i+1,j} - f_{i-1,j})(x_{i,j+1} - x_{i,j-1})}{(x_{i+1,j} - x_{i-1,j})(y_{i,j+1} - y_{i,j-1}) - (x_{i,j+1} - x_{i,j-1})(y_{i+1,j} - y_{i-1,j})}, \quad (6.97)$$

where $[f]_{ij} = f(x_{ij}, y_{ij})$ is the value of z at a grid point.

For a given set of values $[x]_{i,j}$ and $[y]_{i,j}$ at all grid points, the grid point values of the left-hand sides of eqns (6.94) can now be approximated:

$$[L(x)]_{ij} = [g_{22}x_{\xi\xi}]_{ij} - [2g_{12}x_{\xi\eta}]_{ij} + [g_{11}x_{\eta\eta}]_{ij} - \left[J^2\sqrt{\gamma}\left\{\frac{\partial}{\partial x}\left(\frac{1+(f_y)^2}{\sqrt{\gamma}}\right) - \frac{\partial}{\partial y}\left(\frac{f_x f_y}{\sqrt{\gamma}}\right)\right\}\right]_{ij} \quad (6.98)$$

and

$$[L(y)]_{ij} = [g_{22}y_{\xi\xi}]_{ij} - [2g_{12}y_{\xi\eta}]_{ij} + [g_{11}y_{\eta\eta}]_{ij} - \left[J^2\sqrt{\gamma}\left\{-\frac{\partial}{\partial x}\left(\frac{f_x f_y}{\sqrt{\gamma}}\right) + \frac{\partial}{\partial y}\left(\frac{1+(f_x)^2}{\sqrt{\gamma}}\right)\right\}\right]_{ij}. \quad (6.99)$$

The derivatives involving $\partial/\partial x$ and $\partial/\partial y$ in the last terms of these expressions may be calculated using equations of the form (6.96) and (6.97).

A possible iterative solution scheme suggested by the terms $-2[g_{22}]_{ij}x_{i,j} - 2[g_{11}]_{ij}x_{i,j}$ appearing on the right-hand side of eqn (6.98) and $-2[g_{22}]_{ij}y_{i,j} - 2[g_{11}]_{ij}y_{i,j}$ in eqn (6.99) is

$$x_{i,j}^{m+1} = x_{i,j}^{m} + \sigma\frac{[L(x)]_{ij}}{2[g_{22} + g_{11}]_{ij}} \quad (6.100)$$

and

$$y_{i,j}^{m+1} = y_{i,j}^{m} + \sigma\frac{[L(y)]_{ij}}{2[g_{22} + g_{11}]_{ij}}, \quad (6.101)$$

where σ is an iteration parameter lying between 0 and 1. Here the quantities on the right-hand sides are all evaluated at the mth iteration step. The iteration will start with a guessed field of x and y values, and the above equations can be used to calculate a new set of values. The iteration will stop when some measure of convergence has been achieved.

6.6 Website programs

6.6.1 Subdirectory: Book/var.gds

This contains five files:

1. Area.trid.f
   This program solves the Euler-Lagrange equations of the form (6.51) for the area functional with weight function equal to a constant, so that the matrices in (6.48) are given by eqn (6.53) with $\mathbf{S} = \mathbf{0}$. Figure 6.3 shows an example of the resulting grid in a trapezium.

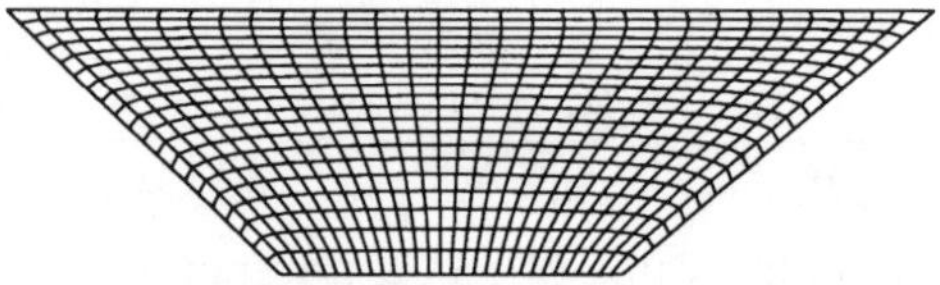

Fig. 6.3 Area functional.

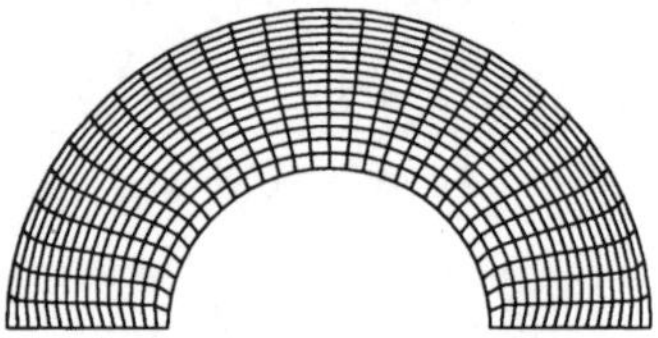

Fig. 6.4 Area functional.

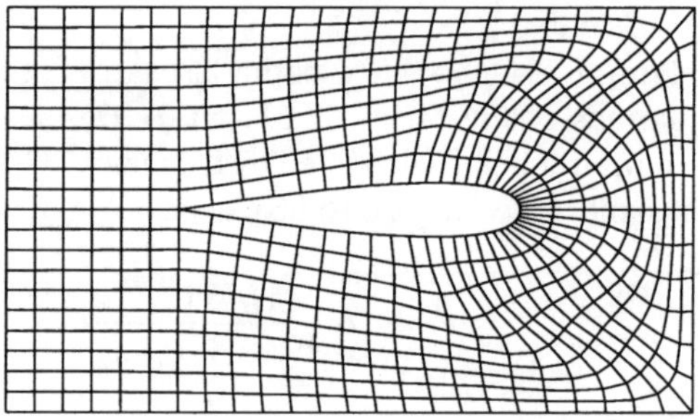

Fig. 6.5 Area-orthogonality.

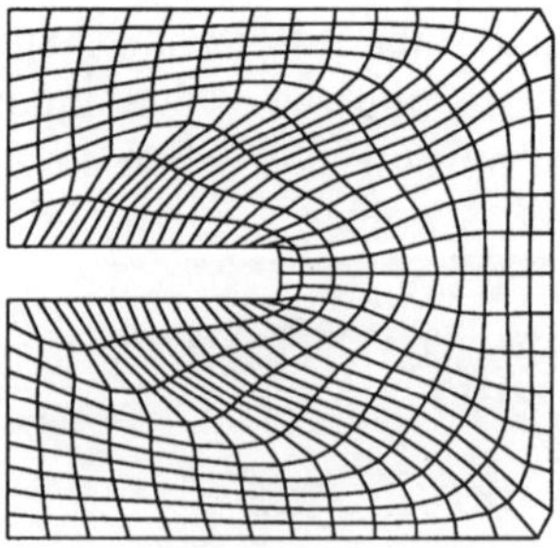

Fig. 6.6 Area-orthogonality.

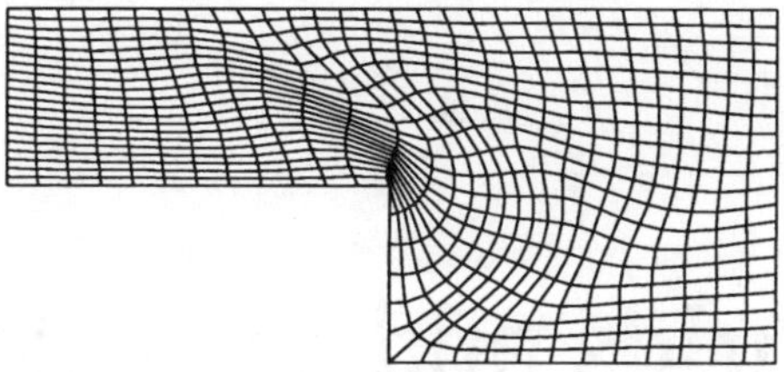

Fig. 6.7 Area-orthogonality.

2. Var1.ao.f
3. Var2.ao.f
4. SOR.ao.f
   These programs solve the Euler-Lagrange equations based on the Area-Orthogonality functional, with matrices given by eqns (6.61). The weight function φ is again a constant, so that $\mathbf{S} = \mathbf{0}$. Examples of the resulting grids for domains of various shapes are shown in Figs 6.4–6.7.
5. Length.conjugate.f

This solves the Euler-Lagrange equations based on the Length-functional (6.40). Thus it solves the pair of Laplace's eqns (6.39). The numerical scheme used here is the Conjugate Gradient method.

6.6.2 Subdirectory: Book/one.d.gds

This contains one file relevant to this chapter, as discussed in Section 6.3.3, namely

1. curve.SOR.f

7

Moving grids and time-dependent co-ordinate systems

7.1 Time-dependent co-ordinate transformations

The numerical solution of time-dependent (otherwise known as transient, evolution, or unsteady) transport equations may require time-dependent 'moving' grids in physical space. For example, transonic flow problems and similar applications involving the propagation of shocks demand moving grids, with a continuing re-distribution of grid-nodes, to capture a shock. Some problems will involve fixed boundaries in physical space, with internal grid-nodes moving in response to the flow developed. Other problems involve moving boundaries, for example the internal combustion chamber. In either case a time variable t enters the hosted equations and the grid generation equations as an independent parameter. Any boundary-conforming grid at any time t with curvilinear co-ordinates $x^i, i = 1, 2, 3$ (in three dimensions), however, may still be mapped to a *fixed* uniform rectangular grid in x^i-computational space. The transformation involved will now be time-dependent and of the form

$$y_i = y_i(x^1, x^2, x^3, t), \quad i = 1, 2, 3, \tag{7.1}$$

with inverse

$$x^i = x^i(y_1, y_2, y_3, t), \quad i = 1, 2, 3, \tag{7.2}$$

where y_i are the cartesian co-ordinates with respect to some rectangular cartesian reference system of a point in physical space which at time t has curvilinear co-ordinates x^i. We assume that the Jacobian

$$J = \det(\partial y_i / \partial x^j)$$

of the transformation remains non-zero at all times.

An arbitrary function $f(\mathbf{r}, t)$ of space (the physical domain) and time has differential (small increment to first order)

$$\mathrm{d}f = \left(\frac{\partial f}{\partial y_i}\right)_t \mathrm{d}y_i + \left(\frac{\partial f}{\partial t}\right)_{y_i} \mathrm{d}t,$$

where now subscripts outside brackets indicate (where clarity is needed) which variables are being held constant when partial differentiation is performed. But the transformation (7.1) also gives

$$\mathrm{d}y_i = \left(\frac{\partial y_i}{\partial x^j}\right)_t \mathrm{d}x^j + \left(\frac{\partial y_i}{\partial t}\right)_{x^j} \mathrm{d}t,$$

so that

$$\mathrm{d}f = \left(\frac{\partial f}{\partial y_i}\right)_t \left(\frac{\partial y_i}{\partial x^j}\right)_t \mathrm{d}x^j + \left[\left(\frac{\partial f}{\partial y_i}\right)_t \left(\frac{\partial y_i}{\partial t}\right)_{x^j} + \left(\frac{\partial f}{\partial t}\right)_{y_i}\right] \mathrm{d}t,$$

which gives the Chain Rules

$$\left(\frac{\partial f}{\partial x^j}\right)_t = \left(\frac{\partial f}{\partial y_i}\right)_t \left(\frac{\partial y_i}{\partial x^j}\right)_t, \tag{7.3}$$

$$\left(\frac{\partial f}{\partial t}\right)_{x^j} = \left(\frac{\partial f}{\partial y_i}\right)_t \left(\frac{\partial y_i}{\partial t}\right)_{x^j} + \left(\frac{\partial f}{\partial t}\right)_{y_i}. \tag{7.4}$$

Note that in eqn (7.4) the left-hand side time derivative is carried out at fixed x^j, i.e. at a fixed point in computational space, which we may imagine corresponding to a given (moving) grid point in physical space, whereas time derivatives evaluated at fixed y_i on the right-hand side are taken at fixed positions of physical space. This equation may be written as

$$\left(\frac{\partial f}{\partial t}\right)_{x^j} = \nabla f \cdot \left(\frac{\partial \mathbf{r}}{\partial t}\right)_{x^j} + \left(\frac{\partial f}{\partial t}\right)_{y_i} = \nabla f \cdot \mathbf{W} + \left(\frac{\partial f}{\partial t}\right)_{y_i}, \tag{7.5}$$

where

$$\mathbf{W} = \left(\frac{\partial \mathbf{r}}{\partial t}\right)_{x^j} \tag{7.6}$$

is the rate of change of position of a given grid point in the physical domain, and may be called the *grid point velocity*. Thus the time-derivative of the quantity f at a fixed point of the physical domain is related to its time-derivative at a fixed point of the computational domain by the equation

$$\left(\frac{\partial f}{\partial t}\right)_{y_i} = \left(\frac{\partial f}{\partial t}\right)_{x^j} - \mathbf{W} \cdot \nabla f, \tag{7.7}$$

with ∇f evaluated in the physical domain.

7.2 Time-dependent base vectors

As the curvilinear co-ordinate system moves in physical space, the covariant base vectors $\mathbf{g}_i$ must be functions of time t. In fact we have

$$\mathbf{g}_i = \frac{\partial \mathbf{r}}{\partial x^i}, \quad i = 1, 2, 3,$$

with cartesian components obtained from differentiating eqn (7.1):

$$(\mathbf{g}_j)_i = \frac{\partial y_i}{\partial x^j}$$

and time-derivatives

$$\frac{\partial}{\partial t}(\mathbf{g}_j)_i = \frac{\partial}{\partial t}\left(\frac{\partial y_i}{\partial x^j}\right) = \frac{\partial}{\partial x^j}\left(\frac{\partial y_i}{\partial t}\right),$$

giving the vector equations

$$\frac{\partial \mathbf{g}_j}{\partial t} = \frac{\partial \mathbf{W}}{\partial x^j}, \quad j = 1, 2, 3. \tag{7.8}$$

This variation of the base vectors with time as well as space complicates the transformation of vector equations (such as the time-dependent Navier-Stokes equations) to a curvilinear co-ordinate system. The variation of base vectors in space was considered in Chapter 1 by the introduction of Christoffel symbols, and a similar formalism is convenient for dealing with time-variation.

Here we put $t = x^0$ and regard this as a fourth generalized co-ordinate. We can also define an associated covariant base vector

$$\mathbf{g}_0 = \left(\frac{\partial \mathbf{r}}{\partial x^0}\right)_{x^j} = \left(\frac{\partial \mathbf{r}}{\partial t}\right)_{x^j} \tag{7.9}$$

and thus we can write

$$\mathbf{g}_i = \frac{\partial \mathbf{r}}{\partial x^i}, \quad i = 0, 1, 2, 3, \tag{7.10}$$

with

$$\frac{\partial \mathbf{g}_i}{\partial x^j} = \frac{\partial^2 \mathbf{r}}{\partial x^j \partial x^i} = \frac{\partial \mathbf{g}_j}{\partial x^i}. \tag{7.11}$$

By the definition of **W** in Section 7.1, we also have

$$\mathbf{g}_0 = \mathbf{W}. \tag{7.12}$$

The covariant metric tensor can also be extended to include the time variable, so that

$$g_{ij} = \mathbf{g}_i \cdot \mathbf{g}_j$$

with the suffixes i, j ranging over the values 0 to 3, so that $g_{00} = \mathbf{W} \cdot \mathbf{W} = |\mathbf{W}|^2$ and

$$g_{0j} = \mathbf{W} \cdot \mathbf{g}_j = W_j. \tag{7.13}$$

Similarly we have the Christoffel symbol $[ij, k]$, which is given, as in eqn (1.108), by

$$[ij, k] = \frac{1}{2}\left(\frac{\partial g_{jk}}{\partial x^i} + \frac{\partial g_{ik}}{\partial x^j} - \frac{\partial g_{ij}}{\partial x^k}\right). \tag{7.14}$$

Here, however, we allow i and j but not k to take the value 0 as well as 1, 2, 3, so that the equation

$$\frac{\partial \mathbf{g}_i}{\partial x^j} = [ij, k]\mathbf{g}^k, \quad i, j = 0, 1, 2, 3, \tag{7.15}$$

as in eqn (1.97), is valid, with k summed over the values 1 to 3. Thus it is not necessary to extend the set of spatial contravariant base vectors $\mathbf{g}^k$, $k = 1, 2, 3$.

With suffixes restricted to the values 1, 2, 3, we now have

$$\frac{\partial \mathbf{g}_i}{\partial x^j} = [ij, k]\mathbf{g}^k, \quad \frac{\partial \mathbf{g}_i}{\partial t} = [i0, k]\mathbf{g}^k,$$

$$\frac{\partial \mathbf{g}_i}{\partial x^j} = \Gamma^k_{ij}\mathbf{g}_k, \quad \frac{\partial \mathbf{g}_i}{\partial t} = \Gamma^k_{i0}\mathbf{g}_k, \tag{7.16}$$

where Christoffel symbols of the second kind are

$$\Gamma^k_{ij} = g^{kl}[ij, k], \quad \Gamma^k_{i0} = g^{kl}[i0, k]. \tag{7.17}$$

Note that substituting $f = x^i$ in eqn (7.5) gives

$$0 = \nabla x^i \cdot \mathbf{W} + \left(\frac{\partial x^i}{\partial t}\right)_{y_j} = \mathbf{g}^i \cdot \mathbf{W} + \left(\frac{\partial x^i}{\partial t}\right)_{y_j} = W^i + \left(\frac{\partial x^i}{\partial t}\right)_{y_j},$$

using eqns (1.11) and (1.48), so that we have the result

$$\left(\frac{\partial x^i}{\partial t}\right)_{y_j} = -W^i, \quad i = 1, 2, 3, \tag{7.18}$$

which states that the time-derivatives of the curvilinear co-ordinates at a fixed spatial point of the physical domain as the co-ordinate system moves are equal to the negative of the contravariant components of the grid point velocity.

Another useful formula gives the time-derivative at fixed x^i of $\sqrt{g}$, where by eqn (1.31)

$$\sqrt{g} = \mathbf{g}_1 \cdot (\mathbf{g}_2 \times \mathbf{g}_3).$$

We have, after re-arrangement of scalar and vector products, and with use of eqns (7.8), (1.8), and (1.134),

$$\begin{aligned}
\frac{1}{\sqrt{g}}\frac{\partial}{\partial t}(\sqrt{g}) &= \frac{1}{\sqrt{g}}\left(\frac{\partial \mathbf{g}_1}{\partial t} \cdot (\mathbf{g}_2 \times \mathbf{g}_3) + \frac{\partial \mathbf{g}_2}{\partial t} \cdot (\mathbf{g}_3 \times \mathbf{g}_1) + \frac{\partial \mathbf{g}_3}{\partial t} \cdot (\mathbf{g}_1 \times \mathbf{g}_2)\right) \\
&= \frac{1}{\sqrt{g}}\left(\frac{\partial \mathbf{W}}{\partial x^1} \cdot (\mathbf{g}_2 \times \mathbf{g}_3) + \frac{\partial \mathbf{W}}{\partial x^2} \cdot (\mathbf{g}_3 \times \mathbf{g}_1) + \frac{\partial \mathbf{W}}{\partial x^3} \cdot (\mathbf{g}_1 \times \mathbf{g}_2)\right) \\
&= \frac{\partial \mathbf{W}}{\partial x^1} \cdot \mathbf{g}^1 + \frac{\partial \mathbf{W}}{\partial x^2} \cdot \mathbf{g}^2 + \frac{\partial \mathbf{W}}{\partial x^3} \cdot \mathbf{g}^3 = \nabla \cdot \mathbf{W}.
\end{aligned} \tag{7.19}$$

This is a fundamental identity in time-dependent co-ordinate theory; it is called the *Geometric Conservation Law*. In Computational Fluid Dynamics it provides an additional equation which has to be solved alongside the usual transport equations when we have moving grids.

From eqn (7.12) we also have

$$\nabla \cdot \mathbf{W} = \nabla \cdot \mathbf{g}_0 = \mathbf{g}^k \cdot \frac{\partial \mathbf{g}_0}{\partial x^k} = \mathbf{g}^k \cdot \Gamma^j_{0k}\mathbf{g}_j = \delta^k_j \Gamma^j_{0k} = \Gamma^j_{0j},$$

using (7.16). Hence

$$\Gamma^j_{0j} = \frac{1}{\sqrt{g}}\frac{\partial}{\partial t}(\sqrt{g}). \tag{7.20}$$

7.3 Transformation of generic convective terms

Here we take a typical convective expression in the form

$$\mathcal{C} = \left(\frac{\partial \varphi}{\partial t}\right)_{y_i} + \nabla \cdot (\mathbf{v}\varphi) \tag{7.21}$$

for some field quantity $\varphi = \varphi(y_1, y_2, y_3, t)$, where $\mathbf{v}$ is the local fluid (or continuum) velocity, and we transform it to a time-dependent curvilinear co-ordinate system x^i. From eqn (7.7) we have immediately

$$\mathcal{C} = \left(\frac{\partial \varphi}{\partial t}\right)_{x^i} - \mathbf{W} \cdot \nabla\varphi + \nabla \cdot (\mathbf{v}\varphi) = \left(\frac{\partial \varphi}{\partial t}\right)_{x^i} + \varphi\nabla \cdot \mathbf{W} + \nabla \cdot [(\mathbf{v} - W)\varphi]. \tag{7.22}$$

Alternatively, using eqns (7.19) and (1.138), we can write

$$\begin{aligned}\mathcal{C} &= \left(\frac{\partial \varphi}{\partial t}\right)_{x^i} + \frac{\varphi}{\sqrt{g}}\frac{\partial \sqrt{g}}{\partial t} + \frac{1}{\sqrt{g}}\frac{\partial}{\partial x^i}(\sqrt{g}\mathbf{g}^i \cdot (\mathbf{v} - \mathbf{W})\varphi) \\ &= \frac{1}{\sqrt{g}}\left\{\left(\frac{\partial}{\partial t}(\sqrt{g}\varphi)\right)_{x^i} + \frac{\partial}{\partial x^i}(\sqrt{g}\mathbf{g}^i \cdot (\mathbf{v} - \mathbf{W})\varphi)\right\}\end{aligned} \tag{7.23}$$

in conservative form. It may be convenient to put

$$\mathbf{U} = \mathbf{v} - \mathbf{W}, \tag{7.24}$$

which clearly represents the continuum velocity *relative* to the moving grid points.

Exercise 1. Show that

$$\sqrt{g}\mathcal{C} = \frac{\partial}{\partial t}(\sqrt{g}\varphi) + \frac{\partial}{\partial x^i}(\sqrt{g}\varphi U^i). \tag{7.25}$$

It follows that if the hosted equations are given by

$$\frac{\partial \varphi}{\partial t} + \nabla \cdot (\mathbf{v}\varphi) + \nabla \cdot (\nu\nabla\varphi) + S = 0, \tag{7.26}$$

rather than the time-independent eqns (5.119), the transformed equations (on a fixed grid in computational space) may be written in conservative form as

$$\frac{\partial}{\partial t}(\sqrt{g}\varphi) + \frac{\partial}{\partial x^i}\left[\sqrt{g}\left(U^i\varphi + \nu g^{ij}\frac{\partial \varphi}{\partial x^j}\right)\right] + \sqrt{g}S = 0, \tag{7.27}$$

with appropriate summation over i and j.

In place of eqn (5.121) we also have the non-conservative form

$$\frac{\partial \varphi}{\partial t} + [U^i + \nu(\nabla^2 x^i)]\frac{\partial \varphi}{\partial x^i} + g^{ij}\frac{\partial}{\partial x^i}\left(\nu\frac{\partial \varphi}{\partial x^j}\right) + \varphi\mathbf{g}^i \cdot \frac{\partial \mathbf{v}}{\partial x^i} + S = 0. \tag{7.28}$$

7.4 Transformation of continuity and momentum equations

In this section we show the transformation of both the continuity and momentum equations of continuum mechanics to a general time-dependent curvilinear co-ordinate system. As indicated in the above sections, an important role is played by the grid-point velocity vector **W**, and both the cartesian and the contravariant components of **W** will be required at the grid nodes at any instant. With a differential model of grid generation, a partial differential equation will have to be solved at each time-step (a Dirichlet boundary-value problem) to obtain the new grid.

7.4.1 Continuity equation

The continuity equation of Continuum Mechanics can be written in terms of the density function $\rho(y_i, t)$ as

$$\left(\frac{\partial \rho}{\partial t}\right)_{y_i} + \nabla \cdot (\mathbf{v}\rho) = 0. \tag{7.29}$$

This can be written with respect to a moving grid, using eqn (7.22), as

$$\left(\frac{\partial \rho}{\partial t}\right)_{x^i} + \rho \nabla \cdot \mathbf{W} + \nabla \cdot [(\mathbf{v} - \mathbf{W})\rho] = 0, \tag{7.30}$$

or, by (7.27), as

$$\frac{\partial}{\partial t}(\sqrt{g}\rho) + \frac{\partial}{\partial x^i}(\sqrt{g}\rho U^i) = \frac{\partial}{\partial t}(\sqrt{g}\rho) + \frac{\partial}{\partial x^i}\left(\sqrt{g}\rho\left[v^i + \left(\frac{\partial x^i}{\partial t}\right)_{y_j}\right]\right) = 0, \tag{7.31}$$

using eqn (7.18).

7.4.2 Momentum equations

We start with the standard (non-conservative) *cartesian* form of the Momentum Equations for a viscous fluid with viscosity μ:

$$\rho\left(\frac{\partial v_i}{\partial t}\right)_{y_i} + \rho v_j \frac{\partial v_i}{\partial y_j} = \frac{\partial \sigma_{ij}}{\partial y_j}, \tag{7.32}$$

with the stress tensor σ_{ij} given in terms of pressure p and viscous stress T_{ij} by

$$\sigma_{ij} = -p\delta_{ij} + T_{ij},$$

where T_{ij} is related to the strain-rate ε_{ij} through the equations

$$T_{ij} = 2\mu\left(\varepsilon_{ij} - \frac{1}{3}\frac{\partial v_k}{\partial y_k}\delta_{ij}\right)$$

and $\varepsilon_{ij} = 1/2(\partial v_i/\partial y_j + \partial v_j/\partial y_i)$.

Now eqn (7.32) may be expressed in vector form as

$$\rho\left(\frac{\partial \mathbf{v}}{\partial t}\right)_{y_i} + \rho(\mathbf{v}\cdot\nabla)\mathbf{v} = \nabla\cdot\sigma = -\nabla p + \nabla\cdot\mathbf{T}$$

and using eqn (7.7) gives

$$\rho\left(\frac{\partial \mathbf{v}}{\partial t}\right)_{x^i} - \rho(\mathbf{W}\cdot\nabla)\mathbf{v} + \rho(\mathbf{v}\cdot\nabla)\mathbf{v} = -\nabla p + \nabla\cdot\mathbf{T}. \tag{7.33}$$

Dropping the subscript x^i for the moment, we have

$$\frac{\partial \mathbf{v}}{\partial t} = \frac{\partial(v^k \mathbf{g}_k)}{\partial t} = \frac{\partial v^k}{\partial t}\mathbf{g}_k + v^k\frac{\partial \mathbf{g}_k}{\partial t} = \left(\frac{\partial v^k}{\partial t} + v^j\Gamma^k_{j0}\right)\mathbf{g}_k. \tag{7.34}$$

The second and third terms of eqn (7.33) may be combined to give

$$\rho[(\mathbf{v}-\mathbf{W})\cdot\nabla]\mathbf{v} = \rho(v^j - W^j)\frac{\partial \mathbf{v}}{\partial x^j} = \rho(v^j - W^j)v^k_{,j}\mathbf{g}_k, \tag{7.35}$$

where $v^k_{,j}$ is a covariant derivative. The right-hand side of eqn (7.33) may be written, using eqns (1.12), (1.51), and (1.155), as

$$-g^{ik}\frac{\partial p}{\partial x^i}\mathbf{g}_k + \left[\frac{1}{\sqrt{g}}\frac{\partial}{\partial x^j}(\sqrt{g}T^{kj}) + T^{ij}\Gamma^k_{ij}\right]\mathbf{g}_k. \tag{7.36}$$

Putting eqns (7.34), (7.35), and (7.36) together into eqn (7.33), and observing that the resulting contravariant coefficients of $\mathbf{g}_k$ must be identical on both sides, gives

$$\rho\left(\frac{\partial v^k}{\partial t} + v^j\Gamma^k_{j0}\right) + \rho(v^j - W^j)v^k_{,j} = -g^{ik}\frac{\partial p}{\partial x^i} + \left[\frac{1}{\sqrt{g}}\frac{\partial}{\partial x^j}(\sqrt{g}T^{kj}) + T^{ij}\Gamma^k_{ij}\right]. \tag{7.37}$$

Exercise 2. Making use of eqns (1.155) and (1.156), show that eqn (7.37) may be expressed as

$$\begin{aligned}\rho\left(\frac{\partial v^k}{\partial t} + v^j\Gamma^k_{j0}\right) + \rho(v^j - W^j)\frac{\partial v^k}{\partial x^j} + \frac{1}{\sqrt{g}}\frac{\partial}{\partial x^j}[\sqrt{g}(pg^{kj} - T^{kj})]\\ + [\rho v^i(v^j - W^j) + pg^{ij} - T^{ij}]\Gamma^k_{ij} = 0.\end{aligned} \tag{7.38}$$

In the last two equations there is summation over the i and j suffixes from 1 to 3, and k can take any value from 1 to 3. We can also express T^{ij} in generalized contravariant form as

$$T^{ij} = 2\mu\left(\varepsilon^{ij} - \tfrac{1}{3}g^{ij}\nabla\cdot\mathbf{v}\right) \tag{7.39}$$

with $\varepsilon^{ij} = g^{ik}g^{jl}\varepsilon_{kl} = 1/2g^{ik}g^{jl}\left(v_{k,l} + v_{l,k}\right)$.

Equations (7.38) are the momentum equations for a general time-dependent curvilinear co-ordinate system in non-conservative form.

7.5 Application to a moving boundary problem

The 'in-cylinder' calculation for an internal combustion engine gives us an application of a time-dependent co-ordinate system. The physical domain in this case is the volume between cylinder head and piston face. The piston moves between the B.D.C. (bottom dead centre) position and the T.D.C. (top dead centre) position, and so the physical domain has a moving boundary. Here we take a two-dimensional model of the situation, as shown in Fig. 7.1.

Our objective is to obtain a set of co-ordinates such that the irregularly-shaped time-dependent physical domain is mapped to a fixed rectangular computational domain at all times. In Fig. 7.1 the shape of the cylinder head is given by the time-independent function $y = y_b(x)$, while the shape of the moving piston is given by $y = y_a(x, t)$.

A possible time-dependent co-ordinate transformation is

$$\xi = x/l$$
$$\eta = \frac{y - y_a(x,t)}{y_b(x) - y_a(x,t)}. \tag{7.40}$$

where l is the width of the cylinder head. This clearly maps the physical domain onto a unit square in the computational plane, in which the transport equations are to be solved, once they have been transformed to ξ, η co-ordinates. To carry out the transformation of transport equations, we require the various components of the metric tensors and Christoffel symbols.

We begin with

$$\xi_x = \frac{1}{l}, \qquad \xi_y = 0.$$

Exercise 3. Show that

$$\eta_x = -\left[(1-\eta)\frac{\partial y_a}{\partial x} + \eta\frac{dy_b}{dx}\right](y_b - y_a)^{-1}, \qquad \eta_y = (y_b - y_a)^{-1}. \tag{7.41}$$

From eqns (1.160) and (1.162) the Jacobian of the transformation is

$$\sqrt{g} = x_\xi y_\eta - x_\eta y_\xi = g(\xi_x\eta_y - \xi_y\eta_x) = \frac{g}{l}(y_b - y_a)^{-1}.$$

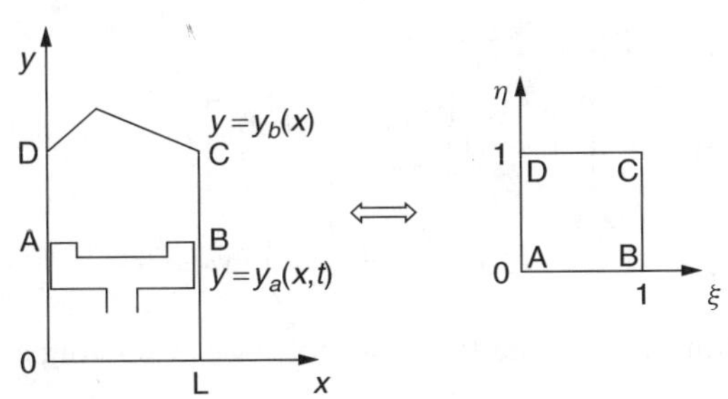

Fig. 7.1 Piston-cylinder assembly with irregular cylinder head and piston face.

Hence

$$\sqrt{g} = l(y_b - y_a). \tag{7.42}$$

Moreover,

$$x_\xi = l, \quad x_\eta = 0, \quad y_\xi = l\left[(1-\eta)\frac{\partial y_a}{\partial x} + \eta\frac{\mathrm{d}y_b}{\mathrm{d}x}\right], \quad y_\eta = (y_b - y_a). \tag{7.43}$$

It follows from eqn (1.158) that

$$\begin{aligned} g_{11} &= l^2\left\{1 + \left[(1-\eta)\frac{\partial y_a}{\partial x} + \eta\frac{\mathrm{d}y_b}{\mathrm{d}x}\right]^2\right\}, \\ g_{22} &= (y_b - y_a)^2 = g/l^2, \\ g_{12} &= l\left[(1-\eta)\frac{\partial y_a}{\partial x} + \eta\frac{\mathrm{d}y_b}{\mathrm{d}x}\right](y_b - y_a). \end{aligned} \tag{7.44}$$

where η is given in terms of x, y, and t by eqn (7.40).

Also, from (1.163),

$$\begin{aligned} g^{11} &= 1/l^2, \\ g^{22} &= \left\{1 + \left[(1-\eta)\frac{\partial y_a}{\partial x} + \eta\frac{\mathrm{d}y_b}{\mathrm{d}x}\right]^2\right\}(y_b - y_a)^{-2}, \\ g^{12} &= -\frac{1}{l}\left[(1-\eta)\frac{\partial y_a}{\partial x} + \eta\frac{\mathrm{d}y_b}{\mathrm{d}x}\right](y_b - y_a)^{-1}. \end{aligned} \tag{7.45}$$

In evaluating Christoffel symbols through eqn (1.102) we have to obtain the second partial derivatives

$$\begin{aligned} x_{\xi\xi} &= x_{\xi\eta} = x_{\eta\eta} = 0, \\ y_{\xi\xi} &= l^2\left[(1-\eta)\frac{\partial^2 y_a}{\partial x^2} + \eta\frac{d^2 y_b}{\mathrm{d}x^2}\right], \\ y_{\xi\eta} &= l^2\left(\frac{\mathrm{d}y_b}{\mathrm{d}x} - \frac{\partial y_a}{\partial x}\right), \quad y_{\eta\eta} = 0. \end{aligned} \tag{7.46}$$

Hence we get

$$\begin{aligned} \Gamma^1_{11} &= \Gamma^1_{12} = \Gamma^1_{21} = \Gamma^2_{22} = 0, \\ \Gamma^2_{11} &= l^2\left[(1-\eta)\frac{\partial^2 y_a}{\partial x^2} + \eta\frac{d^2 y_b}{\mathrm{d}x^2}\right](y_b - y_a)^{-1}, \\ \Gamma^2_{12} &= \Gamma^2_{21} = l^2\left(\frac{\mathrm{d}y_b}{\mathrm{d}x} - \frac{\partial y_a}{\partial x}\right)(y_b - y_a)^{-1}. \end{aligned} \tag{7.47}$$

Turning our attention now to the time variable, we can deduce from eqn (7.16) that

$$\Gamma^k_{i0} = \frac{\partial \mathbf{g}_i}{\partial t}\cdot\mathbf{g}^k = \frac{\partial^2 y_j}{\partial t\,\partial x^i}\frac{\partial x^k}{\partial y_j}, \tag{7.48}$$

which is similar to eqn (1.102). The required second derivatives are

$$\frac{\partial^2 x}{\partial t \partial \xi} = \frac{\partial^2 x}{\partial t \partial \eta} = 0,$$

$$\frac{\partial^2 y}{\partial t \partial \xi} = l(1-\eta)\frac{\partial^2 y_a}{\partial t \partial x}, \quad \frac{\partial^2 y}{\partial t \partial \eta} = -\frac{\partial y_a}{\partial t}. \tag{7.49}$$

Hence from eqn (7.48) we obtain

$$\Gamma^1_{10} = \Gamma^1_{01} = \Gamma^1_{20} = \Gamma^1_{02} = 0,$$

$$\Gamma^2_{10} = \Gamma^2_{01} = l(y_b - y_a)^{-1}(1-\eta)\frac{\partial^2 y_a}{\partial t \partial x}, \tag{7.50}$$

$$\Gamma^2_{20} = \Gamma^2_{02} = -(y_b - y_a)^{-1}\frac{\partial y_a}{\partial t}.$$

In this example the moving grid has been generated essentially by algebraic interpolation from the known moving boundaries.

8

Unstructured grid generation

8.1 Introduction

In structured grid generation, as we have seen, grids are constructed in the solution domain (of the partial differential equations to be solved) in such a way that grid points can be regarded as the points of intersection of curvilinear co-ordinate curves (in two dimensions) or surfaces (in three dimensions). The manner in which grid points are connected to each other (their *connectivity*) anywhere in the solution domain is thus dependent on the overall generation scheme used. In two dimensions a grid point can be specified by a pair of integers (i, j), and neighbouring points by $(i + 1, j)$, etc.; the cartesian co-ordinates $x(i, j)$, $y(i, j)$ of a point can conveniently be stored as the elements of matrices.

In recent years, however, methods of unstructured grid generation have been developed, first in structural and solid mechanics and then in computational fluid dynamics, in which the pattern of connections between grid points can vary from point to point. The connectivity of grid points then has to be described explicitly by an appropriate data structure, and this task will tend to make solution algorithms for partial differential equations using unstructured grids more expensive than for structured grids. This extra cost may be justified when the solution domains are of considerable geometric complexity, when the additional geometric flexibility of unstructured grids compared with structured grids can help to improve overall solution accuracy. Moreover, unstructured grids have been found useful when employed in an 'adaptive' manner in the numerical solution of transient flows or moving boundary problems.

The use of triangular grids in particular has grown rapidly in recent years, given that in computing flow-fields around complex geometries it may be easier to create triangular grids than quadrilateral grids (even if a quadrilateral grid can be created at all). Unstructured grid generation methods based on Delaunay triangulation are frequently used, and have been found particularly suited to adaptive solution strategies, because of their capacity to allow the addition of new grid points to an existing triangulation without affecting the grid as a whole. In other words, the addition of a new grid node to a Delaunay grid alters the grid locally, but not globally.

In this chapter we introduce the two basic approaches to the generation of unstructured grids, namely (1) Delaunay triangulation, and (2) the Advancing Front method. The presentation is confined to two dimensions here, but the extension of these methods to three dimensions is straightforward in principle. Mathematical proofs are mostly omitted. In most cases results for triangles in two dimensions can be generalized to tetrahedra in three dimensions.

8.2 Delaunay triangulation

8.2.1 Basic geometric properties

The principal objective here is to represent the two-dimensional solution domain of a problem by a set of triangles. The method discussed here, Delaunay triangulation, was first presented by Dirichlet in terms of connecting an arbitrary set of points together, thus producing a set of triangles, in such a way that the resulting triangulation was as near uniformly equilateral as possible.

Figure 8.1 shows a typical set of initial points $N_1, N_2, \ldots, N_8$, which give rise to a so-called *Dirichlet*, or *Voronoi, tessellation* of the plane into convex polygons T_1, $T_2, \ldots, T_8$, called *tiles*(or *Voronoi regions*), as shown. The tiles have the property that any point in the interior of the tile T_j is closer to the point N_j than to any other point N_k, $j \neq k$. Hence the boundaries of the tiles consist of segments of the perpendicular bisectors of lines joining the points N_1, $N_2, \ldots, N_8$ to each other. Tiles having a common edge are called *contiguous*. A Delaunay triangulation is then obtained by connecting together those initial points belonging to contiguous tiles (see Figure 8.2). The convex polygon ($N_1N_2N_3N_4N_5N_6$ in Fig. 8.2) which is the outer boundary of the triangulation is called the *convex hull* of the set of initial points.

A typical vertex of a Dirichlet tessellation (or Voronoi polygon) is clearly equidistant from three initial points, and these points form a Delaunay triangle of which the given vertex of the tessellation is the circumcentre (the centre of the circumcircle of the triangle). Fig. 8.3 shows one such circumcircle. Thus to each Delaunay triangle there corresponds a unique vertex of a Voronoi polygon lying at its circumcentre.

An important feature of a Delaunay triangulation is the *Circumcircle Property*: this guarantees that in a Delaunay triangulation none of the points (vertices) of a triangle can lie within the circumcircle of any other triangle. The triangles ABD and BCD in Fig. 8.4, for example, cannot represent a Delaunay triangulation, since the circumcircle of the triangle BCD contains the point A. However, changing the diagonal of the quadrilateral ABCD to AC instead of BD produces the triangles ABC and ADC, which

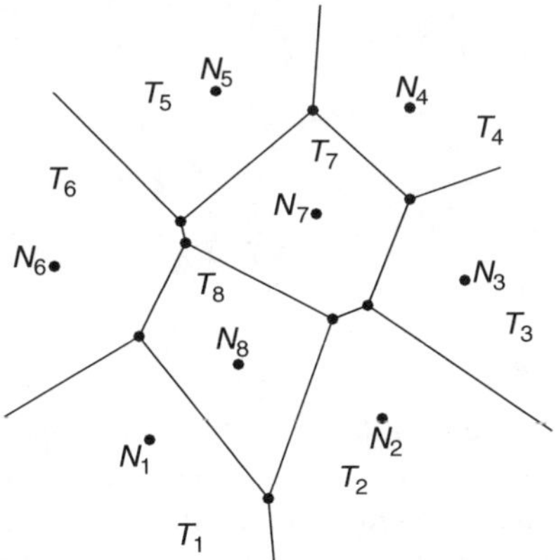

Fig. 8.1 Dirichlet tessellation with Voronoi regions.

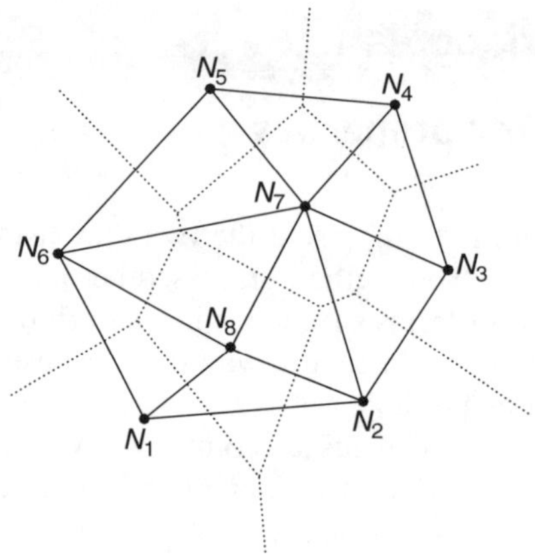

Fig. 8.2 Delaunay triangulation.

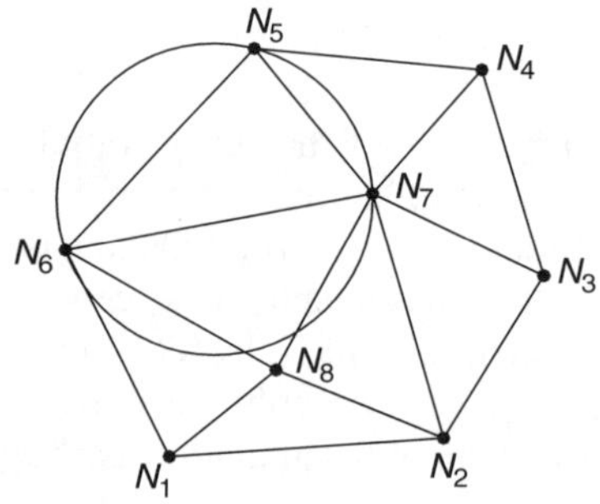

Fig. 8.3 Circumcircle for N_5, N_6, N_7.

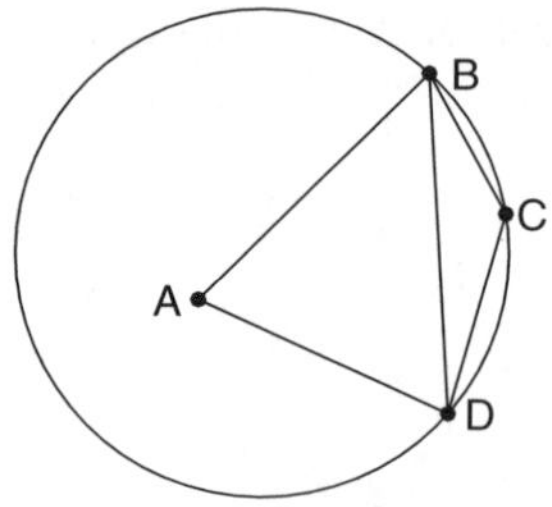

Fig. 8.4 Circumcircle property.

do represent a Delaunay triangulation. (This procedure is called, naturally, *diagonal swapping*.) Of the two possible choices of diagonal, it can be shown that the one corresponding to the Delaunay triangulation maximizes the minimum of the six angles in the resulting two triangles, and thus can be said to make the triangles as close to equilateral as possible. This choice can also be shown to be the one for which the sum of opposite interior angles ABC and ADC is less than 180 degrees (as opposed to the sum of the angles BAD and BCD, which is greater than 180 degrees). This gives a simple method for checking if the way in which four data points are connected to form two triangles is Delaunay or not.

8.2.2 The Bowyer-Watson algorithm

Among various methods for implementing Delaunay triangulation by means of a sequence of local operations, the Bowyer-Watson algorithm, due to Bowyer (1981) and Watson (1981), has been found to be very convenient and efficient. We suppose that there already exists a set of data points with a Delaunay triangulation, and seek to incorporate further points, one at a time. To initiate the iterative procedure, we could start if necessary with a sufficiently large equilateral triangle, large enough to contain all the data points to be triangulated, using the three vertices as the initial points. When all data points have been incorporated and triangulated, triangles containing the original three points of the equilateral triangle could then be removed. A sufficiently large square (divided by a diagonal into two triangles), with four initial vertices, could also be used.

Suppose the existing Delaunay triangulation contains i data points $\mathbf{x}_1, \mathbf{x}_2, \ldots, \mathbf{x}_i$, where the **x**s represent position vectors. Let the union of the triangles in the triangulation be T_i and the union of all the circumcircles of the triangles be B_i. We now wish to insert a new point $\mathbf{x}_{i+1}$ into the triangulation. There are three main possibilities:

1. $\mathbf{x}_{i+1} \in T_i$;
2. $\mathbf{x}_{i+1} \notin T_i$ and $\mathbf{x}_{i+1} \in B_i$;
3. $\mathbf{x}_{i+1} \notin B_i$.

In case 1, the new point lies in one of the triangles of T_i. It is necessary to determine the set S which is the union of those triangles belonging to T_i whose circumcircles contain the new point. The next step of the algorithm is to remove the internal edges of the triangles composing S, thus creating a 'hole' within the triangulation. Finally, a new set of edges is obtained by connecting the new point to each vertex of S. This process always produces a new Delaunay triangulation T_{i+1}.

Figure 8.5 shows an example of a Delaunay triangulation with five points $\mathbf{x}_1, \mathbf{x}_2, \ldots, \mathbf{x}_5$, with a new point $\mathbf{x}_6$ lying in one triangle and also within two circumcircles. In this case S is the quadrilateral with vertices $\mathbf{x}_2, \mathbf{x}_3, \mathbf{x}_4, \mathbf{x}_5$. The only internal edge in S is that connecting $\mathbf{x}_2$ and $\mathbf{x}_4$. Removing this creates a hole in the

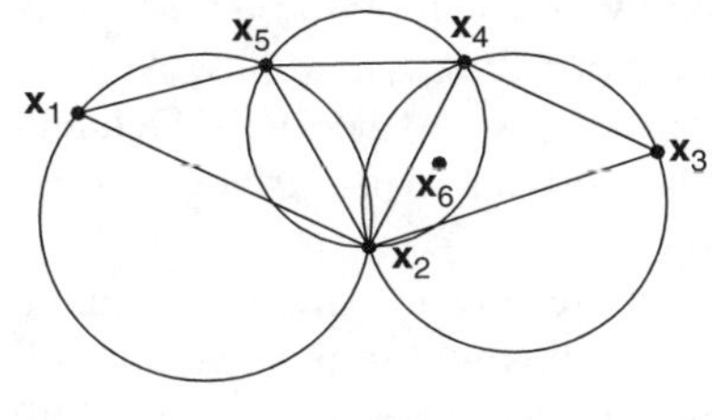

Fig. 8.5 Delaunay triangulation and new point $\boldsymbol{X}_6$.

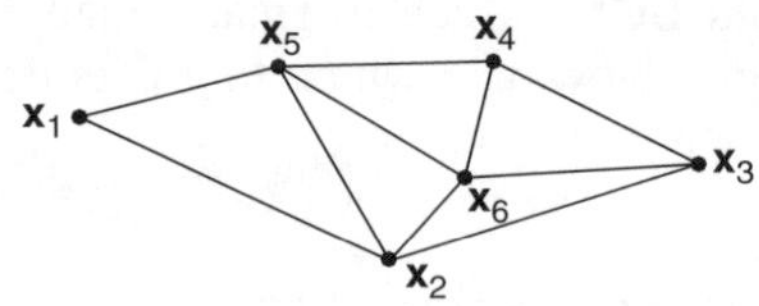

Fig. 8.6 New Delaunay triangulation.

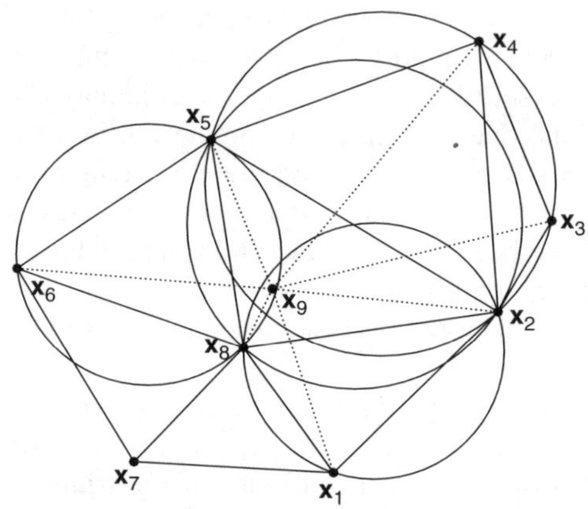

Fig. 8.7 New Delaunay triangulation with removal of four edges and creation of seven edges.

tessellation, and new edges are then created by connecting $\mathbf{x}_6$ to the four vertices of S, as shown in Fig. 8.6.

Another example of the Bowyer-Watson procedure is shown in Fig. 8.7, where we start from a Delaunay triangulation of eight points $\mathbf{x}_1, \mathbf{x}_2, \ldots, \mathbf{x}_8$. To reduce the complexity of the diagram this example has been given a degree of degeneracy, in that $\mathbf{x}_5$ lies on the circumcircle of triangle $\mathbf{x}_2\mathbf{x}_3\mathbf{x}_4$. In other words the circumcircles of triangles $\mathbf{x}_2\mathbf{x}_3\mathbf{x}_4$ and $\mathbf{x}_2\mathbf{x}_4\mathbf{x}_5$ coincide. The new data point $\mathbf{x}_9$ is now found to lie within the circumcircles of the five triangles $\mathbf{x}_2\mathbf{x}_3\mathbf{x}_4$, $\mathbf{x}_2\mathbf{x}_4\mathbf{x}_5$ (which coincide), $\mathbf{x}_1\mathbf{x}_2\mathbf{x}_8$, $\mathbf{x}_5\mathbf{x}_6\mathbf{x}_8$, and $\mathbf{x}_5\mathbf{x}_2\mathbf{x}_8$. The set S thus consists of the (non-convex) polygon $\mathbf{x}_1\mathbf{x}_2\mathbf{x}_3\mathbf{x}_4\mathbf{x}_5\mathbf{x}_6\mathbf{x}_8$, and removal of internal edges implies removal of the connections $\mathbf{x}_2\mathbf{x}_8$, $\mathbf{x}_2\mathbf{x}_4$, $\mathbf{x}_2\mathbf{x}_5$, and $\mathbf{x}_5\mathbf{x}_8$. The large hole created in the triangulation is then filled with connections from $\mathbf{x}_9$ to the points $\mathbf{x}_1$, $\mathbf{x}_2$, $\mathbf{x}_3$, $\mathbf{x}_4$, $\mathbf{x}_5$, $\mathbf{x}_6$, and $\mathbf{x}_8$ (dotted lines as shown), giving a new Delaunay triangulation.

In case 2, the new point $\mathbf{x}_{i+1}$ lies outside the existing Delaunay triangulation but within at least one circumcircle. We can define the set S as before as the union of the triangles whose circumcircles contain $\mathbf{x}_{i+1}$, but now we remove those edges in S which are nearest to (and 'visible' from) $\mathbf{x}_{i+1}$. The new Delaunay triangulation is obtained by connecting $\mathbf{x}_{i+1}$ to all the vertices of the triangles composing S and to any other vertex of T_i visible from $\mathbf{x}_{i+1}$.

Figure 8.8 shows an example of a Delaunay triangulation of five points $\mathbf{x}_1$, $\mathbf{x}_2, \ldots, \mathbf{x}_5$. A new data point $\mathbf{x}_6$ lies within the circumcircle of triangle $\mathbf{x}_1\mathbf{x}_2\mathbf{x}_5$, which is the set S. The edge connecting $\mathbf{x}_1$ to $\mathbf{x}_2$ is the one visible to $\mathbf{x}_6$, and this is deleted. The next Delaunay triangulation is obtained by connecting $\mathbf{x}_6$ to $\mathbf{x}_1$, $\mathbf{x}_5$, $\mathbf{x}_2$, and also $\mathbf{x}_3$.

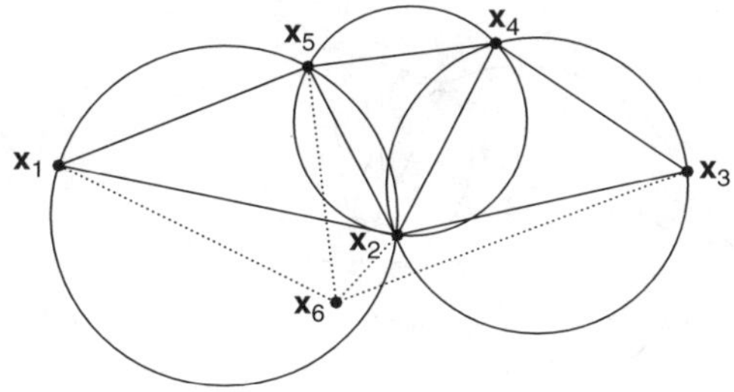

Fig. 8.8 New point $\mathbf{x}_6$ lying outside triangulation and inside one circumcircle.

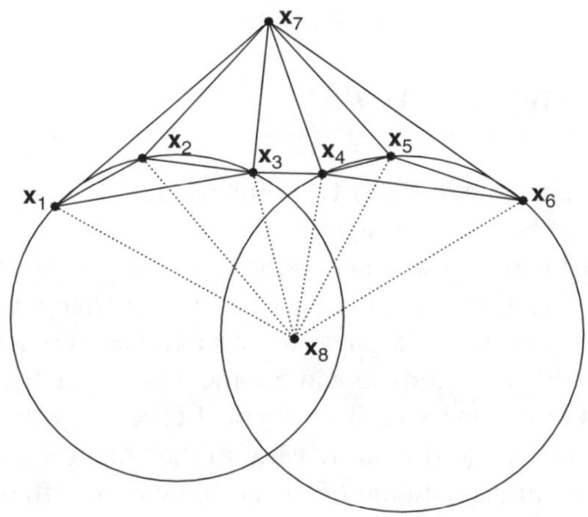

Fig. 8.9 New point lying outside triangulation and inside two circumcircles.

An example of a new point lying outside T_i but within two circumcircles is shown in Fig. 8.9. Here we start with a Delaunay triangulation of seven points $\mathbf{x}_1, \mathbf{x}_2, \ldots, \mathbf{x}_7$; the new point $\mathbf{x}_8$ lies within the circumcircles of triangles $\mathbf{x}_1\mathbf{x}_2\mathbf{x}_3$ and $\mathbf{x}_4\mathbf{x}_5\mathbf{x}_6$, which together make up the set S. We therefore remove edges $\mathbf{x}_1\mathbf{x}_3$ and $\mathbf{x}_4\mathbf{x}_6$, and connect $\mathbf{x}_8$ to the points $\mathbf{x}_1, \mathbf{x}_2, \mathbf{x}_3, \mathbf{x}_4, \mathbf{x}_5, \mathbf{x}_6$. The remaining vertex $\mathbf{x}_7$ is not 'visible' from $\mathbf{x}_8$.

In the final case 3, the new point $\mathbf{x}_{i+1}$ lies outside the existing triangulation T_i *and* the union of circumcircles. No deletion of edges in T_i is now necessary. All that is required is to locate the external edges of T_i which are visible to $\mathbf{x}_{i+1}$. The triangulation is completed by connecting $\mathbf{x}_{i+1}$ to each of the vertices on these external edges. An example is shown in Fig. 8.10, which starts with a Delaunay triangulation of five points $\mathbf{x}_1, \mathbf{x}_2, \ldots, \mathbf{x}_5$. Here the outer edges of T_i visible to $\mathbf{x}_6$ (which is not contained in any of the three circumcircles) are $\mathbf{x}_1\mathbf{x}_2$ and $\mathbf{x}_2\mathbf{x}_3$. So $\mathbf{x}_6$ must be connected to each vertex $\mathbf{x}_1, \mathbf{x}_2$, and $\mathbf{x}_3$ for the new Delaunay triangulation to be complete. Clearly the circumcircles of each of the new triangles $\mathbf{x}_1\mathbf{x}_2\mathbf{x}_6$ and $\mathbf{x}_2\mathbf{x}_3\mathbf{x}_6$ do not contain any of the three remaining points, and the Circumcircle Property is satisfied.

As might be expected, the final Delaunay triangulation of a set of data points is unaffected by the order in which individual points are inserted into a pre-existing triangulation (using the Bowyer-Watson algorithm).

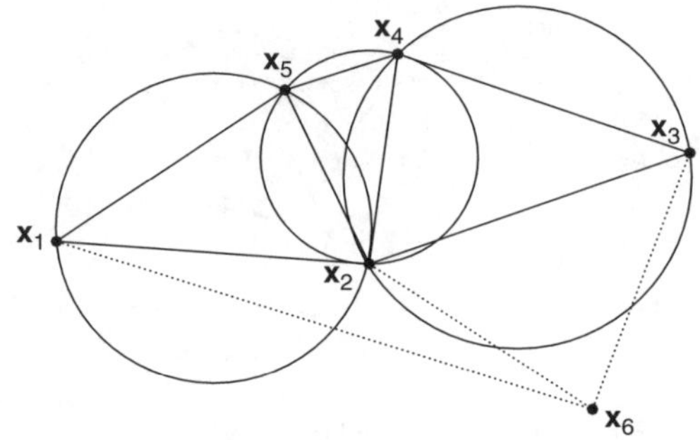

Fig. 8.10 New point lying outside triangulation and outside all circumcircles.

8.2.3 Point insertion strategies

The initial Delaunay triangulation of a two-dimensional domain, based on a selection of boundary points together with the application of the Bowyer-Watson algorithm, can often yield an unsatisfactory grid. Improvement of the grid must then be sought by inserting new points into the domain, and the resulting triangulations will be judged according to various geometric and physical criteria. The geometric criteria will normally be that the triangulation (grid) should be smooth and that the triangles (grid-cells) should be of uniform shape and size. The physical criterion will be that of adaptivity, which normally requires the grid-point density in the solution domain to vary so that areas of sharp variation in the solution of the hosted partial differential equations have greater grid-point density (and areas of low variation smaller grid-point density).

In Delaunay triangulation (in two dimensions) there are two main approaches to the insertion of new points. The first Shenton and Cendes (1985) and Holmes and Snyder (1988) is to place new interior points at the circumcentres of the triangles in an existing triangulation. The second is known as *Voronoi-segment point insertion* Rebay (1993), according to which new points are placed along so-called *Voronoi segments*.

One difficulty associated with Delaunay triangulation is that, since a given set of data points has a unique Delaunay triangulation, there is no guarantee that edges connecting boundary points will be preserved. In other words, the triangulation may not be boundary-conforming. It is possible to carry out a check on a triangulation to detect all possible missing boundary edges, and, by inserting new nodes at the mid-points of missing boundary edges, boundary-conforming triangulations can be produced.

Point insertion at triangle circumcentres

New points to be inserted should not be too close to existing grid nodes. Insertion at the circumcentre of an existing triangle at least results in the point being equidistant from the three vertices. This strategy thus involves insertion of a new point at the circumcentre of a triangle, followed by re-triangulation of the whole domain. The only remaining decisions to be made are which triangles to select for the insertion of new points. The plan here is to devise a set of rules for identifying 'unsuitable' triangles in the existing triangulation. These will be referred to as *forming triangles*. Points will then be placed at their circumcentres and re-triangulation carried out until the grid is satisfactory.

The following rules for the selection of forming triangles may be used:

1. The triangle with the largest area is selected. This rule works well when the initial triangulation is based on a set of boundary data points that are distributed in a fairly uniform manner. However, when this does not apply, the rule when used on its own does not generally produce good grids.
2. The triangle with the largest circumcircle radius is selected. This rule has the desirable feature that it will eliminate triangles with obtuse angles, together with triangles that are 'skinny' and have small area (so that they would be missed by Rule 1).
3. Obtuse-angled and skinny triangles are targeted directly. To this end a criterion for recognizing skinny triangles is required. One such is provided by consideration of the *aspect ratio* of a triangle.

The aspect ratio $A.R.$ of a triangle ABC (Fig. 8.11) is defined as $R/2r$, where R is the radius of the circumcircle and r is the radius of the inscribed circle. A standard formula for the area of a triangle is

$$area = \sqrt{s(s-a)(s-b)(s-c)} \tag{8.1}$$

where

$$s = \tfrac{1}{2}(a+b+c). \tag{8.2}$$

The following formulas are also straightforward to establish:

$$area = \tfrac{1}{2}(a+b+c)r = sr \tag{8.3}$$

and (making use of the Sine Rule for triangles)

$$area = \frac{abc}{4R}. \tag{8.4}$$

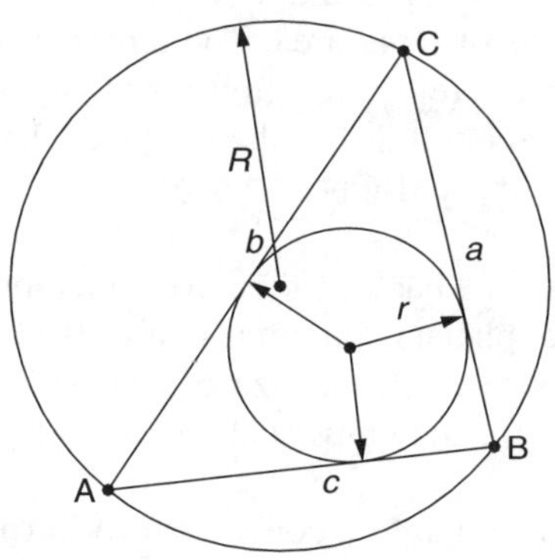

Fig. 8.11 Circumcircle and inscribed circle of triangle ABC.

Equating these expressions for area, it is easy to show that the aspect ratio of the triangle may be expressed in terms of the lengths of the sides as

$$A.R. = \frac{abc}{8(s-a)(s-b)(s-c)}. \tag{8.5}$$

For an equilateral triangle it follows that $A.R. = 1$. Clearly, the nearer the three vertices of a triangle approach to lying on a straight line, the closer will s approach to the values a, b, or c, and the larger the value of the aspect ratio will become. In that sense the aspect ratio is a measure of the 'skinniness' of a triangle.

The following algorithm makes use of the above ideas:

Step 1 – Generate an initial triangular grid using boundary data points and the Bowyer-Watson algorithm.

Step 2 – Apply Rule 1 to eliminate all triangles having an area larger than 1.5 times the area of the equilateral triangle formed by the largest boundary segment between two data sets. This step will eliminate all large triangles.

Step 3 – Apply a combination of Rules 2 and 3 to eliminate triangles with large circumcircle radius and high aspect ratio. It may be convenient to leave alone triangles near the boundary which are small but have a high aspect ratio (by applying Rule 2 only). The criterion for high aspect ratio is empirical. Often a critical aspect ratio of 1.5 is used; this would be the aspect ratio of isosceles triangles with apex angles of about 24 or 104 degrees. This criterion can be applied to eliminate all triangles of high aspect ratio, or until all triangles with high aspect ratio have an area of less than twice the area of an equilateral triangle with sides equal to the minimum boundary-point spacing.

However, when this algorithm is adopted, three situations may arise when point insertion at circumcentres of the selected triangles must be rejected, namely:

Case 1 – The circumcentre of the selected triangle does not lie within the solution domain.

Case 2 – The circumcentre of the selected triangle is too close to the boundary of the solution domain.

Figure 8.12 shows an example of a solution domain with inner and outer boundaries and a selection of boundary points (circled). The insertion of point B leads to a triangle ABC of high aspect ratio. However, the circumcentre P of ABC lies outside the solution domain and must be rejected. As an example of Case 2, insertion of point D leads to a triangle EDA which may be of large area. The circumcentre Q is, however, close to the boundary. The criterion for rejection often taken is that the distance of the point from the boundary is less than one-third of the length of the nearest boundary segment. In this case rejection would apply if QN $< \frac{1}{3}$ EA.

Control of grid-density A drawback of the above method is that the grid-density of triangulation over the whole (planar) domain is controlled by the grid-density along the boundary of the domain. As a result, the size of triangles far from the boundary may turn out to be unsatisfactory for the numerical solution of partial differential equations in the domain.

To overcome this problem, it may be convenient to introduce a function of position $f(\mathbf{x})$ which specifies the required value of a characteristic dimension of a triangle, say its circumradius, at a location $\mathbf{x}$ in the domain. For a triangle with circumradius R and

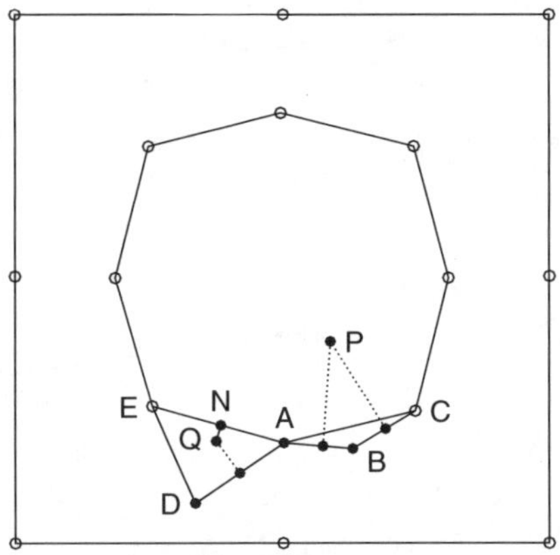

Fig. 8.12 Constructing new points in a doubly-connected region.

circumcentre $\mathbf{X}$ we can then define the parameter $\alpha = R/f(\mathbf{X})$. New grid-points may now be inserted at triangles having the largest value of α. After a number of iterations it should be possible to arrive at a grid in which the maximum value of α for any triangle is less than or equal to unity, which means that all triangles have reached their target size.

Voronoi-segment point insertion method

An alternative point insertion strategy, as proposed by Rebay, is to place a new point along a segment of a Voronoi polygon inside a triangle instead of at its circumcentre. The locations at which new points are inserted in the circumcentre approach described above are fixed (at the circumcentres of the chosen triangle), and the required size of the grid cells (triangles) is achieved after a number of iterations. In the Voronoi-segment approach, however, one can in principle aim to produce grid cells of the final specified size as soon as the procedure is invoked.

For any given Delaunay triangulation of the domain, the triangles are divided into *external* triangles, which have at least one side consisting of a boundary segment, and *internal* triangles, which do not. (An initial triangulation based on boundary points would consist of external triangles only.) The internal triangles are divided into so-called *accepted* and *non-accepted* triangles, accepted triangles being those defined as having circumradius less than 1.5 times the user-specified target value given by local values of the function $f(\mathbf{X})$ discussed in the previous section. The remaining triangles are non-accepted.

The algorithm starts by considering non-accepted triangles which have an edge in common with an accepted triangle and choosing the one with largest circumradius. Figure 8.13 shows a typical situation, in which the triangle ABD is non-accepted, with circumradius R_{ABD}, and ABC an accepted triangle. The common edge AB is called an *active* edge. The Voronoi segment for the two triangles is the line EF connecting the circumcentres E and F, which is the perpendicular bisector of AB, the point of intersection being M. The idea now is to choose a new point X somewhere on the segment EF such that the triangle ABD will be replaced by an accepted triangle ABX, with a new Delaunay triangulation. The local target circumradius may be taken as the value of $f(\mathbf{X})$ at M, which we denote by f_M.

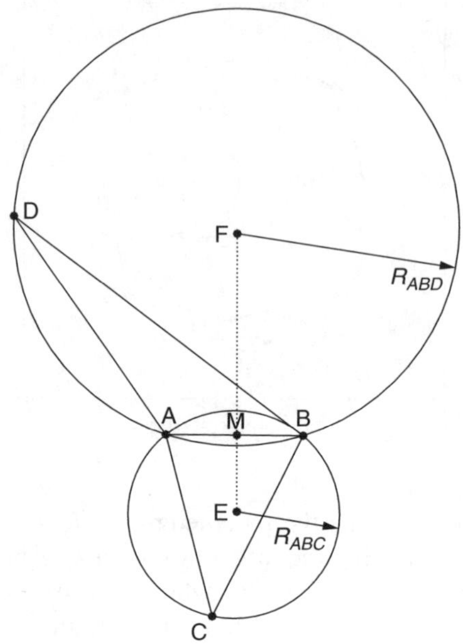

Fig. 8.13 Non-accepted and accepted triangles.

Writing lengths $\overline{AM} = p$ and $\overline{MF} = q$, it is straightforward to show that insertion of a new point at the triangle circumcentre F (by choosing X to coincide with F) would give a triangle AFB with circumradius given by $(p^2+q^2)/2q$. Since the smallest radius of any circle passing through the points A and B is clearly p (the circle with centre M), we must have

$$(p^2+q^2)/2q \geqslant p$$

(which is also true for simple algebraic reasons). If $f_M \leqslant p$, the best choice for X is such that the circumradius R_{ABX} of the triangle ABX is equal to p. This occurs when $\overline{MX} = p$ and the angle AXB is a right-angle. If $f_M > p$, however, we locate X so that

$$R_{ABX} = \min(f_M, (p^2+q^2)/2q)$$

which will give a position for X between the previously stated position and F. In this case we have

$$\overline{MX} = R_{ABX} + \sqrt{(R_{ABX})^2 - p^2}. \tag{8.6}$$

In either case we can write as in Liseikin (1999)

$$R_{ABX} = \min\{\max(f_M, p), (p^2+q^2)/2q\}, \tag{8.7}$$

with the position of X still given by (8.6).

Suppose that $f_M < p < 1.5 f_M$ and q is large compared with p. Then the point X is chosen such that the angle AXB is a right-angle; moreover, the triangle AXB satisfies the criterion to be *accepted*. The edges XA and XB will now be candidates for the next active edge, assuming that the triangle DAX is non-accepted. Here we consider

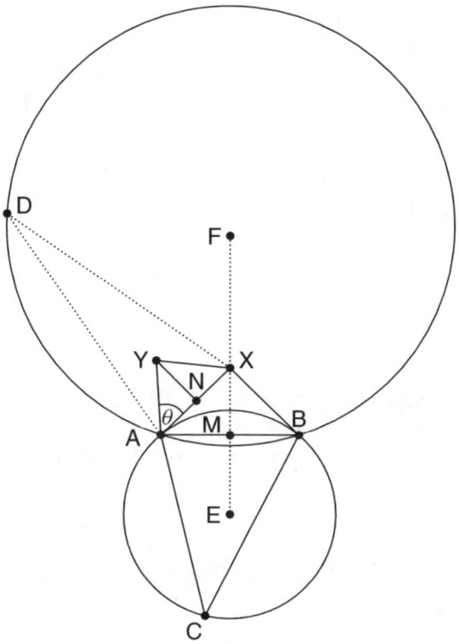

Fig. 8.14 Voronoi-segment point insertion.

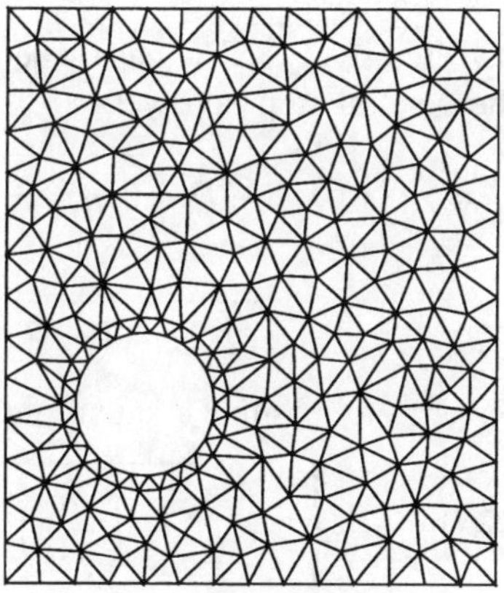

Fig. 8.15 Delaunay triangulation. Voronoi-segment algorithm.

XA. We look for a new vertex Y on the perpendicular bisector of AX (Fig. 8.14). We can re-label p as p_0, and in the new construction, where the mid-point of AX is N, we put $\overline{AN} = p_1$. Clearly, $p_1 = p_0/\sqrt{2}$.

Now if $p_1 > f_N$, where f_N is the value of the target circumradius at N, then the previous step is repeated, and Y will be chosen such that the angle AYX is a right-angle. But if $p_1 < f_N$, we expect that, given that q is large compared with p_0, the circumradius $R_{AYX} = f_N$ according to eqn (8.7), and, by eqn (8.6),

$$\overline{NY} = d_1 = f_N + \sqrt{(f_N)^2 - (p_1)^2}. \tag{8.8}$$

This means that since for the angle θ in Fig. 8.14

$$\tan\theta = \frac{d_1}{p_1} = \frac{f_N}{p_1} + \sqrt{\left(\frac{f_N}{p_1}\right)^2 - 1},$$

provided that the value of f is essentially the same at M and N (and equal to f_N), so that $p_1 < f_N < \sqrt{2}p_1$, we have

$$1 < \tan\theta < \sqrt{2} + 1,$$

which implies that

$$45^\circ < \theta < 67.5^\circ.$$

If AY is taken as the next active edge, we will have

$$\overline{AY}^2 = (2p_2)^2 = (p_1)^2 + (d_1)^2,$$

and, using eqn (8.8), we obtain

$$(p_2)^2 = \frac{1}{2}\left\{(f_N)^2 + f_N\sqrt{(f_N)^2 - (p_1)^2}\right\}. \tag{8.9}$$

Since we are assuming that $p_1 < f_N < \sqrt{2}p_1$, it follows that

$$\frac{1}{\sqrt{2}} < \frac{p_2}{p_1} < \sqrt{1 + \frac{1}{\sqrt{2}}}.$$

If this procedure is repeated iteratively in the case where the value of f is assumed to be effectively constant, the general step involves

$$(p_{n+1})^2 = \frac{1}{2}\left\{(f_N)^2 + f_N\sqrt{(f_N)^2 - (p_n)^2}\right\} \tag{8.10}$$

and

$$d_n = f_N + \sqrt{(f_N)^2 - (p_n)^2}. \tag{8.11}$$

The values of p_n then converge to a value p satisfying, according to eqn (8.10),

$$\left(\frac{p}{f_N}\right)^2 = \frac{1}{2}\left\{1 + \sqrt{1 - \left(\frac{p}{f_N}\right)^2}\right\}. \tag{8.12}$$

It is straightforward to show that the solution of this equation is

$$p = \frac{\sqrt{3}}{2} f_N, \tag{8.13}$$

and, furthermore, the corresponding value of d, according to eqn (8.11), is

$$d = \tfrac{3}{2} f_N. \tag{8.14}$$

These are characteristic measurements for an equilateral triangle of circumradius f_N. Thus this algorithm tends to produce approximately equilateral triangles of the target circumradius after a few iterations.

Examples of triangulations in domains of various shapes using Voronoi point-insertion are shown in Figs 8.15–8.18.

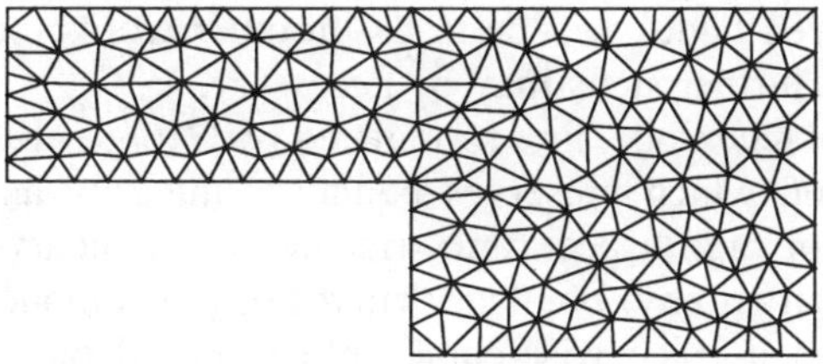

Fig. 8.16 Delaunay triangulation. Voronoi-segment algorithm.

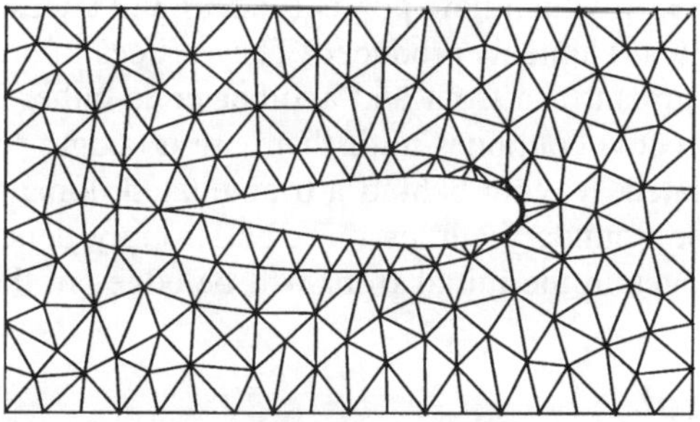

Fig. 8.17 Delaunay triangulation for an airfoil. Voronoi-segment algorithm.

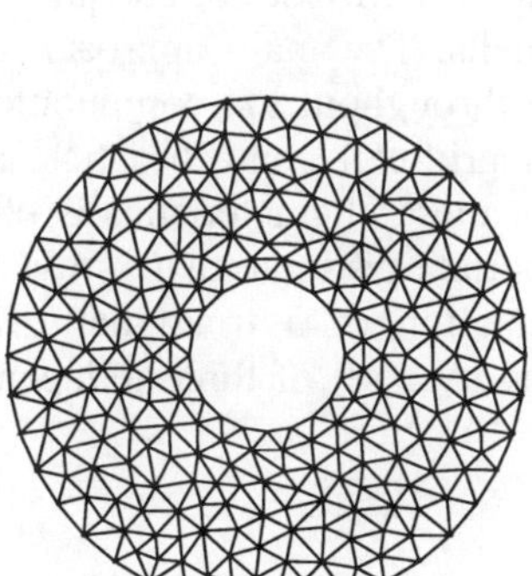

Fig. 8.18 Delaunay triangulation in an annulus. Voronoi-segment algorithm.

8.3 Advancing front technique (AFT)

8.3.1 Introduction

In certain problems, for example the computation of viscous flow solutions, the use of Delaunay triangulation for generation of unstructured grids may not be satisfactory. This could be due to the need to create triangular elements with high aspect ratio in boundary-layer regions, which would be difficult with Delaunay Triangulation alone. Another problem with Delaunay triangulation, as mentioned above, is that, even though boundary nodes will be vertices in the final triangulation, there is no guarantee that boundary edges between nodes will be sides. In other words, boundary integrity may not be preserved, and further steps may be necessary.

The advancing front technique, first formulated by George (1971), is an unstructured grid generation method which preserves boundary integrity and has the capacity to create the clustering of high aspect-ratio triangles in boundary-layer regions. In this method, outer and inner (if any) boundary curves of the computational domain, which are commonly defined as piecewise-cubic splines based on a user-specified set of points, are discretized by being divided into straight-line segments which correspond to a chosen node distribution of the boundary of the domain. These sets of straight edges compose the initial 'fronts'. The fronts then move into the interior of the domain in a 'marching' process, in which new points (nodes) and edges are created, old edges are deleted, and triangular elements produced. The vertices of a new triangle consist of the two nodes of a segment of a front and another node either already in the front or newly created. This process continues until there are no edges left in the front, i.e. the front has been annihilated, leaving behind a triangulated domain. Note that the initial choice of nodes on the boundary curves must be strongly dependent on the required grid-cell size, since edges in the initial front will be edges in the final triangulation.

8.3.2 Grid control

Any grid-generation method should provide for adequate grid control regarding acceptable size and shape of grid cells. The main approach to control of grid-cell size in the AFT (in two dimensions) throughout the computational domain involves first the definition of certain required grid-cell characteristics, and then the generation of a *background grid*. Control over these characteristics is obtained by the specification of a spatial distribution of appropriate grid parameters over the background grid.

The size, shape, and orientation of a triangular grid-cell (see triangle ABC in Fig. 8.19) is roughly described by a set of three independent parameters:

- a size parameter δ;
- a 'stretching' parameter s;
- the orientation of the cell ϕ, which is associated with two mutually orthogonal vectors $\mathbf{s}$, $\mathbf{n}$.

To define a grid-cell a user can input four grid generator parameters (δ, s, n_x, n_y), where n_x, n_y specify the components of $\mathbf{n}$ with respect to the axes of the global cartesian

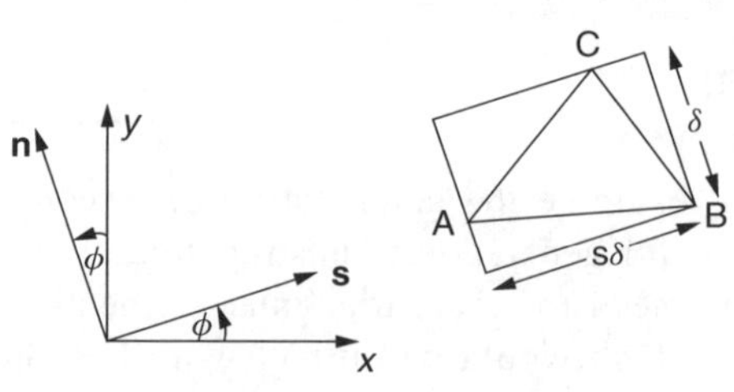

Fig. 8.19 Descriptive parameters for triangular elements.

co-ordinate system *Oxy*. For control purposes, required values of these parameters are specified at each node of the background grid. The initial background grid is usually generated manually by the user, and can be quite coarse, even for complex domains. For example, a background grid consisting of a single element, or two elements (triangles), can be used to impose the requirement of linear variation of parameters, or of constant spacing and stretching throughout the domain.

The background grid does not have to align with the boundary of the domain to be triangulated. In cases where no initial background grid is supplied, a default background grid is generated, consisting of two triangular elements with a uniform grid-density requirement, based on empirical rules. The default grid-density value of δ is taken to be five per cent of the length of the diagonal of the background grid. When adaptive methods are required, the first grid generated could become the background grid for the next grid, and then a more detailed specification of the spatial variation of grid generator parameters can be achieved.

An alternative, or additional, method for grid control, particularly when complex geometries are involved and there is a need to specify grid parameters in certain regions, such as the leading and trailing edges of aircraft wings, is provided by a so-called *distribution of sources*. In this approach a spatial distribution of grid-cell size is specified as a function of the distance from a given point to a 'source', which could be a point or a line. The distribution is 'isotropic' if it depends only on the distance x measured in *any* direction from the source. The isotropic source function for a point source at S is taken to be

$$\delta(x) = \begin{cases} \delta_1 & \text{if } 0 < x < x_c \\ \delta_1 \exp\left(\left\{ \dfrac{x - x_c}{D - x_c} \right\} \ln 2\right) & \text{if } x \geqslant x_c, \end{cases} \tag{8.15}$$

where δ_1, D, and x_c are user-specified parameters that can be tuned to control the variation of triangle size δ about S. A typical graph of $\delta(x)$ is shown in Fig. 8.20.

8.3.3 Searching algorithm

To be able to interpolate grid parameters from the background grid (in two dimensions), it is necessary to locate the triangle of the background grid in which a given point in the domain lies. This can be achieved by computing the so-called *area co-ordinates* of the point. Suppose we have a triangle with vertices labelled 1, 2, 3. The area of this

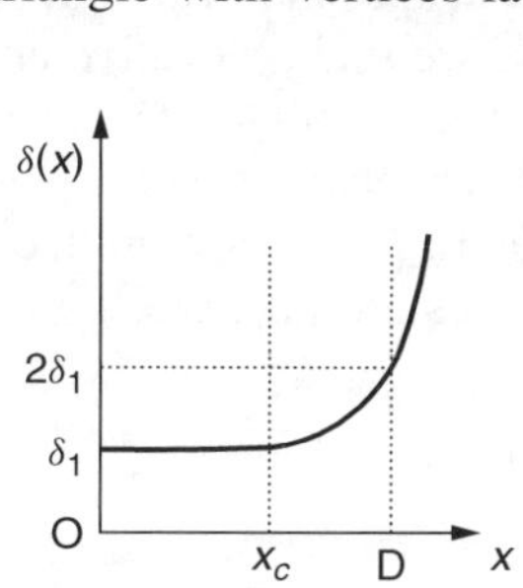

Fig. 8.20 A possible source function.

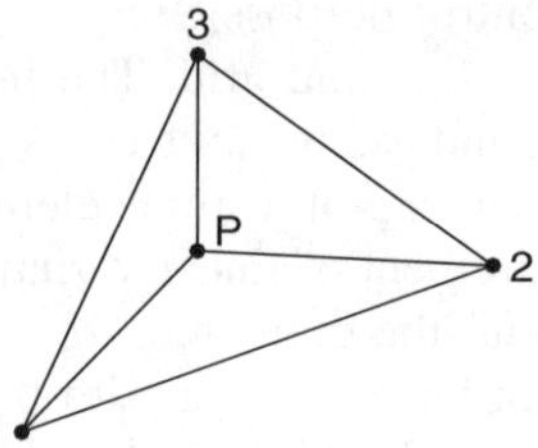

Fig. 8.21 Points 1, 2, 3 forming anti-clockwise sequence.

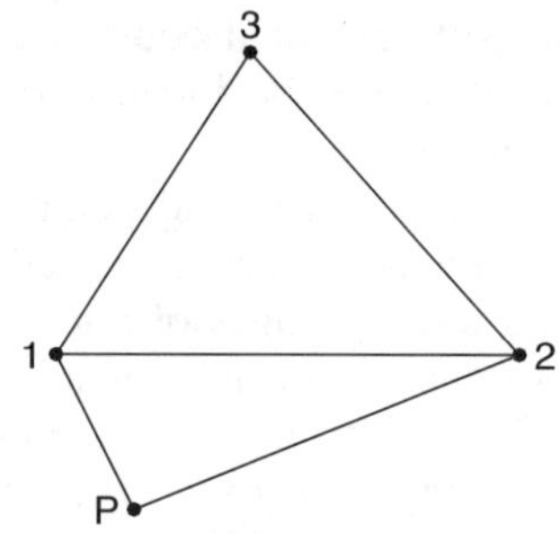

Fig. 8.22 Points 1, 2, P forming a clockwise sequence.

triangle, denoted by $\Delta 123$, will be taken to be positive if the order 1, 2, 3 follows an anti-clockwise sequence of points (as in Fig. 8.21), but negative otherwise. The area co-ordinates of a point P are then given as ratios of areas:

$$l_1 = \frac{\Delta 23P}{\Delta 123}, \quad l_2 = \frac{\Delta 31P}{\Delta 123}, \quad l_3 = \frac{\Delta 12P}{\Delta 123}, \tag{8.16}$$

where areas are taken positive or negative following the same anti-clockwise or clockwise convention. Thus if P lies within a triangle with positive area, as in Fig. 8.21, the area co-ordinates are all positive. However, for point P located as in Fig. 8.22, we have $l_3 < 0$, while $l_1 > 0$, $l_2 > 0$. It is clear that, when P lies outside the triangle 123, at least one area co-ordinate is negative.

So, for a given point $P(x_P, y_P)$, we take a triangle (for example, the last triangle that has been generated) of the background grid, and calculate the area co-ordinates of P with respect to that triangle. If we find a negative area co-ordinate, this will indicate the direction of search for the next triangle to be tested. For example, in Fig. 8.22, since l_3 is found to be negative, we next test the triangle opposite to the vertex 3, i.e. the triangle that shares the edge 12. We can continue this procedure until we reach a triangle where all area co-ordinates are positive.

8.3.4 AFT algorithm

A principal feature of the AFT is the simultaneous generation of grid nodes and triangular grid-cells. The validity of each new generated triangle is checked locally as soon as it is created. Here we give the algorithmic steps of the AFT followed by a test-case example.

The basic steps of the algorithm are as follows:

1. Set up the initial grid generation front, a set of oriented line-segments connecting selected nodes on the boundary (in the direction of the orientation). Interpolate grid-cell parameters for all the nodes in the front. Any nodes in the current position of the front are called *active*.
2. Select the shortest side in the front, with length l. A value of δ is obtained by averaging the interpolated values of δ corresponding to the two nodes.
3. Compute the position of an 'ideal' point K_{ideal} on the perpendicular bisector of this side such that an equilateral triangle is formed with K_{ideal} as vertex.
4. Construct a circle with centre at K_{ideal} and radius r given by the empirical formula $r = 0.8 * \delta'$, where

$$\delta' = \begin{cases} 0.55 * l & \text{if } \delta < 0.55 * l \\ \delta & \text{if } 0.55 * l \leqslant \delta \leqslant 2.0 * l \\ 2.0 * l & \text{if } \delta > 2.0 * l, \end{cases}$$

   and δ takes its local value. This formula helps to avoid the creation of highly distorted triangles and to ensure geometrical compatibility.
5. Find the active nodes that lie within this circle, and list them in terms of their distance from K_{ideal}. The closest would be the best candidate for the third vertex of the next triangular element.
6. If there are no such active nodes, the validity of the equilateral triangle with K_{ideal} as a vertex must be checked. This requires that:

   - The point K_{ideal} does not lie inside another triangular element.
   - The sides of the new element do not intersect any of the existing sides of the active front.

   If the triangle is not valid, look for an active node giving a triangle with the best shape.
7. If step 6 fails, re-order the front, take the second shortest active side, and go to step 3.
8. If step 5 or step 6 succeeds, generate a new triangular cell and update the front (from which the chosen shortest side has been deleted). In the case of step 6 being successful with K_{ideal} as a vertex (which then becomes one of the active nodes of the new front), interpolate the grid generation parameter for this new point.
9. If the updated front is 'non-empty', go to step 3 and repeat the procedure. Continue until there are no edges left in the front. Thus all edges have become *passive* instead of active, and the computational domain has been completely triangulated.

As a simple example, we consider the simply-connected two-dimensional domain shown in Fig. 8.23, consisting of a square ABDC from which a semi-circle has been removed. The background grid consists of the two triangles ACD, ABD, and grid generation parameters on the background grid are specified to be $\delta = 1$, $s = 1$, $n_x = 1$, $n_y = 0$, at each point A, B, C, D. Here we show a simple step-by-step method of grid generation.

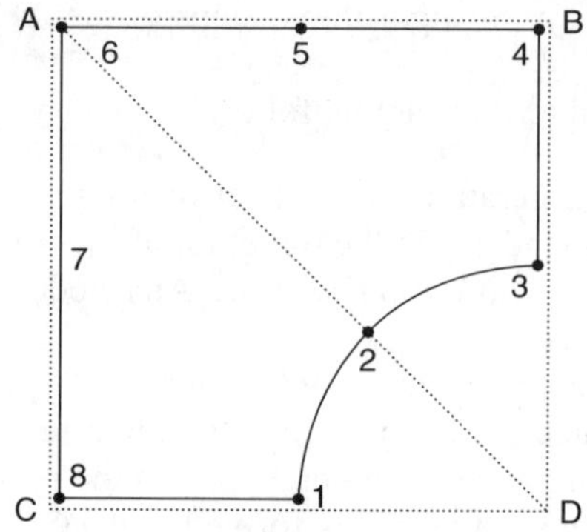

Fig. 8.23 Domain for AFT with background grid.

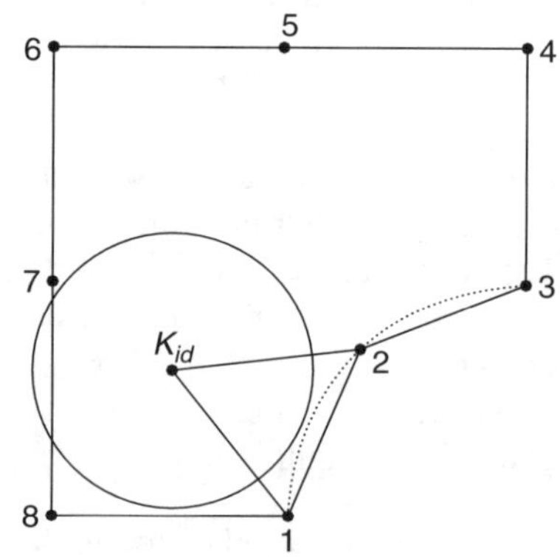

Fig. 8.24 First front with ideal point.

Nodes 1 to 8 are chosen, and the first front is shown in Fig. 8.24. One of the two shortest faces (1-2) is chosen and an ideal point is constructed as shown in the same figure. The circle of radius δ' is also indicated. This contains no nodes, and the equilateral triangle created is valid. Hence the ideal point is acceptable, and can be taken as a new node. The front is now updated. As shown in Fig. 8.25 we have:

- active nodes: 1, 9, 2, 3, 4, 5, 6, 7, 8
- active sides: 1-9, 9-2, 2-3, 3-4, 4-5, 5-6, 6-7, 7-8, 8-1
- passive nodes: none
- passive faces; 1-2
- npoint (total number of points): 9
- nelem (total number of elements created): 1
- nptr (number of points remaining, i.e. active nodes): 9

Next the face 2-3 is chosen. This leads to a similar construction to that in the previous step, as shown in Fig. 8.26, where again the ideal point is acceptable and taken as the new node. The updated front is shown in Fig. 8.27, with

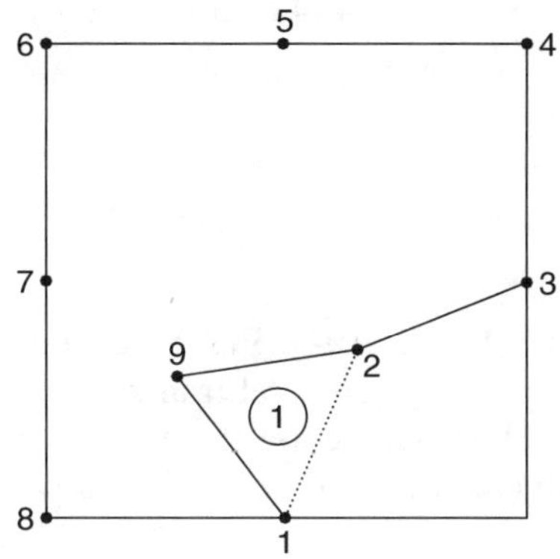

Fig. 8.25 Second front.

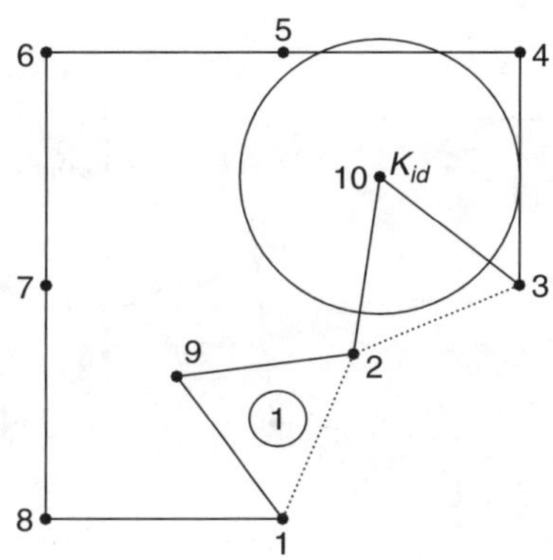

Fig. 8.26 Ideal point for side 2-3.

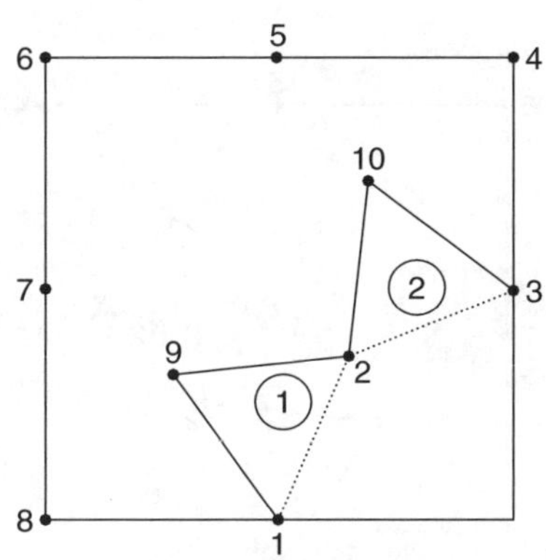

Fig. 8.27 Third front.

- active nodes: 1, 9, 2, 10, 3, 4, 5, 6, 7, 8
- active sides: 1-9, 9-2, 2-10, 10-3, 3-4, 4-5, 5-6, 6-7, 7-8, 8-1
- passive nodes: none
- passive sides 1-2, 2-3
- npoint: 10
- nelem: 2
- nptr: 10

In the next step the chosen face is 10-3 (Fig. 8.28), but now K_{ideal} does not lie in the interior of the domain, and the equilateral triangle produced is not valid. Thus the ideal point is not acceptable. Instead, we use the existing active node 4, and take 10-4 as a new active side. The updated front (Fig. 8.29) has properties as follows:

- active nodes: 1, 9, 2, 10, 4, 5, 6, 7, 8
- active sides: 1-9, 9-2, 2-10, 10-4, 4-5, 5-6, 6-7, 7-8, 8-1
- passive nodes: 3

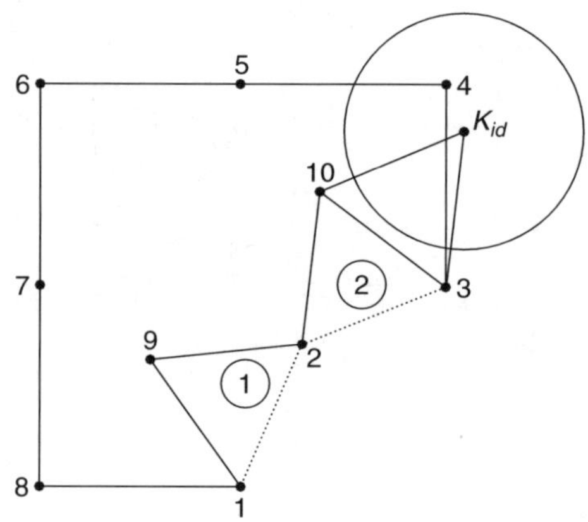

Fig. 8.28 Side 3-10; ideal point.

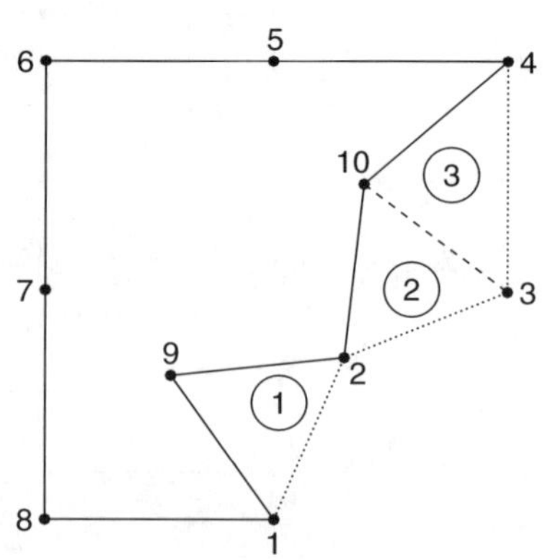

Fig. 8.29 Fourth front.

- passive sides: 1-2, 2-3, 3-4
- npoint: 10
- nelem: 3
- nptr: 9

Next, the chosen side is 2-10, but the circle with centre K_{ideal} now contains node 9 (Fig. 8.30) and so this is used as the vertex of the next element. The updated front is shown in Fig. 8.31 with properties:

- active nodes: 1, 9, 10, 4, 5, 6, 7, 8
- active sides: 1-9, 9-10, 10-4, 4-5, 5-6, 6-7, 7-8, 8-1
- passive nodes: 2, 3
- passive sides: 1-2, 2-3, 3-4
- npoint: 10
- nelem: 4
- nptr: 8

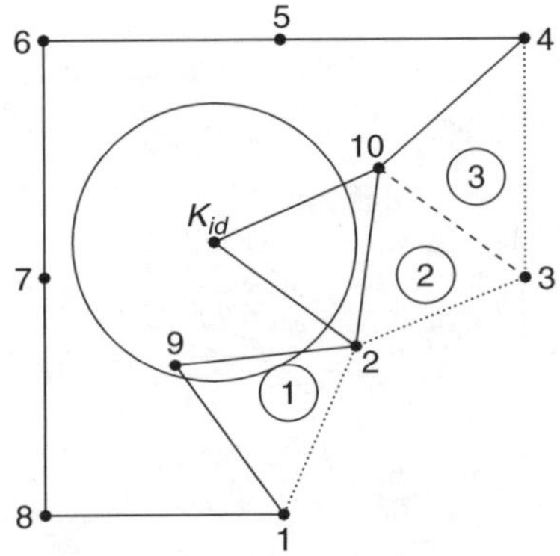

Fig. 8.30 Side 2-10; ideal point.

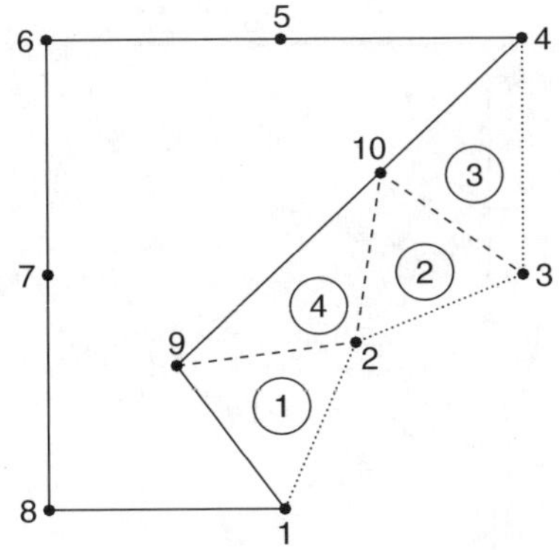

Fig. 8.31 Fifth front.

Next, the side 1-9 is chosen. This leads to a procedure symmetrical to that involved in the choice 10-3 above. The resulting updated front is shown in Fig. 8.32, and has the properties:

- active nodes: 9, 10, 4, 5, 6, 7, 8
- active sides: 9-10, 10-4, 4-5, 5-6, 6-7, 7-8, 8-9
- passive nodes: 1, 2, 3
- passive sides: 8-1, 1-2, 2-3, 3-4
- npoints: 10
- nelem: 5
- nptr: 7

The next chosen side is 8-9. This time K_{ideal} is found to lie outside the computational domain (as shown in Fig. 8.33) and is rejected since a side of the new equilateral triangle cuts the current face. Point 7 is chosen instead, giving the updated front shown in Fig. 8.34 with properties:

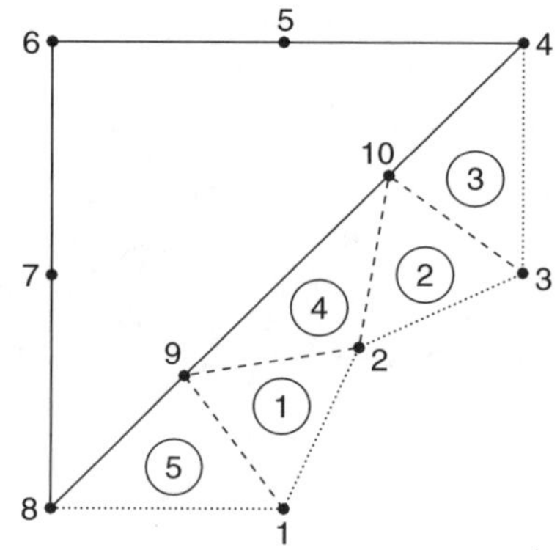

Fig. 8.32 Sixth front.

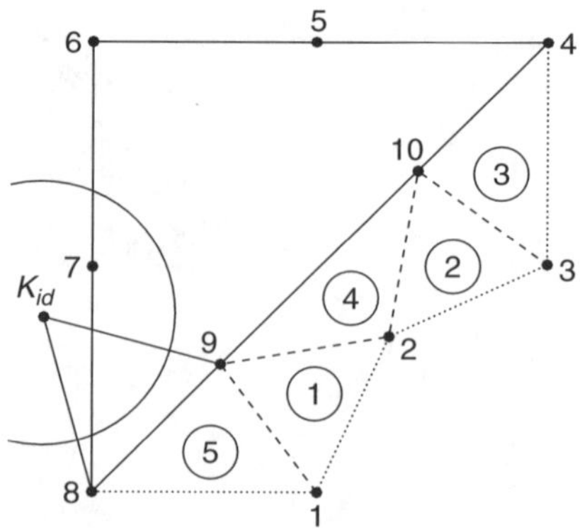

Fig. 8.33 Side 8-9; ideal point.

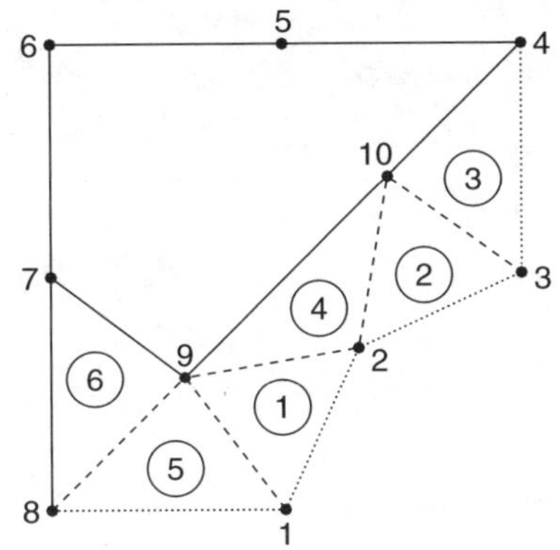

Fig. 8.34 Seventh front.

- active nodes: 9, 10, 4, 5, 6, 7
- active sides: 9-10, 10-4, 4-5, 5-6, 6-7, 7-9
- passive nodes: 8, 1, 2, 3
- passive sides: 7-8, 8-1, 1-2, 2-3, 3-4
- npoint: 10
- nelem: 6
- nptr: 6

Now the side 7-9 is chosen, and the new equilateral triangle with centre K_{ideal} (Fig. 8.35) is valid. Thus K_{ideal} becomes the new point 11, and the updated front is as shown in Fig. 8.36, with properties:

- active nodes: 9, 10, 4, 5, 6, 7, 11
- active sides: 9-10, 10-4, 4-5, 5-6, 6-7, 7-11, 11-9
- passive nodes: 8, 1, 2, 3
- passive sides: 7-8, 8-1, 1-2, 2-3, 3-4
- npoint: 11

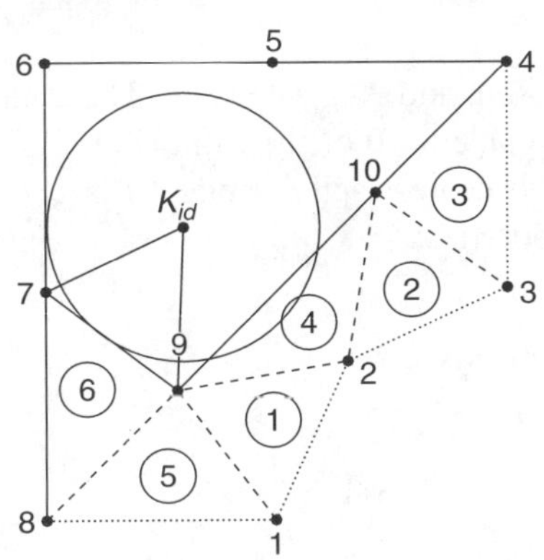

Fig. 8.35 Side 7-9; ideal point.

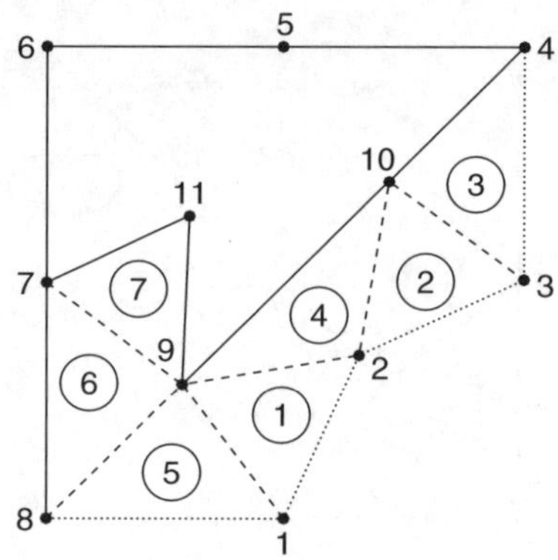

Fig. 8.36 Eighth front.

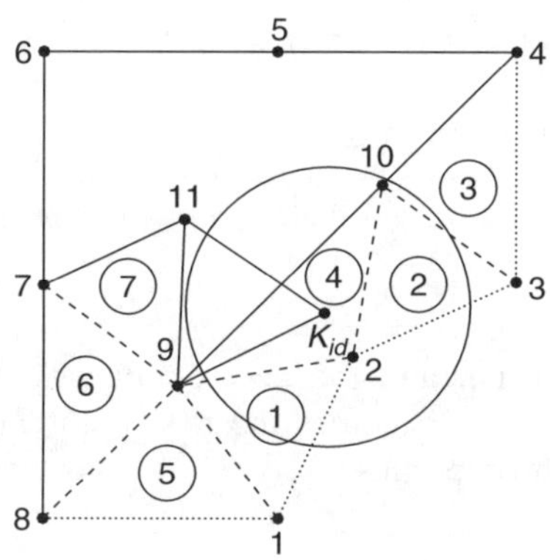

Fig. 8.37 Side 11-9; ideal point.

- nelem: 7
- nptr: 7

Next the side 11-9 is chosen, and K_{ideal} located. The equilateral triangle constructed has one side which cuts the side 9-10 of the front (Fig. 8.37), and so K_{ideal} is rejected as a new point. Instead, the nearest active node 10 is chosen, and the updated front shown in Fig. 8.38 has properties:

- active nodes: 7, 11, 10, 4, 5, 6
- active sides: 7-11, 11-10, 10-4, 4-5, 5-6, 6-7
- passive nodes: 8, 1, 2, 3
- passive sides: 7-8, 8-1, 1-2, 2-3, 3-4
- npoint: 11
- nelem: 8
- nptr: 6

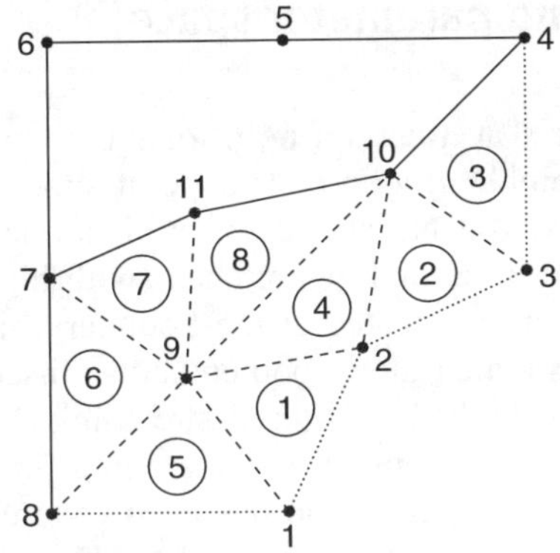

Fig. 8.38 Ninth front.

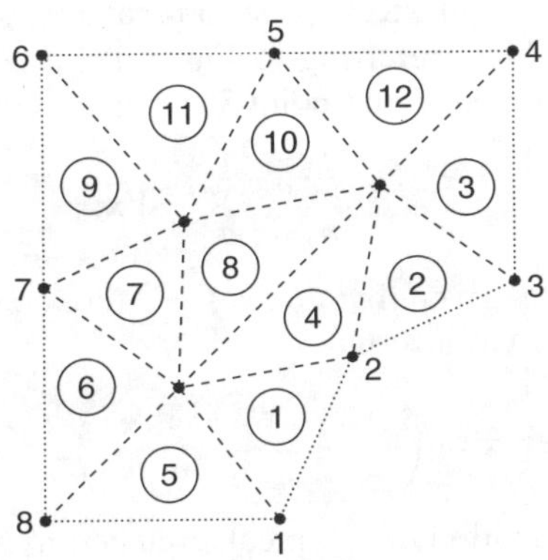

Fig. 8.39 Completed triangulation.

If the algorithm is applied for a further four stages, the shortest face being chosen at each stage, we arrive at the triangular grid shown in Fig. 8.39, in which the front has finally disappeared and the following properties can be deduced:

- active nodes: none
- active sides: none
- passive nodes: 1, 2, 3, 4, 5, 6, 7, 8
- passive sides: 1-2, 2-3, 3-4, 4-5, 5-6, 6-7, 7-8, 8-1
- npoint: 11
- nelem: 12
- nptr: none

Note that, in general, when a new point is found to be very close to an active node on the front, it is replaced by this node. This avoids having to deal with a triangular cell with a very small edge at some later stage.

8.3.5 Adaptation and parameter space

If the stretching parameter s at a certain point of the computational domain is equal to unity, the grid in the neighbourhood of that point should consist of approximately equilateral triangles. However, as has already been mentioned, grid cells near boundary segments are often required to be highly stretched (with high aspect ratio). A uniform distribution of such cells can be created at the boundary, new points (vertices of the stretched triangles) being situated along perpendicular bisectors of the initial boundary segments in accordance with the required stretching. Stretched cells can then be extended into the interior of the domain in a marching process, advancing one layer at a time. When the boundary-layer region has been covered with stretched cells, the rest of the domain can be covered according to the AFT as described above.

It may be convenient to generate stretched cells by means of a mathematical transformation, such that grid generation takes place in a parameter (or 'normalized') space in which the stretched triangles are transformed to approximately equilateral triangles. The transformation from physical space to parameter space involves rotating and shrinking an element, in which, referring to Fig. 8.19, a point with position vector $\mathbf{x}$ in physical space is transformed to the point $\tilde{\mathbf{x}}$, where

$$\tilde{\mathbf{x}} = \begin{pmatrix} \frac{1}{s}s_x & \frac{1}{s}s_y \\ n_x & n_y \end{pmatrix} \mathbf{x}, \tag{8.17}$$

the components of $\mathbf{s}$ being given by $(s_x, s_y) = (\cos\phi, \sin\phi)$ and those of $\mathbf{n}$ by $(n_x, n_y) = (-\cos\phi, \sin\phi)$. We also have

$$\tilde{\mathbf{x}} = \begin{pmatrix} \tilde{x} \\ \tilde{y} \end{pmatrix} = \begin{pmatrix} \frac{1}{s}\mathbf{s}\cdot\mathbf{x} \\ \mathbf{n}\cdot\mathbf{x} \end{pmatrix}. \tag{8.18}$$

The effect of the transformation on a typical stretched triangle is shown in Fig. 8.40. It follows that approximately equilateral triangles may be generated in parameter space and may then be transformed back into stretched triangles in physical space by applying the inverse transformation to that given in eqn (8.17).

8.3.6 Grid quality improvement

After an unstructured grid has been generated, two procedures may be applied in order to improve it. These procedures do not change the total number of triangles and nodes in the grid.

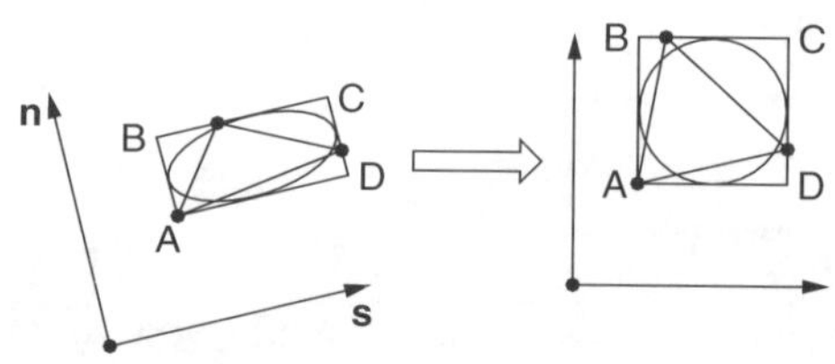

Fig. 8.40 Transformation from physical space to parameter space.

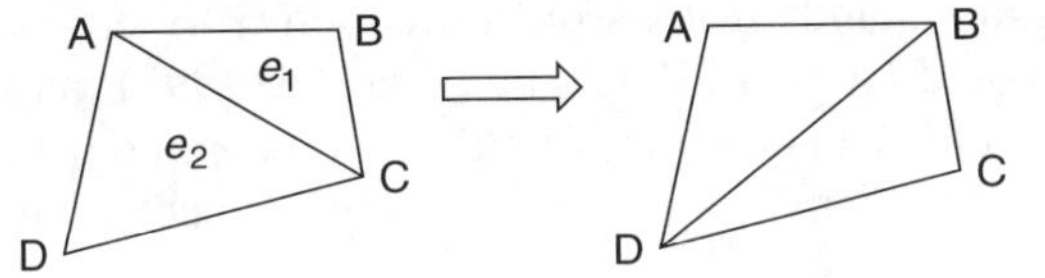

Fig. 8.41 Diagonal swapping.

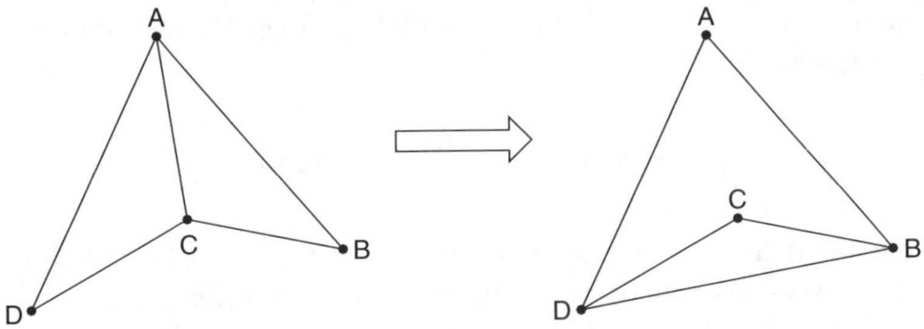

Fig. 8.42 No diagonal swapping in this case.

Diagonal swapping

This procedure, already referred to in Section 8.2.1, does not change the position of the nodes, but may change their connectivities. A loop (i.e. a DO loop) over all the triangle sides, excluding those elements on the boundary, is set up, in which for a typical side such as AC in Fig. 8.41, which is common to the two triangles ABC, ACD, the possibility of replacing AC by BD is considered. This diagonal swapping is performed if the resulting configuration, consisting of the two triangles ABD and BCD, is preferable according to some specified criterion. One such criterion would be that the minimum angle in the two triangles in the changed configuration is larger than the one in the original one.

No diagonal swapping is carried out if the original triangles compose a quadrilateral which is not convex (Fig. 8.42).

Grid smoothing

In this procedure the positions of interior nodes are altered without changing any connectivities. The idea is to regard the triangle sides as linear springs with identical stiffnesses and tensions proportional to the lengths of the springs. The overall equilibrium position of the nodes is then sought by iteration. In each iteration a loop over the interior nodes is carried out in which each node is moved to the centroid of the three nodes to which it is connected. Between three and five iterations are generally required to arrive at a satisfactory smoothed grid.

8.4 Solving hosted equations using finite elements

Here we present a very brief introduction to the solution of field equations (the hosted equations) in a domain which has been triangulated, using linear triangular elements.

For further information, standard texts on finite element methods such as Hinton and Owen (1979), Taylor and Hughes (1981), and George (1991) may be consulted.

Suppose that the field equation for a field quantity ϕ, subject to certain boundary conditions, in a planar domain D (assumed simply connected here) is

$$L(\phi) = 0 \tag{8.19}$$

where L is a second order partial differential operator. Suppose also that the domain has been triangulated, with n triangles and m points (nodes).

In the Method of Weighted Residuals, we seek an approximate solution for ϕ from the family of functions

$$\tilde{\phi} = f_0(x, y) + \sum_1^N \phi_i f_i(x, y), \tag{8.20}$$

where usually the function f_0 is chosen to satisfy the given boundary conditions and the f_is satisfy homogeneous boundary conditions in order to make $\tilde{\phi}$ satisfy the boundary conditions for any choice of the constant coefficients ϕ_i. In addition, we choose a set of N weighting functions $W_i(x, y)$ and impose the conditions

$$\iint_D W_i L(\tilde{\phi})\,\mathrm{d}x\,\mathrm{d}y = 0, \quad i = 1, 2, \ldots, N, \tag{8.21}$$

to make $L(\tilde{\phi})$ as close to zero as possible in some sense over the domain D. These equations will generate a set of N equations for the N unknown constants ϕ_i and hence the best (in some sense) approximate solution from our original set.

In the special case of the Galerkin Method, the weighting functions are taken to be identical to the approximating functions, which gives the set of equations

$$\iint_D f_i L(\tilde{\phi})\,\mathrm{d}x\,\mathrm{d}y = 0, \quad i = 1, 2, \ldots, N, \tag{8.22}$$

with $\tilde{\phi}$ given by eqn (8.20).

In the particular form of the finite-element method which is appropriate here for a plane triangulated region, we look for approximate solutions in the form

$$\tilde{\phi} = \sum_{i=1}^m \phi_i R_i(x, y), \tag{8.23}$$

where the ϕ_is are constant coefficients to be determined and the known $R_i(x, y)$ functions are called *roof functions*, which are continuous and piecewise-linear in both x and y, such that R_i takes the value unity at the ith node and zero at all other nodes. Each node of the triangulation is assigned a global number between 1 and m. Moreover, a node constitutes one of three vertices of a triangular element, and is assigned a local number between 1 and 3, these numbers being ordered in an anti-clockwise sense. This assigning of numbers, together with the numbering of triangular elements and the listing of cartesian co-ordinates of all nodes, is carried out in the Delaunay grid generation program *delaunay1.f* listed below.

Roof functions may be built up from linear *shape* functions $N_i^e(x, y)$ defined over each triangular element. Here the suffix refers to the local number of a vertex; $N_i^e(x, y)$

takes the value unity at the ith vertex and zero at the other two. Over a particular triangle with vertices numbered 1, 2, 3 in anti-clockwise manner, having known cartesian co-ordinates (x_1, y_1), (x_2, y_2), (x_3, y_3), respectively, it is easy to show that a shape function linear in x and y and taking the value 1 at the vertex 1 and zero at the vertices 2 and 3 is given by

$$N_1^e(x, y) = \frac{1}{2A^e}(a_1 + b_1 x + c_1 y), \tag{8.24}$$

where $a_1 = x_2 y_3 - x_3 y_2$, $b_1 = y_2 - y_3$, $c_1 = x_3 - x_2$, and A^e is the area of the triangle. Expressions for similar shape functions N_2^e and N_3^e which take the value 1 at the vertices 2, 3, respectively, and zero at the other vertices may be immediately written down by cyclic permutation of the suffixes in eqn (8.24). Over a typical triangular element of the triangulation, we can then write the approximating solution as

$$\phi^e = \phi_1^e N_1^e + \phi_2^e N_2^e + \phi_3^e N_3^e, \tag{8.25}$$

where now the coefficients ϕ_1^e, ϕ_2^e, ϕ_3^e represent the values of ϕ^e at the vertices.

As an example of a field equation, we take Poisson's equation

$$\kappa\left(\frac{\partial^2\phi}{\partial x^2} + \frac{\partial^2\phi}{\partial y^2}\right) + Q\,(x, y) = 0 \tag{8.26}$$

where κ is a constant and in heat transfer problems $Q(x, y)$ could be a heat source function. The boundary conditions are taken to be that ϕ is specified on a part C_ϕ of the boundary and the normal derivative $\partial\phi/\partial n$ (in the direction of the outward normal) takes the value q (possibly a function of position) on the remaining part C_q. This means that at nodes on that part of the boundary of the triangulation which approximates C_ϕ the value of the coefficients ϕ_i in eqn (8.23) is effectively specified.

Suppose that N of the nodes are in the interior of the triangulation or on the boundary C_q and that nodes with global vertex numbers $N+1, N+2, \ldots, m$ lie on that part of the boundary which approximates C_ϕ. Then instead of eqn (8.23) we can put

$$\tilde{\phi} = \sum_{i=N+1}^{m} \phi_i R_i(x, y) + \sum_{i=1}^{N} \phi_i R_i(x, y) = \tilde{\phi}_0(x, y) + \sum_{i=1}^{N} \phi_i R_i(x, y), \tag{8.27}$$

where in the first term the coefficients ϕ_i are known from the boundary conditions. This representation of the family of approximating functions has the same form as eqn (8.20), since the roof functions $R_i(x, y)$ take the value zero at the boundary nodes $N+1, N+2, \ldots, m$.

The Galerkin approach gives the integral form

$$\iint_D R_i L(\tilde{\phi})\,\mathrm{d}x\,\mathrm{d}y = \iint_D R_i \left\{\kappa\left(\frac{\partial^2\tilde{\phi}}{\partial x^2} + \frac{\partial^2\tilde{\phi}}{\partial y^2}\right) + Q\,(x, y)\right\} \mathrm{d}x\,\mathrm{d}y = 0,$$

$$i = 1, \ldots, N, \tag{8.28}$$

but to be able to deal with approximating functions with discontinuous slopes across triangular edges, we need to transform the double integral by integration by parts (the

Divergence Theorem, or Green's formula) to

$$-\iint_D \left\{ \kappa \left(\frac{\partial R_i}{\partial x}\frac{\partial \tilde{\phi}}{\partial x} + \frac{\partial R_i}{\partial y}\frac{\partial \tilde{\phi}}{\partial y} \right) - R_i Q \right\} \mathrm{d}x\,\mathrm{d}y + \oint_C \kappa R_i \frac{\partial \tilde{\phi}}{\partial n} ds = 0, \qquad (8.29)$$

where the contour integral is taken anti-clockwise around the boundary C of D. If we replace $\partial\tilde{\phi}/\partial n$ by q on C_q, and R_i by zero on C_ϕ, we obtain

$$\iint_D \kappa \left(\frac{\partial R_i}{\partial x}\frac{\partial \tilde{\phi}}{\partial x} + \frac{\partial R_i}{\partial y}\frac{\partial \tilde{\phi}}{\partial y} \right) \mathrm{d}x\,\mathrm{d}y = \iint_D R_i Q\,\mathrm{d}x\,\mathrm{d}y + \oint_{C_q} \kappa R_i q\,\mathrm{d}s,\ i = 1, \ldots, N. \qquad (8.30)$$

Using eqn (8.27), this becomes, for $i = 1, \ldots, N$,

$$\sum_{j=1}^{N} \left\{ \iint_D \kappa \left(\frac{\partial R_i}{\partial x}\frac{\partial R_j}{\partial x} + \frac{\partial R_i}{\partial y}\frac{\partial R_j}{\partial y} \right) \mathrm{d}x\,\mathrm{d}y \right\} \phi_j$$

$$= \iint_D R_i Q\,\mathrm{d}x\,\mathrm{d}y + \oint_{C_q} \kappa R_i q\,\mathrm{d}s - \iint_D \kappa \left(\frac{\partial R_i}{\partial x}\frac{\partial \tilde{\phi}_0}{\partial x} + \frac{\partial R_i}{\partial y}\frac{\partial \tilde{\phi}_0}{\partial y} \right) \mathrm{d}x\,\mathrm{d}y. \qquad (8.31)$$

The contribution of a triangular element to the terms

$$\iint_D \kappa \left(\frac{\partial R_i}{\partial x}\frac{\partial R_j}{\partial x} + \frac{\partial R_i}{\partial y}\frac{\partial R_j}{\partial y} \right) \mathrm{d}x\,\mathrm{d}y$$

in eqn (8.31) is then, using eqn (8.25),

$$\sum_{j=1}^{3} \left\{ \iint_e \kappa \left(\frac{\partial N_i^e}{\partial x}\frac{\partial N_j^e}{\partial x} + \frac{\partial N_i^e}{\partial y}\frac{\partial N_j^e}{\partial y} \right) \mathrm{d}x\,\mathrm{d}y \right\} \phi_j^e = \sum_{j=1}^{3} K_{ij}^e \phi_j^e, \text{ say,} \qquad (8.32)$$

where K_{ij}^e is clearly symmetric and can be regarded as the element stiffness matrix. By eqn (8.24) we can write, in the present case,

$$K_{ij}^e = \iint_e \frac{\kappa}{(2A^e)^2}(b_i b_j + c_i c_j)\,\mathrm{d}x\,\mathrm{d}y = \frac{\kappa}{4A^e}(b_i b_j + c_i c_j). \qquad (8.33)$$

For a given triangulation, reducing eqns (8.31) to a global matrix equation

$$\sum_{j=1}^{N} K_{ij}\phi_j = F_i \qquad (8.34)$$

involves first the evaluation of the stiffness matrices for all elements and then assembling and adding them into a large $N \times N$ matrix K_{ij}. This is usually a straightforward task once a *connectivity table* has been established listing the global and local vertex numbers for each element in the triangulation. This enables global numbers to be made to correspond to the row (and column) numbers of K_{ij} so that the 3×3 entries K_{ij}^e can be added in the correct places. The 'force' terms F_i can be calculated from the RHS of eqn (8.31) using the values of Q and q at appropriate nodes. Finally a numerical scheme must be used to solve the global matrix equation for the unknown coefficients ϕ_i.

Similar procedures can be used to solve non-linear hosted equations, such as the Navier-Stokes equations, using iterative methods based on an initial approximate solution. For example, see Rao, S.S. (1982) and Lewis, Morgan, Thomas, and Seetharamu (1996).

8.5 Website programs

8.5.1 Subdirectory: book/Delaunay

This subdirectory contains two files, called Delaunay1.f and Delaunay2.f. The codes involved are rather lengthy, and we do not give every detail here. However, they are annotated to make them easier to follow.

1. Delaunay1.f
   The 'main' routine contained in Delaunay1.f was written by S.W. Sloan (Sloan (1987)). This is a small program, described as 'A fast algorithm for constructing Delaunay triangulation in the plane'. It triangulates domains starting from sets of boundary data, and generally produces completely unacceptable triangulations. For domains with an internal boundary, such as the region around an airfoil, not only is the quality of the triangles in the domain unacceptable, but the inner circular region is also filled with unacceptable triangles. To make the grid acceptable, (a) the unwanted triangles must be removed from the inner region, and (b) the remaining triangles must be processed so that the final triangulation does not contain triangles that are excessively large or skinny.
   Delaunay1.f contains a subroutine called 'Add-point' which carries out the lengthy process of refining the initial triangulation. The first task is the removal of unwanted triangles from within any inner boundaries, and this is followed by the re-numbering of the remaining triangles. Then the processing of these triangles is carried out. The criteria used in this processing are those based on area and on aspect ratio, as discussed in Section 8.2.3. Target values for area and aspect ratio, called here 'amax' and 'armax' respectively, are specified by the user.
   All the triangles go through the 'Add-point' subroutine one at a time. From the cartesian co-ordinates of the vertices of a triangle the lengths of the sides can be calculated, and then eqns (8.1) and (8.5) permit the calculation of area and aspect ratio. If either of these is greater than 'amax' or 'armax' respectively, then the triangle qualifies for processing. (Otherwise we proceed to the next triangle for consideration.)
   If a triangle qualifies for processing, it is classified by the program according to whether one of its sides is parallel to the x-axis (in which case it is called 'horizontal'), perpendicular to the x-axis (when it is called 'vertical'), or neither of these possible cases (when it is called 'general'). Furthermore, given that in the initial triangulation the vertices of a triangle are organized in an anti-clockwise manner, and labelling them here as 1, 2, 3, we can label side 1-2 as side 1, side 2-3 as side 2, and side 3-1 as side 3. This gives three further categories for classification. Finally

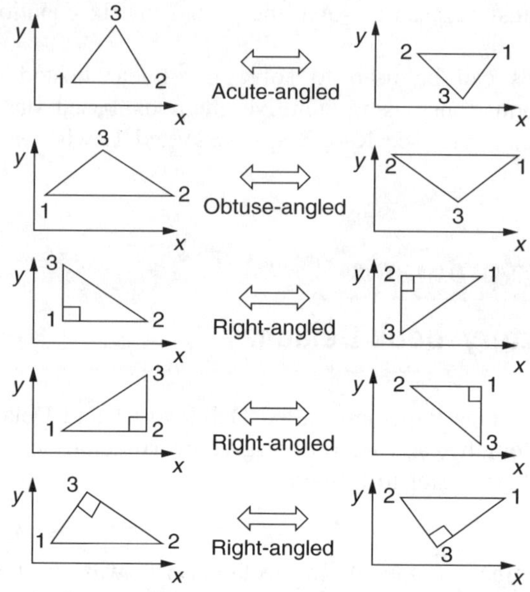

Fig. 8.43 Triangles with side 1 parallel to the x-axis.

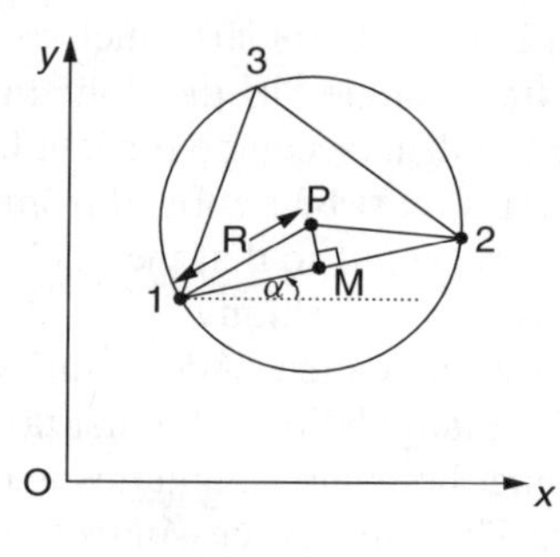

Fig. 8.44 Locating the circumcentre of a triangle.

the triangle is classified according to whether it is acute-angled, obtuse-angled, or right-angled.

For example, suppose that side 1 is parallel to the x-axis. Fig. 8.43 shows the various cases into which the triangle can fall. Subroutine 'Add-point' contains six subroutines of its own, called H_1, H_2, H_3, V_1, V_2, V_3, where, for example, H_1 considers triangles whose side 1 is parallel to the x-axis and V_1 triangles with side 1 perpendicular to the x-axis. 'General' triangles are processed by subroutine 'Add-point' itself, and are classified as (a) acute-angled triangles, (b) obtuse-angled triangles in which the longest side is *either* side 1 *or* not side 1, or (c) right-angled triangles.

The main task of the subroutines is to calculate the position of the circumcentre of the triangle being processed. When the co-ordinates of the circumcentre have been calculated, tests are carried out to determine whether the circumcentre lies inside

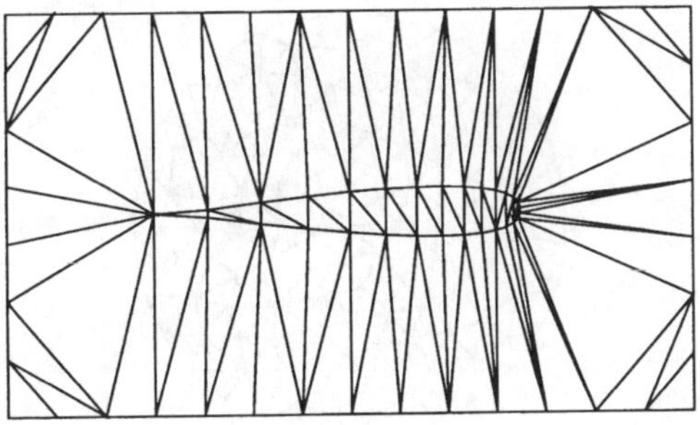

Fig. 8.45 Initial Delaunay triangulation for an airfoil.

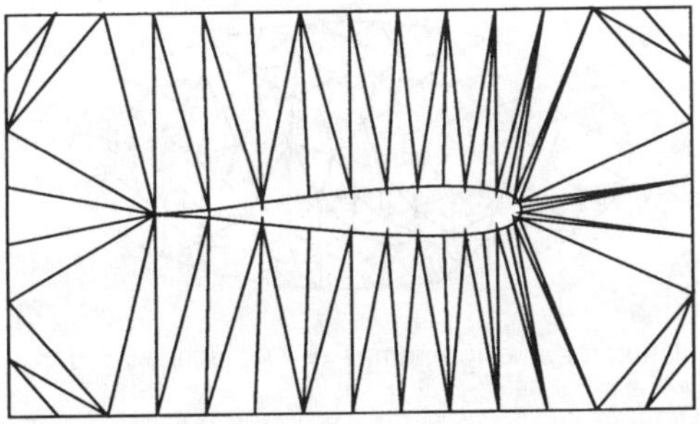

Fig. 8.46 Second stage Delaunay triangulation for an airfoil.

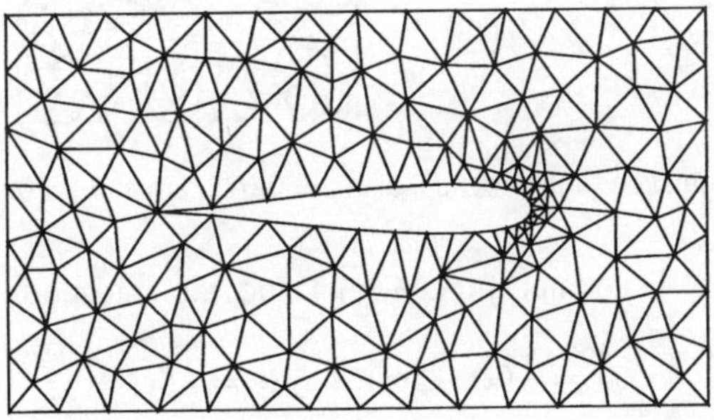

Fig. 8.47 Delaunay triangulation of a geometrically constructed aerofoil, amax = 0.04, armax = 2.0.

any inner boundary or outside any outer boundary of the domain. If so, then the circumcentre is rejected as a new point. Otherwise it is accepted, added to the list of data points, and a new triangulation carried out. This procedure is continued until all the 'bad' triangles have been removed.

Calculation of the co-ordinates of the circumcentre is illustrated in Fig. 8.44 for a general acute-angled triangle. Given the co-ordinates of the vertices, the circum-radius R may be found by using the combination of eqns (8.1), (8.2), and (8.4).

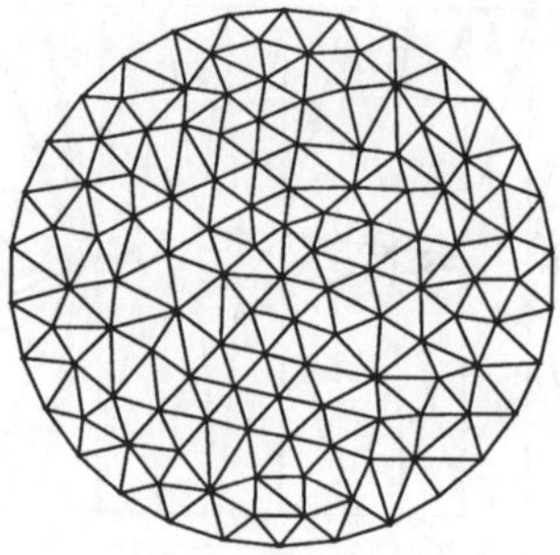

Fig. 8.48 Delaunay triangulation for a circle, amax = 2.2, armax = 5.5.

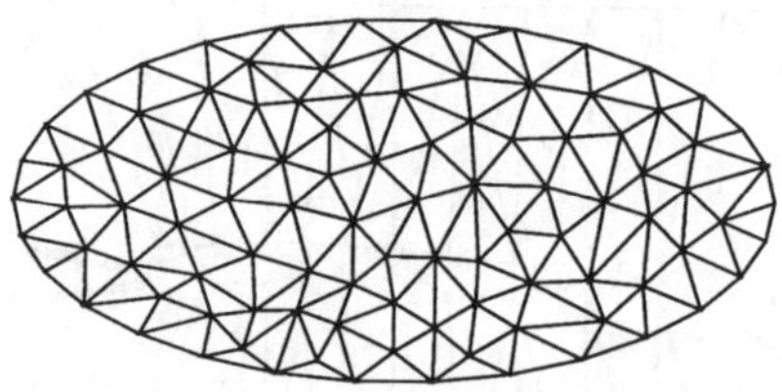

Fig. 8.49 Delaunay triangulation for an ellipse, amax = 1.4, amax = 5.5.

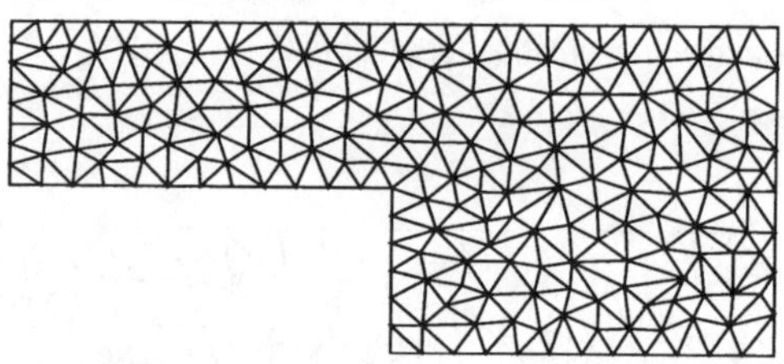

Fig. 8.50 Delaunay triangulation for a backstep, amax = 0.025, armax = 5.0.

The circumcentre P is located at (x_P, y_P) and the mid-point M of side 1-2 is at (x_M, y_M), where

$$x_M = \tfrac{1}{2}(x_1 + x_2), \quad y_M = \tfrac{1}{2}(y_1 + y_2)$$

in terms of the co-ordinates of the vertices 1, 2. We have

$$\overline{PM} = \sqrt{R^2 - \left(\tfrac{1}{2}c\right)^2},$$

where c is the length of side 1-2. Then if 1-2 makes an angle α with the x-axis (positive if anti-clockwise), where

$$\sin\alpha = \frac{1}{c}(y_2 - y_1)$$

$$\cos\alpha = \frac{1}{c}(x_2 - x_1),$$

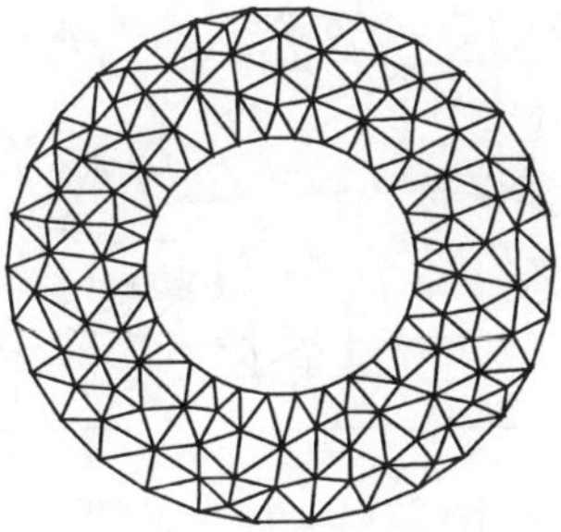

Fig. 8.51 Delaunay triangulation for two concentric circles, amax = 1.6, armax = 3.9.

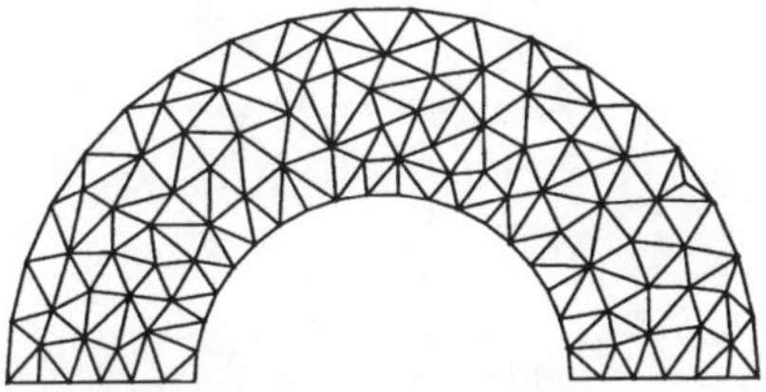

Fig. 8.52 Delaunay triangulation for two semi-circles, amax = 1.0, armax = 4.0.

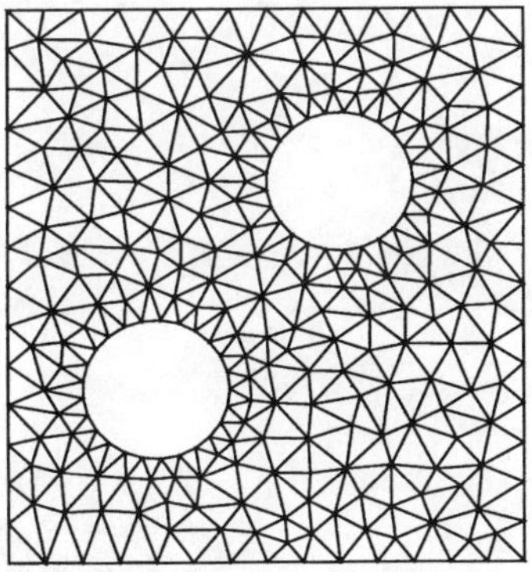

Fig. 8.53 Delaunay triangulation for two circles and one square, amax = 0.2, armax = 1.5.

we have

$$x_P = x_M - \overline{PM} \sin\alpha$$

$$y_P = y_M + \overline{PM} \cos\alpha. \tag{8.35}$$

Figures 8.45–8.47 illustrate three stages of this program in operation as applied to the region around an airfoil. More resulting triangulations for domains of various geometries are shown in Figs 8.48–8.53.

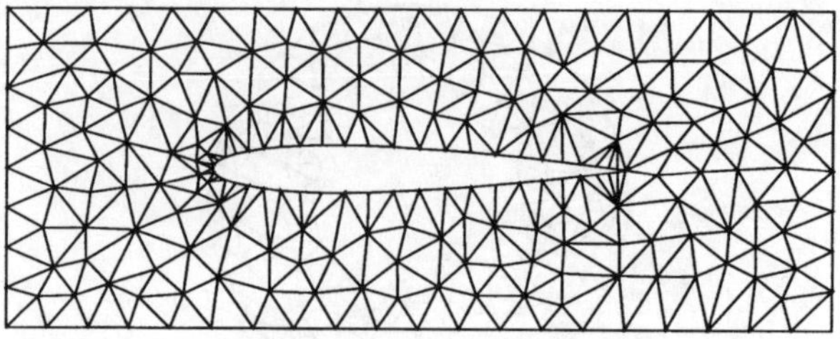

Fig. 8.54 Delaunay triangulation for NASA-0012, amax = 0.007, armax = 3.0.

2. Delaunay2.f
   This program generates an unstructured grid around a NASA 0012 airfoil using actual NASA boundary data. An example of a resulting triangulation is shown in Fig. 8.54.

Bibliography

Ahlfors, L.V. (1996) *Lectures on Quasiconformal Mappings*, Van Nostrand, New York.

Allwright, S.E. (1988) Techniques in multiblock domain decomposition and surface grid generation. In: Sengupta, S., Thompson, J.F., Eiseman, P.R. and Hauser, J. (eds.) *Numerical Grid Generation in Computational Fluid Mechanics '88*. Pineridge Press, Miami, FL, 559–568.

Aris, R. (1962) *Vectors, Tensors, and the Basic Equations of Fluid Mechanics*, Prentice-Hall, Englewood Cliffs, NJ.

Boor, C. (1974) Good approximation by splines with variable knots. *Lect. Notes Math.* **363**, 12–20.

Bowyer, A. (1981) Computing Dirichlet tessellations. *Comput. J.* **24**(2), 162–166.

Carey, G.F. (1997) *Computational Grids – Generation, Adaptation, and Solution Strategies*, Taylor and Francis.

do Carmo, M.P. (1976) *Differential Geometry of Curves and Surfaces*, Prentice-Hall.

Eiseman, P.R. (1979) A multi-surface method of co-ordinate generation. *J.Comput.Phys.* **33**, 118–150.

Eiseman, P.R., Cheng, Z. and Hauser, J. (1994) Application of multiblock grid generation with automatic zoning. In: Weatherill, N.P., Eiseman, P.R., Hauser, J. and Thompson, J.F. (eds.) *Numerical Grid Generation in Computational Fluid Mechanics and Related Studies* – Proceedings of the 4th International Conference. The Cromwell Press Ltd., Melksham, Wiltshire.

Erikkson, L.-E. (1982) Generation of boundary-conforming grids around wing-body configurations using transfinite interpolation. *AIAA Journal* **20**, 1313–1320.

Farrashkhalvat, M. and Miles, J.P. (1990) *Tensor Methods for Engineers*, Ellis Horwood, Chichester.

George, A.J. (1971) *Computer Implementation of the Finite Element Method*, Stanford University Department of Computer Science, STAN-CS-71–208.

George, P.L. (1991) *Automatic Mesh Generation: Application to Finite Element Methods*, Wiley, New York.

Hinton, E. and Owen, D.R.J. (1979) *An Introduction to Finite Element Computations*, Pineridge Press.

Holmes, D.G. and Snyder, D.D. (1988) The generation of unstructured triangular meshes using Delaunay triangulation. In: Sengupta, S., Hauser, J., Eiseman, P.R., Thompson, J.F. (eds.) *Numerical Grid Generation in Computational Fluid Dynamics*, Pineridge, Swansea.

Kay, D.C. (1988) *Tensor Calculus*, Schaum's Outline Series, McGraw-Hill.

Knupp, P. and Steinberg, S. (1993) *The Fundamentals of Grid Generation*, CRC Press, Boca Raton, Florida.

Kreyzig, E. (1968) *Differential Geometry,* Dover Publications.

Lewis, R.W., Morgan, K., Thomas, H.L. and Seetharamu, K.N. (1996) *The Finite Element Method in Heat Transfer Analysis,* John Wiley & Sons Ltd., Chichester, England.

Liao, G. and Liu, H. (1993) Existence and $C^{(0,\alpha)}$ regularity of a minimum of a functional related to grid generation problems. *Num. Math. PDEs.* **9**, 3.

Liseikin, V.D. (1999) *Grid Generation Methods,* Springer-Verlag, Berlin, Heidelberg.

Mastin, C.W. (1991) Elliptic grid generation and conformal mapping. In: Castillo, J.E. (ed.) *Mathematical Aspects of Numerical Grid Generation.* SIAM, Philadelphia.

Mastin, C.W. and Thompson, J.F. (1984) Quasiconformal mappings and grid generation, *SIAM J. Sci. Stat.Comput.,* **5**, 305–310.

Nehari, Z. (1975) *Conformal Mapping*, Dover Publications.

Parakash, N. (1981) *Differential Geometry,* McGraw-Hill, New Delhi.

Rao, S.S. (1982) *The Finite Element in Engineering,* Pergamon Press, Oxford, England.

Rebay, S. (1993) Efficient unstructured mesh generation by means of Delaunay triangulation and Bowyer-Watson algorithm. *J. Comput. Phys.* **106**, 125–138.

Roberts, G.O. (1971) *Proceedings of the Second International Conference on Numerical Methods in Fluid Dynamics,* Springer, Berlin, Heidelberg.

Ryskin, G. and Leal, L.G. (1983) Orthogonal Mapping. *J. Comput. Physics* **50**, 71–100.

Shenton, D.N. and Cendes, Z.J. (1985) Three-dimensional finite element mesh generation using Delaunay tessellation. *IEEE Trans. Magnetics* **MAG-21**, 2535–2538.

Sloan, S.W. (1987) A fast algorithm for constructing Delaunay triangulations in the plane. *Adv. Eng. Software* **9**, 1.

Spain, B. (1953) *Tensor Calculus,* Oliver and Boyd, Edinburgh and London.

Steger, J.L. and Chaussee, D.S. (1980) Generation of body fitted co-ordinates using hyperbolic differential equations. *SIAM J. Sci. Stat. Comput.* **1**(4), 431–437.

Stoker, J.J. (1968) *Differential Geometry,* Wiley-Interscience.

Struik, D.J. (1950) *Lectures on Classical Differential Geometry,* Dover Publications.

Taylor, C. and Hughes, T.G. (1981) *Finite Element Programming of the Navier-Stokes Equations,* Pineridge Press.

Thompson, J.F., Soni, B.K. and Weatherill, N.P. (eds.) (1999) *Handbook of Grid Generation,* CRC Press.

Thompson, J.F., Thames, F.C. and Mastin, C.W. (1974) Automatic numerical generation of body-fitted curvilinear co-ordinate system for field containing any number of arbitrary two-dimensional bodies. *J. Computat. Phys.,* **15**, 299–319.

Thompson, J.F., Warsi, Z.U.A. and Mastin, C.W. (1985) *Numerical Grid Generation: Foundations and Applications*, North Holland.

Warsi, Z.U.A. and Thompson, J.F. (1980) A non-iterative method for the generation of two-dimensional orthogonal curvilinear co-ordinates in a Euclidean space: in *Numerical Grid Generation Techniques* NASA CP2166.

Watson, D. (1981) Computing the n-dimensional Delaunay tessellation with applications to Voronoi polytopes. *Comput. J.* **24**(2), 167–172.

Winslow, A.M. (1967) Equipotential zoning of two-dimensional meshes, *J. Computat. Physics,* **1,** 149–172.

Index